BIOTECHNOLOGY-2

INCLUDING
CELL BIOLOGY, GENTICS
MICROBIOLOGY

Rajeshwari S Setty
Professor
Department of Biotechnology
SSMRV College, Bangalore
Karnataka

V Sreekrishna
Formerly
Principal, City College
Bangalore

Publishing Globally

NEW AGE INTERNATIONAL (P) LIMITED, PUBLISHERS
LONDON • NEW DELHI • NAIROBI
Bangalore • Chennai • Cochin • Guwahati • Hyderabad • Kolkata • Lucknow • Mumbai
Visit us at www.newagepublishers.com

Published by New Age International (P) Ltd., Publishers
First Edition: 2003
Reprint: 2018

GLOBAL OFFICES

- **New Delhi** **NEW AGE INTERNATIONAL (P) LIMITED, PUBLISHERS**
7/30 A, Daryaganj, New Delhi-110002, (INDIA)
Tel.: (011) 23253771, 23253472, **Telefax:** 23267437, 43551305
E-mail: contactus@newagepublishers.com • Visit us at www.newagepublishers.com
- **London** **NEW AGE INTERNATIONAL (UK) LTD.**
27 Old Gloucester Street, London, WC1N 3AX, UK
E-mail: info@newacademicscience.co.uk • Visit us at www.newacademicscience.co.uk
- **Nairobi** **NEW AGE GOLDEN (EAST AFRICA) LTD.**
Ground Floor, Westlands Arcade, Chiromo Road (Next to Naivas Supermarket)
Westlands, Nairobi, KENYA, **Tel.:** 00-254-713848772, 00-254-725700286
E-mail: kenya@newagepublishers.com

BRANCHES

- **Bangalore** 37/10, 8th Cross (Near Hanuman Temple), Azad Nagar, Chamarajpet, Bangalore- 560 018
Tel.: (080) 26756823, **Telefax:** 26756820, **E-mail: bangalore@newagepublishers.com**
- **Chennai** 26, Damodaran Street, T. Nagar, Chennai-600 017, **Tel.:** (044) 24353401
Telefax: 24351463, **E-mail: chennai@newagepublishers.com**
- **Cochin** CC-39/1016, Carrier Station Road, Ernakulam South, Cochin-682 016
Tel.: (0484) 2377303, **Telefax:** 4051304, **E-mail: cochin@newagepublishers.com**
- **Guwahati** Hemsen Complex, Mohd. Shah Road, Paltan Bazar, Near Starline Hotel
Guwahati-781 008, **Tel.:** (0361) 2513881, **Telefax:** 2543669
E-mail: guwahati@newagepublishers.com
- **Hyderabad** 105, 1st Floor, Madhiray Kaveri Tower, 3-2-19, Azam Jahi Road, Near Kumar Theater
Nimboliadda Kachiguda, Hyderabad-500 027, **Tel.:** (040) 24652456, **Telefax:** 24652457
E-mail: hyderabad@newagepublishers.com
- **Kolkata** RDB Chambers (Formerly Lotus Cinema) 106A, 1st Floor, S N Banerjee Road
Kolkata-700 014, **Tel.:** (033) 22273773, **Telefax:** 22275247
E-mail: kolkata@newagepublishers.com
- **Lucknow** 16-A, Jopling Road, Lucknow-226 001, **Tel.:** (0522) 2209578, 4045297, **Telefax:** 2204098
E-mail: lucknow@newagepublishers.com
- **Mumbai** 142C, Victor House, Ground Floor, N.M. Joshi Marg, Lower Parel, Mumbai-400 013
Tel.: (022) 24927869, **Telefax:** 24915415, **E-mail: mumbai@newagepublishers.com**
- **New Delhi** 22, Golden House, Daryaganj, New Delhi-110 002, **Tel.:** (011) 23262368, 23262370
Telefax: 43551305, **E-mail: sales@newagepublishers.com**

ISBN: 978-81-224-1417-2
C-17-06-10523

Printed in India at Krishna Printers & Binders, Delhi.

NEW AGE INTERNATIONAL (P) LIMITED, PUBLISHERS
7/30 A, Daryaganj, New Delhi-110002
Visit us at **www.newagepublishers.com**
(CIN: U74899DL1966PTC004618)

Dedicated to

Mr. S.S. Sanjai Kumar
(Husband of Rajeshwari Setty)
And
Dr. S. Vijaya Sree
(Wife of Dr. V. Sreekrishna)

Preface

This book has been written for Biotechnology students of B.Sc. All aspects of Cell Biology, Genetics and Microbiology have been discussed exhaustively as per Biotechnology curriculum. The material is presented in a simple manner that should make the understanding of the subject easy for the students.

We have received a warm response for my Biotechnology- I book from students and teachers. We do hope that this book would also generate a similar response from the students and the teachers.

We are thankful to New Age International (P) Ltd., Publishers, New Delhi, Mr. Saumya Gupta, Managing Director for his cooperation and encouragement in the development of the text. We are also thankful to all our friends and family members for their encouragements.

Suggestions for further improvement of the book and for correction of errors if any, would be gratefully acknowledged and incorporated

RAJESHWARI S. SETTY
DR. V. SREEKRISHNA

Contents

Part A
Cell Biology

1

Cells as a Basic Unit of Living Organisms: The Cell Theory

The cell can be understood by studying its complexity and functions. In the beginning the term itself was coined by the invention of the simple microscope by Anton Von Lewenhock and Jenssenis and the subsequent invention of the microscope by Robert Hooke in 1665; he introduced the term Cell while describing the microscopic texture of the cork.

The term Cell comes from the Latin word *Cellula*, which means small compartments. The improvement in microscopes and the techniques of staining helped us examining the structure and content of the cell.

The structural living unit other than the virus is the cell, which consists of smaller structures called Organelles, namely:

Cell Wall.

Plasma Membrane.

Cytosol, Cytoplasm

Golgi Bodies

Endoplasmic Reticulum,

Ribosomes.

Cytoskeletal structure.

Mitochondria.

Chloroplasts.

Lysosomes and Peroxisomes and

Nucleus.

Each organelle has a specialised function within the cell. A Cell containing these organelles and a well defined nucleus is called a Eukaryotic Cell.

The cells of bacteria and blue green algae lack a definite nucleus and organelles like eukaryotes, and are called prokaryotes. Many plants and animals exist as single cells; hence, the cell is the basic unit.

Three requirements have to be fulfilled for a cellular organisation. These are:

(i) a selectively permeable boundary separating the cellular components from the outside.

(ii) a metabolic system which would allow a series of enzyme catalysed reactions, and

(iii) a genetic information system which would store and transmit hereditary information for future generations.

However, viruses are exceptional as they possess a molecule containing the information required for reproduction and a capsular coat around this molecule. Viruses, however, cannot reproduce without a larger host which has an efficient biological machinery. For this reason they do not conform to a truly living system, and may be called semi living particles. Viruses have simple life histories. While the capsular coat is made up of protein, the information molecule is a nucleic acid DNA or RNA.

EUKARYOTES

They are highly evolved cells, possessing specialised organelles viz., endoplasmic reticulum; the endoplasmic reticulum may be granular or agranular. Ribosomes are deposited on the granular endoplasmic reticulum. Golgi complex, consisting of stacks of lamellae and vesicles of different shapes and sizes act as an intracellular pump to regulate fluid movement, expulsion of excretory products.

Lysosomes, phagosomes and pinocytosomes appear as irregular vesicular bodies. Lysosomes are rich in hydrolytic enzymes which carry on cellular digestion. Phagosomes contain semisolid or solid nutritive materials which are engulfed by the plasma membrane. Pinocytosomes are microscopic vesicles, pinched off from the plasma membrane like phagosomes; they too contain nutritive material but in liquid phase.

Mitochondria are most important organelles; they may be capsule or rodlike in shape. They are enveloped by a double membrane. The inner membrane, invaginates to form compartments. The mitochondria are responsible for cellular respiration. They also function as energy transducing organelles; contain enzymes of electron transport chain; sites for oxidation or carbohydrates, lipids and amino acids.

Plastids, which are characteristic of plant cell are another cell organelle; plastids without pigment are known as leukoplasts and those possessing colored pigments are known as chromoplasts. The plastids possessing green chlorophyll pigments are known as chloroplasts, which takes part in the photosynthetic activity of the cell. The leucoplasts serve for storage of starch, oil droplets and proteins.

The cytoplasm of the cell also contains vacuoles, which are large in plant cells but in animal cells they are small. The vacuoles are filled with liquid matter.

Another major component of eukaryotic cells is membrane bound neuclear complex. The nucleus consists of the nucleoplasm in which chromosomes and nucleolus are present. The nuclear membrane is double layered, with minute pores and nuclear membrane has direct contact with granular endoplasmic reticulum. Deoxyribonucleic acid (DNA) and the basic protein called histone are present in the chromosomes.

The Centriole is another organelle, that participates in cell division. They are spherical in shape located near the nucleus.

THE CELL THEORY

In 1665 Robert Hooke suggested the term "cell" while describing the microscopic texture of the cork. This was followed by illustrations conforming to cellular nature of living organisms by a number of workers during the next two centuries.

Schleiden and Schwann are generally accredited as the exponents of the "cell theory", which states that both plants and animals are made up of aggregations of basic units called cells.

Schleiden and Schwann writing in the 19th century, introduced the cell theory, which can be summarised as follows:

(i) Organisational cells are differentiated into distinct cell types.

(ii) Nucleus is a common feature of all cells with the exception of viruses and red blood cells.

(iii) The protoplasm is the living content of the cell, and determines the activity of the cell and of the whole organism.

(iv) New cells originate from pre-existing cells through division.

Life can be considered as a unique and complex interaction of matter and energy that assumes a hierarchy of structural order. The biological continuity, thus, consists of different levels of biological organisation. Atoms, the smallest units of matter, interact to form molecules that may be variously joined to form more elaborate macromolecules characteristic of living organisms. The interacting macromolecules often result into structures called organelles and the next higher level of organisation is the cell. The cell is the smallest unit of interacting matter and energy that exhibits properties of life.

However, the chemical components of the living cells are qualitativeiy quite different from those of the physical environment. Majority of the molecules in living cells are organic compounds, which occur in great variety and complexity.

For example in *Escherichia coli* bacterium (Fig. 1.1), there are more than 5, 000 types of organic molecules, of which 3,000 are proteins and 1,000 are nucleic acids. *Homo sapiens*, i.e., Man is known to have more than 1,00,000 types of proteins. This immense variety of organic compounds is reducible to a surprising simplicity; all proteins are composed of 20 types of linearly arranged amino acids. Similarly all nucleic acid molecules are composed of four types of nucleotides.

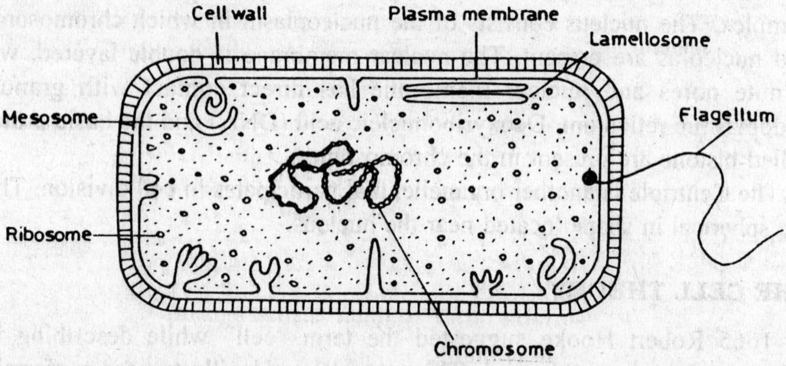

Fig. 1.1 A typical Escherichia coli cell showing prokaryotic organisation

These building blocks, thus, serve as the raw material for macro-molecular synthesis and perform various functions. Amino acids, for example, also serve as the raw material for macromolecular synthesis and perform various functions.

Further, Amino acids also serve as precursors of hormones, alkaloids, pigments, etc. Strange as it may appear, the building blocks are identical in all living organisms. Therefore, the following generalisations can be formulated:

(A) There is an underlying simplicity in the molecular organisation of all living cells.

(B) There is an underlying principle of molecular economy in living cells.

(C) All living organisms have a common ancestor.

PARAMETERS OF LIFE

Though it is not possible to define life, a set of properties can be suggested which characterise life.

(a) *Self Organisation:* Living systems are self organising, whereby specialised structures would have metabolic cooperation in an organised manner.

(b) *Self Regulation Capacity:* Self regulation of metabolic processes should be expressed by living systems. Proteins, which also function as enzymes, are capable of performing polyeric synthesis and degradation

(c) *Mechanism of Energy Transfer:* Living systems obey laws of thermodynamics. They have the capacity to liberate energy which is released through degradation of food materials, and are conserved in the form of ATP. Energy of these phosphate groups is utilised for different metabolic activities.

(d) *Self Reproducing Capacity:* All living organisms reproduce their own kind during their limited lifespan. DNA molecule helps in this process and also in passing genetic information to daughter cells.

(e) *Isothermal Open System:* Living organisms are capable of exchanging matter and energy within the environment; chemical transformations of matter occur at a temperature of around 37°C in majority.

(f) *Adaptation Capacity:* Genetically, living systems are capable of adapting to new environments or improving their efficiency within the same environment and causing evolutionary variations.

Life could be considered as a unique and complex interaction of matter and energy which possesses a sequence of structural order. The biological continuation consists of various levels of biological organisation. Atoms, the smallest units of matter interact to form molecules that may join to form more complicated macromolecules characteristic of living organisms.

Interacting larger molecules often result into structures called organelles and the higher level of organisation is the cell. The cell thus is the smallest unit of interacting matter and energy that exhibits properties of life.

Cells obey the laws of thermodynamics. The first law, which is also known as the law of conservation of energy, states that energy can neither be created nor destroyed; it can, however, be transformed. Living cells could only transform one form of energy into another. Cells absorb the useful form of energy, i.e. free energy, and return an equivalent amount of less useful form of energy to the environment.

The second law states that all physical and chemical processes proceed always with an increase in disorderliness of the universe. This is how cells follow the principle. Living cells neither create nor destroy energy, but transform into a useful form. Cells maintain their orderliness at the cost of their environment which they make more disorderly.

SUMMARY

It is believed that the ancestral organisms both plants and animals gained their energy from chemotrophs and later specialised in synthesising own food through photosynthesis.

Structural life unit other than the virus is the cell. A cell consists of smallest structures called organelles, each having a specialised function within the cell. A cell consisting of cellular organelles and well defined nucleus is called eukaryotic cell.

The cell of blue green algae, and bacteria take a definite nucleus and organelles like eukaryotes; they are thus called prokaryotes. As many plants and animals exist as unicellular organisms.

Several cell theories proposed are discussed.

QUESTIONS

1. Why we should not consider virus as a living entity?
2. What are the three main requirements for a cellular organisation?
3. Explain in detail the cell theory?
4. Write shortnotes on:
   (a) Prokaryotes
   (b) Eukaryotes
   (c) Parameters of life.

2

Precellular Evolution, Artificial Creation of Cell

If we examine the process of evolution which might have originated life, whatever might have been the ancestral form of life, it would have been a primitive form. It must have evolved through highly complex interacting forces abiotically on the primaeval earth.

The planet earth is believed to have originated about 15 billion years ago, according to the Big-Bang hypothesis. It was very much hot to begin with. When it cooled as much as it was necessary, neutrons and protons fused together to form atomic nuclei. Hydrogen and Helium were only produced in the early evolutionary history of earth. Heavier elements were produced after the stars formed, under the influence of gravitational forces. As stars increased in density they got heated up. Consequently,thermonuclear interactions led to the synthesis of heavier elements. After the full-fledged formation of earth, the ecosphere consisted of hydrogen (H), nitrogen (N_2), methane (CH_4) ammonia (NH_3) hydrogen sulphide (H_2S) and water vapour; of course, oxygen was not present.

ORIGIN OF COMPLEX MOLECULES

It is believed that organic molecules might have been synthesised abiotically on the primaeval earth, in the absence of Oxygen, under volcanic eruptions, heavy rains and lightning conditions, which contained Hydrogen, Nitrogen, Hydrogen Sulphide, Ammonia, Carbon-di-oxide, Methane and water. These micromolecules were the precursors for organic molecules. However, the source of energy was by solar radiation, temperature produced by volcanic eruptions and lightning. Once complex organic molecules were formed, mixed with rain water giving rise to diluted form of organic molecules, which is believed to be the" Hot dilute soup".

So as to prove the above mentioned organic molecules evolution, Stanley Miller, Chicago University Student in 1953, synthesised organic compounds under conditions reflecting the primitive atmosphere of the earth was holding water vapours, methane, Hydrogen, and Ammonia which were circulated in a close apparatus by the help of steam with an electric spark of 75,000 volts subjected. As a result the gas mixture was condensed, recirculated and many amino acids were synthesised like alanine, glycerine, glutaric acid, asperitic acid, valine, glycine and many organic compounds namely: hydrogen cyanide, aldehyde and cyano compounds were produced.

On other scientists repeating such experiments purines and pyrimidines were also produced, which can be proved as a step towards synthesis of a living cell.

In 1964 Sydney W. Fox from Miami University, Florida, concluded that proteins were synthesised from amino acids on the primitive earth, due to thermal energy. He could synthesise stable proteins through an experiment and named them as "Proteinoids", which are similar to coccoid bacteria found in ancient geological strata.

The amino acids were proven to be produced by condensation reaction by HCN, NH_3 and RCHO (Aldehyde) in aqueous medium, like

$$HCN + NH_3 = R.CHO \longrightarrow \underset{NH_2}{R.CH} - CN \xrightarrow[2H_2O]{} \underset{\text{Amino acid}}{RCH.COOH} + NH_3$$

Several workers have practically synthesised all biological molecules abiotically in the laboratory during the last three decades.

CAUSE OF PROTOBIONT

In spite of consensus on the abiotic synthesis of organic molecules, no agreement has been reached as to the manner in which these organic molecules organised themselves into living organisms. A primitive life form, "Protobiont" might have appeared in sea during 'primaeval' era, which was capable of showing so conspicuous of life. Three popular models of protobiont have been proposed. They are:

1) Coacervate model of Oparin,
2) Protenoid model of Sydney Fox, and
3) Lipid bilayer model of Richard Gold acre.

COACERVATE MODEL

Coacervate is a droplet spontaneously formed while polar water molecules are oriented around an electrically-charged molecule. Boundary of the water acts as a barrier between the interior of the coacervate and its surroundings. Oparin described them as diverse systems containing enzymes which could synthesise and hydrolyse starch and release maltose to the surroundings. A growing coacervate may fragment into smaller droplets. They are destructed if the medium PH becomes unfavourable.

PROTENOID MODEL

Sydney Fox in 1965 suggested that protenoids are protein-like molecules, which could be synthesised by polymerisation of amino acids at 130° to 160°c. Protenoids are spherical particles with a double-layered outer membrane like that of plasma membrane. These microspheres show weak catalytic property.

LIPID BILAYER MODEL

A bilipid layer model of protobiont was proposed by Robinson, who suggested that different kinds of lipids in aqueous medium may segregate to form double layers enclosed around a central space. As this selectively permeable bilipid layer resembles the plasma membrane, hence this model could be considered as a close possibility.

Thus the formation of a protobiont in abiotic conditions was a unique phenomenon and might have been a forerunner of a true living cell.

EUBIONT ORIGIN

Three requirements are to be fulfilled for a cellular organisation. They are:

i) a partially permeable boundary separating the cellular parts from the outside.

ii) a system involved with series of enzyme catalysed reactions that change food into energy, and

iii) a genetic system to preserve and transfer information of heredity to successive generations.

The protobiont expressed a few qualities of life, but did not contain a genetic system.

PROTEINS AND NUCLEIC ACIDS EVOLUTION

Proteins formation is a major part in the evolution of nucleic acids which led to the formation of revolutionary RNA, which has self-replicating property. RNA molecules have two important properties:

1) Transfer of information through nucleotide sequence of replication, and

2) a specially folded structure that decides its function in response to external and environmental stimulus.

Since RNA are unstable owing to their strands being open to degradation by enzymes, a more stable system with coding strands evolved in the form of DNA double helix. Thus DNA molecules became diversified by storing information in a variety of proteins.

It can also be surmised that during stabilisation of protein synthesis by RNA in the hot dilute soup on earth, a chance trapping of an RNA, along with its protein, into a bilipid layer might have generated a first living and self reproducing unit life – the eubiont.

ORIGIN OF PROKARYOTIC CELLS

If mycoplasma, the simplest of the existing living cells, is compared to the organisation of first cell eubiont, the entire process of evolution narrated appears to be correct. It is correct to assume that a chance collision of RNA, protein and phospholipid molecules spontaneously resulted in the first living primitive cell.

This primitive cell possesses all similarities like mycoplasma, i.e. cell size 0.3 μm; much smaller than lipid bilayervesicle generated experimentally, and nucleic acids of mycoplasma (RNA & DNA) can synthesise about 750 types of proteins, the minimum number required by cells to survive. Thus they bear close resemblance the first evolved cell. The primordial cell could have competed with others and survived by evolving its own metabolic regulating and trapping energy from surroundings. It might have also developed a simple method of reproduction by cell division.

ORIGIN OF EUKARYOTES

Due to the accumulation of oxygen in the environment of earth, several anaerobic cells adopted to aerobic life style, with three alternatives;

1) either they have to perish or
2) find a place where oxygen supply was very low or nil, or
3) switch over to aerobic life.

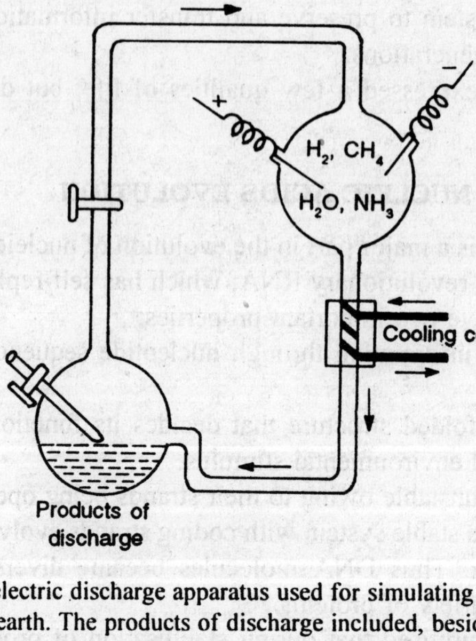

Fig. 2.1 Miller's electric discharge apparatus used for simulating conditions prevailing on primaeval earth. The products of discharge included, besides other organic molecules, a number of amino acids, the building blocks of proteins

Margulis proposed "Symbiotic theory of Eukaryotic Cell Evolution", and stated that eukaryotes evolved 1.5 million years ago. Basic differences expressed by eukaryotes in comparison with prokaryotes are:

1) presence of a rigid cell wall with mobile membranc is system.
2) presence of a distinct nucleus.
3) presence of a photosynthetic organelle.
4) presence of mitochondria.
5) presence of several organelles in the cytoplasmic medium.

SUMMARY

Life existing in this universe is the result of a long process of evolution.

Several assumptions, hypothesis were made to understand—how and when did life originate on the earth? In what form?

It is certain that a primitive form of life must have evolved through complex interaction of the forces of evolution.

The planet earth is believed to have originated 15 billion years ago and, if we immensely hot; on cooling sufficiently, neutrons and protons fused together by random collisions to form atomic nuclei. Further cooling took about 7×10^5 years.

Only hydrogen and helium were produced in the early phases of earth evolutionary history, followed by the formation of heavier elements after the stars were formed. After the total formation of earth, its atmosphere consisted of H_2, N_2, CH_4, NH_3, H_2S and water vapour without oxygen.

Several records are available which prove that the nature of atmosphere, and the hydrosphere in which these chemical reactions took place.

EXERCISE

1. What are coascervates? Were they precursors of life? Explain in detail.
2. Describe Miller's experiment to support Oparian Hypothesis.
3. Write in detail process and origin of life.
4. Explain in detail about evolution of proteins.
5. Write Shortnotes on:
   (a) RNA Model
   (b) General characters of proteins
   (c) Evolution of proteins

3

Broad Classification of Plant Cell and Animal Cell

All living cells have specialised roles to perform. Therefore, the diversity in biological form and function is a universal phenomenon. In animal and plant kingdoms several cell types are observed, each one having a specialised structure adapted to a specific function.

Animal cells are devoid of outer cell wall, capable of ingesting large particles of food and are often motile. Various tissues found in animal kingdom are composed of cells that cooperate among themselves.

There are several types of animal cells which are classified according to structure, chemical composition and function.

Epithelial cell: Found covering external and internal surfaces of the body.

- Epithelial cells found lining various spaces within the body and intestine are referred as *endothelial cells*. When they are arranged in single layers they are simple epithelial cells; when they are in several layers they are stratified epithelial cells. Their main function is protection, absorption, secretion and sensation.

Squamous epithelial cell: They are flat, thin, irregular in shape or (polygonal) compactly arranged without intercellular spaces forming a continuous sheet known as *pavement epithelial cell*. Each cell has a central bulging portion with thinner edges. It is uninucleated and the nucleus is placed exactly in the centre of the cell. These cells are found in alvioli of lungs, Boueman's capsule, skin, oesophagus.

Columnar epithelial cell: The cells are elongated at right angle to the basement membrane.

- They are tall and narrow column like
- The basal ends are generally narrow
- They are uninucleated, and the nucleus is found towards the basal end.

Ciliated epithelial cells: They are a specialised form of columnar epithelial cells.

- The free end of each cell bears numerous hair like short protoplasmic cilia capable of beating rapidly and rhythmically.
- Such type of cells are found in respiratory tract and genital duct.

Hyaline cartilage cell: They contain organic matrix called *chondrin*. These cells are covered by tough fibril membranes called *perichondrin* acting as connective tissue for the passage of nutritive substances. These cells are found in matrix along with lacunae. They are also known as chondrioblast lying in groups of not more than 8 in each lacunae found in larynx, trachea and at the end of bones.

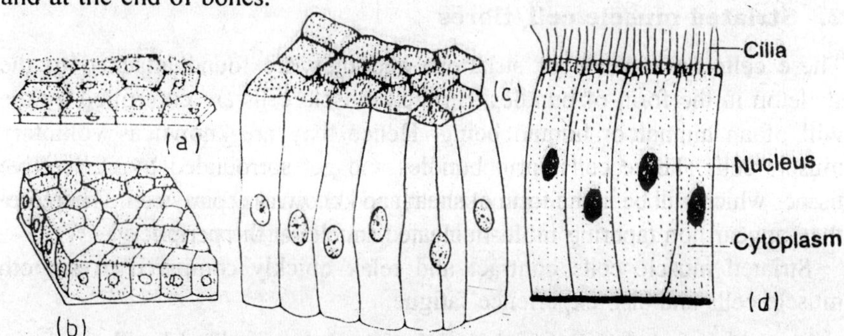

Fig. 3.1. Showing A. Squamous, B. Cuboidal, C. Columnar and D. Citiated epithelial tissues.

Bone cells: They are rigid and harder than cartilage cells. Bone cells are known as osteoplasts and surrounded by Osten secreted by themselves and delicated fibres membrane which is periostium. Oslin is impregnated with calcium salts and calcium phosphate which give hardness to bone, ex. Mammalian bone.

Muscular cells

Muscle cells are devoid of intercellular spaces and are also known as matrix.

- They are elongated and therefore known as muscle fibres.
- They do not divide but increase in size due to growth.
- The protoplasm of these cells are known as sarcoplasm and the membrane is known as sarcolemma.
- They contain contractile elements known as myofibrils or myofilaments or fibrils.
- The filament and fibrils contain protein called as actin and myosin, which are responsible for contraction and relaxation.
- There are 3 types of muscle cells classified according to structure and functional basis.

1. Smooth muscle cells

They are elongated fusi form or spindle shaped having central wider region and tapering ends. Nucleus lies in the centre, but they lack true sarcolemma. They are arranged in sheets around hollow along with connective tissues.

- Muscle cells lack striations or bands. Therefore, they are known as nonstriated muscle cells.
- The smooth muscle cells receive autonomous nerves and, hence, they are involuntary.
- They are capable of prolonged contraindication and are found in stomach, intestine and urinary bladder.

2. Striated muscle cell/fibres

These cells have bands or striations and they are found attached to the skeleton in the form of bundles. Striated muscle cells are controlled by the will of an animal or human being. Hence they are known as voluntary muscle cells. These cells form bundles and get surrounded by connective tissues which will be in the form of sheath and known as epimysium. Therefore they appear non tapering multi-nucleated (nuclei at the periphery).

Striated muscle cells contract and relax quickly compared to smooth muscle cells and also experience fatigue.

3. Cardiac muscle cells

They are involuntary muscle cells:

- They are striped muscle cells present around the heart.
- Each unit lying between the disc has a centrally located nucleus.
- They are united by cross connections to form a contractile network.
- These cells are supplied by autonomous nerves. Therefore, they show rhythmic contraction and relaxation and never experience fatigue.

BLOOD

It is a liquid connective tissue (cells) which has a large number of cells present in inter cellular matrix called plasma. Blood consists of 3 types of cells:

(i) Erythrocytes.
(ii) Leucocytes
(iii) Thrombocytes suspended in all plasma.

It is slightly alkaline having scarlet red colour. The blood plasma contains 90% of water, 1% salts and 9% organic compounds.

Erythrocytes: These are red corpuscles which are biconcave and non nucleated. They consist of stroma mesh network and haemoglobin which imparts red colour and is also an oxygen carrier.

Leucocytes: These are the white blood corpuscles. They are nucleated colourless amoeboid cells and are classified into 3 groups:

1. Agranular
2. Granular
3. Thrombocytes

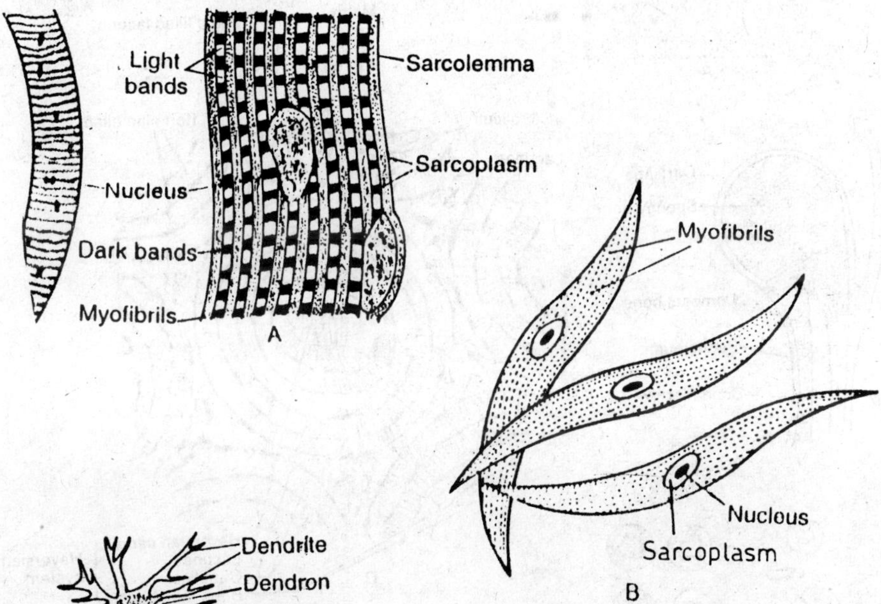

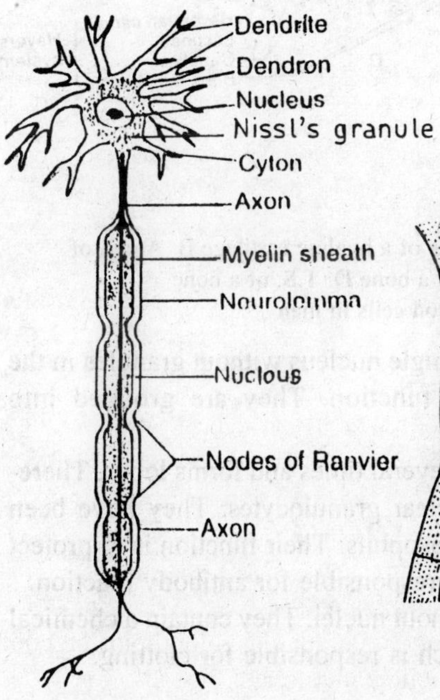

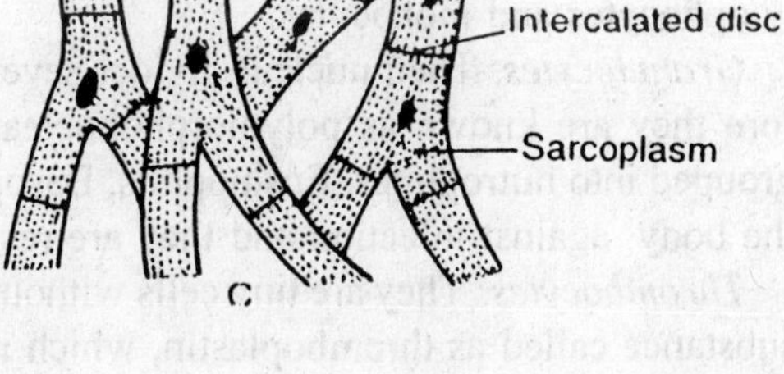

: Detailed structure of a neuron

Fig. 3.2. Muscular tissues, A. Striated muscles
B. Unstriated muscles C. Cardiac muscles

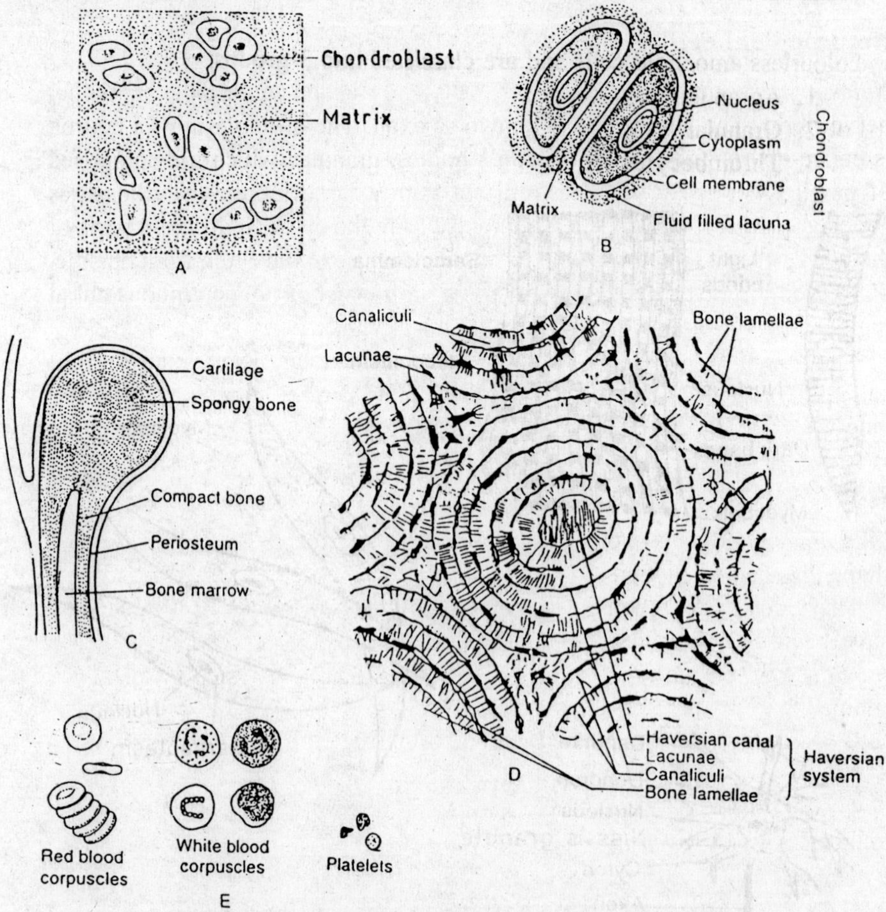

Fig. 3.3. Connective tissues. A. Structure of a hyaline cartilage B. A cell of a cartilage magnified C. L.S. of a bone D. T.S. of a bone E. Three types of blood cells in man

Agranular Leucocytes: They have a single nucleus without granules in the cytoplasm based on their origin and function. They are grouped into lymphocytes and monocytes.

Granulocytes: Their nucleus divides several times and forms lobes. Therefore they are known as polymorphonuclear granulocytes. They have been grouped into nutrophiles, Eosinophils, Basophils. Their function is to protect the body against infection and they are responsible for antibody reaction.

Thrombocytes: They are tiny cells without nuclei. They contain a chemical substance called as thromboplastin, which is responsible for clotting.

Nerve cell

There are two types of nerve cells:

1. Neurological cell
2. Neuron

Neurological cell

Neuron: Each neuron consists of a cell body called cyton and branching nerve fibres. The largest one is known as axon. The protoplasm of a neuron is called neuroplasm. It has Nissile's bodies, granules which are composed of parallel flat sacks of Ergastroplasm with a large number of ribosomes attached to it. Axon carries nerve impulse from the cyton to different parts of the body, whereas dendron conducts nerve impulse to the cyton from outside.

Nerve fibre is surrounded by myelin sheath which is not continuous but at regular intervals comes in contact with axis cylinder.

The region of contact is known as Node of Ranvier. Their function is to carry impulses from the receptor organs to the central nervous system and vice versa.

Plant cell

Like animal cells, plant cells are also diversified. They vary in their origin, shape size and arrangement.

These cells are assembled from various types of tissues. Basically they have been classified into Meristem cells and permanent cells, taking into consideration the stages of development. The dividing cells which have not attained maturity are known as meristem cells. On the other hand, permanent cells would have lost the capacity of cell division and attained matured form.

Meristem cells: These cells are closely packed without intercellular spaces.

The shape of the meristematic cells vary. They may be oval, polygonal, rounded or rectangular in shape.

The cell wall is very thin and cytoplasm is dense without much of reserved food materials. The cell inclusions will be in their precursor stages like – The nuclei are very prominent; plane of division varies, i.e.

(i) Periclinal

(ii) Anticlinal (⊥r)

Classification of Meristem

They have been classified based on development, origin and location.

Meristems are classified into (i) Promeristem and (ii) Marginal meristem cells

Promeristem cells: They constitute new growth cells; new organs are initiated by them. Therefore, they are known as embryonic meristem cells. They are also known as Primordial meristem cells or ur meristem cells. These cells have very thin cell wall, prominent nucleus and non-vacuolated cytoplasm.

Marginal meristem cells: They are found in cortex and epidermis. Based on plane of division they have been classified into mass meristem cell, plate meristem cell and rib meristem cell.

Mass meristem cells: Divide in all planes and result in increasing mass. Ex. sporangia, endosperm and embryo.

Plate meristem cells: They divide in two planes longitudinally and vertically. Ex. leaves.

Rib meristem cells: Divide only in single plane and produce rows or columns of cell thereby length of the organs. Ex. young roots, pith, cortex of young stems.

Based upon their origin: They have been classified into primary and secondary meristem cells.

Primary: They are present from embryonic stages and build the fundamental parts of promeristem cells found in apexes of the plant organ.

Secondary meristem cells: Arise later and originate in permanent tissues and they may be lateral in position. Permanent cells sometimes start dividing and this type of division is known as secondary division meristem cells are found in the cork cambium.

Based upon their location: They are of 3 types:

(i) Apical
(ii) Intercalary
(iii) Lateral

Apical: These are located at the apexes of the root-shoot and other organs. They initiate plant and plant organs growth in length.

It can be initiated by one or more cells.

Apical meristem cells consists of two zones:

(i) Promeristem cells
(ii) Meristem cells

Promeristem cells are dividing cells. They are also known as apical initials.

Meristematic cells: They are of 3 types:

(i) Protoderm
(ii) Proto cambium
(iii) Ground meristem cells

Intercalary meristematic cells: They are the detached portions of the apical meristem cells, separated due to growth of an organ. They are always internodal. In the early stages of development, the whole or part of the internode is meristematic. Subsequently, a portion of it matures and, as a result, patches of the meristem are found embedded in the permanent tissues like leaf base, branch base or internode. Ex. grasses equisetum pinus.

Lateral meristem cells: Present along the sides of the stem. They divide only in one plane and add to the girth of the tissues. Ex. vascular cambium. Vascular cambium produces secondary xylem and phloem brings about secondary growth.

Permanent cells: They are classified into simple and complex cells based upon their composition.

A simple cell is homogeneous and a complex cell is heterogeneous.

Simple cells are classified into three types:

(i) Parenchyma cells
(ii) Collenchyma
(iii) Schlerenchyma

Parenchyma: They are living cells variable in morphology and physiology. They are fundamental. Ontogenetically, they are precursor of other cells. Phylogenetically, they are ancient types found in primitive organisms. But they are for photosynthesis, respiration, assimilation, storage and secretion.

Ground meristem cells: They give rise to parenchyma but in the vascular tissues, parenchyma are formed by the procambium and vascular cambium. These arise from phellogen and phelloderm.

Distribution: They are found in all parts of plants like pith, cortex of the root, stems, mesophylls of leaves, flesh of roots and the endosperms of the seeds.

Cell shape

Based on the function Meristem are classified as:

(i) Protoderm
(ii) Protocambium
(iii) Ground meristem cell

Protoderm develop into epidermic cells.

Protocambium give rise to primary vascular cells.

Ground meristem cell develops into the pith and the cortex. They are thin and isodiametric and give rise to endodermis, hypodermis and pericycle.

Cell shape of permanent cells: They are spherical, polygonal, elongated like column in leaves. Some are polyhedron which may be closely packed or loosely arranged along with the inter-cellular spaces.

Cell structure: They have thin walls and dense cytoplasm with prominent nucleus. Cell becomes prominent when they store abundant proteins, fats and carbohydrates. Parenchyma of the leaves have chloroplast for photosynthesis.

Cell type

Based upon their structural function they have been classified as—

1. Simple parenchyma found in the ground cells.
2. Chlorenchyma found in the photosynthetic cells of the leaves which are of two types
   (i) Pallisade parenchyma 2. Spongy are columnar, filled with chloroplasts and the spongy parenchyma have chloroplasts rich in starch. They are also know erophytic-store water.
   (ii) Aerenchyma-parenchyma with large inter-cellular spaces string air present in hydrophes help in hydrophytes raising in buoyancy.

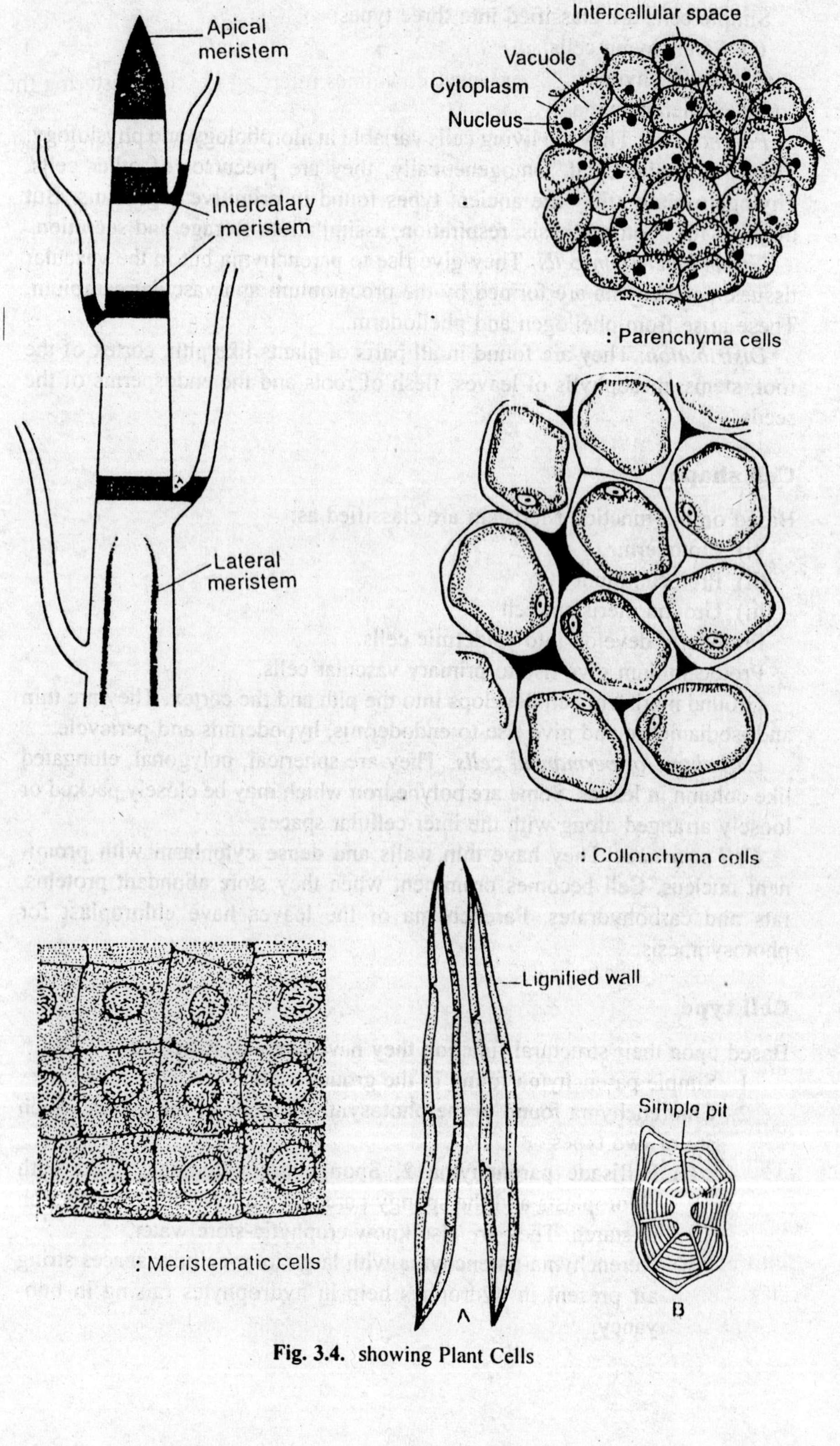

Fig. 3.4. showing Plant Cells

(iii) Collenchyma are living cells. Elongated with thick, primary, non-lignified walls.

Protoplasts are living in Greek colamines referring to thick glistering the walls.

Origin: According to Mojumdar 1941 Chollenchyma originated with vascular tissues from the procambium.

Distribution: They occur in the periphery of the stems, leaves and the floral parts. This may occur in the root cortex only if exposed to light. It is hypodermal situated immediately below the epidermis and separated by parenchyma. These types of cells are usually present in dicotyledons.

Cell shape: They are elongated cells which resembles fibres. The shorter are prismatic and appear polygonal in cross section and appear rectangular in longitudinal section.

Cell Structure and function. Cell wall is the most important feature of chollenchyma. Thickenings are deposited unevenly at the corners where several cells join together depending upon the direction of the cut. Sections of chollenchyma show thin and thick walls made up of cellulose and pectin along with water. Ultra structure show alternations of layer, longitudinally oriented microfibrils and noncellulosic materials.

They have living protoplasts and chloroplasts. The latter are lesser in number in long and narrow cells and more in parenchyma forms.

It is a mechanical cell adopted for support to growing organs with thick walls and close packing make the cells strong, but at the same time of growth, they become flexible and plastic.

Schlerenchyma

They are thick walled cells. Their main function is mechanical. Schlerous means hard enchyma in fusion. Haberland and Foster combined chollenchyma and Schlerenchyma and named it as Steriones which includes the entire mechanical system of the plant. These cells withstand various stresses and strain which may result from stretching, bending, weight and pressure. A mature cell is devoid of protoplasts. Cell wall is thick with uniform deposition of lignin. They are of different forms structure origin and development. But they are classified into fibres and schlerids. Fibres are long cells. Schlerids are short cells.

SUMMARY

There are several types of plants and animal cells which are diversified according to structure, chemical composition and function. They vary in their origin, shape, sizes and arrangement. Plant cells are basically classified into meristem cells and permanent cells, taking into consideration of their stages of development.

Animal cells are basically classified into epithelial cells, muscular cells, nerve cells and blood, while epithelial, muscular and nerve cells are solid tissues, blood is liquid connective tissue possessing a large number of cells present in cellular matrix called plasma.

EXERCISE

1. Write in detail structure and functions including their location of Epithelial Cells.
2. Write in detail about various types of muscular cells and their structure and functions.
3. What do you know about blood? Explain in detail.
4. Classify plant meristem cells and write about their structure, location and functions.
5. What are the permanent cells in plants, describe their structure and functions.
6. Write shortnotes on:
   (a) Bone Cell
   (b) Neuron
   (c) Blood
   (d) Collenchyma
   (e) Sclerenchyma

4

Biochemical Composition of the Cells

By detailed examination of Biochemical composition of a cell, we can know that a living cell is composed of large size complex molecules, small size molecules, inorganic ions and water. As chemical reactions proceed always in aqueous medium, water is most abundant in living cells in the form of individual molecules. Another 5% is in bound form, in association with other molecules. The relatively small amount of water contains chemical substances of different types in soluble or insoluble form. Some chemical substances are present in colloidal suspension. Besides water the living cells possess molecular species of compounds which are involved in various metabolic activities like production, utilisation and storage of energy. Some other classes of compounds are also present, which are involved in storage and transmission of hereditary information, catalytic f :nctions, structure building, repairs and other functions. So the structure and importance of these biochemical molecules have to be understood.

The study of chemical compounds in living cells, systems and the branch of this science is known as Biochemistry. Studies of structure and behaviour of individual molecule is known as Molecular Biology, by which one can study the secret of life (Robert 1986).

The cells of organisms or protoplasm consists of nonliving constituents, Proteins, nucleic acids, fats, carbohydrates, vitamins, etc. The protoplasm is alike because of highly complex organisation of these nonliving substances, which may interact with one another. One can understand in detail the phenomena of life by studying the functions of protoplasm and its constituents.

Many chemical elements are present in the form of atoms, ions and molecules. Among the 46 elements present in the cell, 24 are considered essential for life. But with others, the cell or organism interacts. Among 24 essential elements only 6 play a very important role. They are Carbon,

Hydrogen, Nitrogen, Oxygen, Phosphorus and Sulphur, by which most organic molecules are built. The next 5 essential elements are Calcium, Potassium, Sodium, Chlorine, Magnesium and other elements are called trace elements which are essential for life in minutest quantities, viz., Ferrous, Iodine, Copper, Zinc, Manganese, Selenium, fluorine, Chromium, Silicon, Nickel, Boron, ions.

Ions are important to maintain osmotic pressure. The concentration vary from inside to outside. For example, K and Mg. concentration is high inside the cell; Na and Cl. is high outside the cell. One can find free Ca ions in the blood stream. The ions act as buffer system, and tend to stabilize ph of the blood and cellular fluids. Amino acids are also considered as ions.

Chemical compounds are divided into two groups:

1) Organic and
2) Inorganic.

The most abundant inorganic component is water, which is free or bound and depends on age, habitat and metabolic activities.

MOLECULAR STRUCTURE OF WATER

Water (H_2O) is formed by covalent bond in which atoms are sharing paired electrons. H_2O acts as a solvent by maintaining temperature, transparency and surface tension.

MACROMOLECULES

To describe chemical compounds of relatively large size and large molecular weights, Schrodinger, in 1920, coined the term "Macromolecules". For example, 1, 4 isoprene could be condensed and produce a linear structure polyisoprene which possesses repeated sequence of isoprene units. The repeated unit is termed a monomer and the condensation product or the polymer is called macromolecule. Biologically, significant macromolecules are mostly linear polymers, but branched or cross-linked polymers also exist with a variety of functions. Biological polymers may be homopolymers (composed of identical monomers) or heteropolymers (composed of different monomers).

PROPERTIES OF POLYMERS

1) Many types of macromolecules show elasticity and can undergo deformation. Ex. cell membranes and muscle tissues.
2) They have an ability to provide a definite structure and form even when they are dissolved in a solvent.
3) Macromolecules cannot cross the membrane barrier. Hence they do not get lost to the environment, as they are large in size. Macromolecules include 3 types of compounds viz., carbohydrates, nucleic acid and proteins. Added to these compounds, there are

molecules which can be categorised as lipids. Chemical nature and functions of these macromolecules are discussed hereunder:

CARBOHYDRATES

Carbohydrate compounds contain the elements carbon, hydrogen and oxygen with the carbon atoms linked to one another to form chains of varying lengths. Carbohydrates include sugars and starches which are primary source of chemical energy by every cell. In addition, carbohydrates serve a structural role as components of RNA and DNA.

Carbohydrates are divided into three types which are characterised by the length of carbon chains, viz.,

1) Monosaccharides (simple sugars)
2) Disaccharides (double sugars)
3) Polysaccharides (complex sugars)

MONOSACCHARIDES

These simple sugars have short carbon chains. Most important, simpler sugar is glucose. It is a 6 carbon sugar with the formula $C_6H_{12}O_6$. Because it has six carbon atoms it is called a hexose. Glucose is present in dry state as a straight chain but forms a cyclic compound when dissolved.

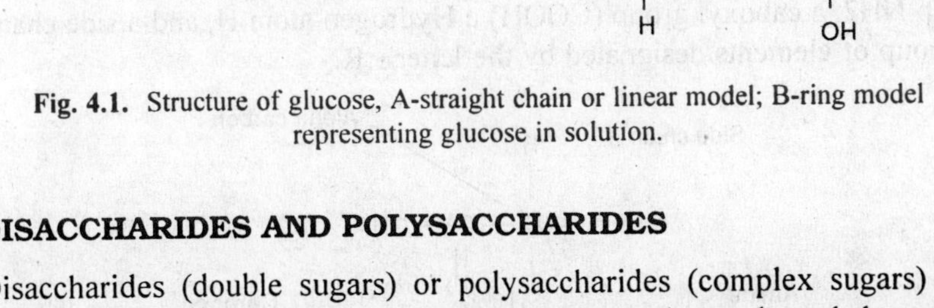

Fig. 4.1. Structure of glucose, A-straight chain or linear model; B-ring model representing glucose in solution.

DISACCHARIDES AND POLYSACCHARIDES

Disaccharides (double sugars) or polysaccharides (complex sugars) are carbohydrates composed of 2 or more simple sugars that are bounded together through a synthesis reaction that involves the removal of water. Sucrose (table sugar) maltose, and lactose are disaccharides. Each consists of two monosaccharides linked together.

Polysaccharides consist of many monosaccharides chemically joined to form straight or branched chains. Once again, water is removed as the many monosaccharide subunits are joined. Any large molecule made up of many identical small molecules is called a polymer. Polysaccharides are polymers of monosaccharides. Glycogen is referred to as an animal starch. It is the main polysaccharide and has an estimated molecular weight of several millions.

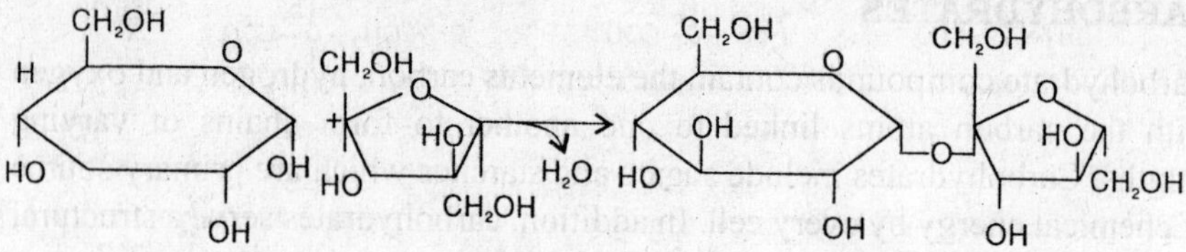

Fig. 4.2. Formation of sucrose, glucose and fructose joined in a synthesis reaction that involves the removal of water.

PROTEINS

Proteins are characterised by the presence of four elements, i.e. carbon, oxygen, hydrogen and nitrogen. More specialised proteins also contain sulphur, iron and phosphorus. Protein molecules are giant sized, i.e. they are macromolecules.

Protein molecules, regardless of size, have a similar basic structure. They are chain-like polymers composed of multiple subunits or building blocks linked end to end. The building blocks of proteins are called amino acids.

AMINO ACIDS

Elements that make up a protein molecule are bounded together to form chemical units called amino acids. Proteins are composed of 20 commonly occurring amino acids, and are present in almost every protein. Among these 20, eight are considered as essential amino acids. Structural formula of amino acid consists of a carbon atom, C, to which are bounded an amino group NH2, a caboxyl group (COOH) a Hydrogen atom H, and a side chain or group of elements designated by the letter - R.

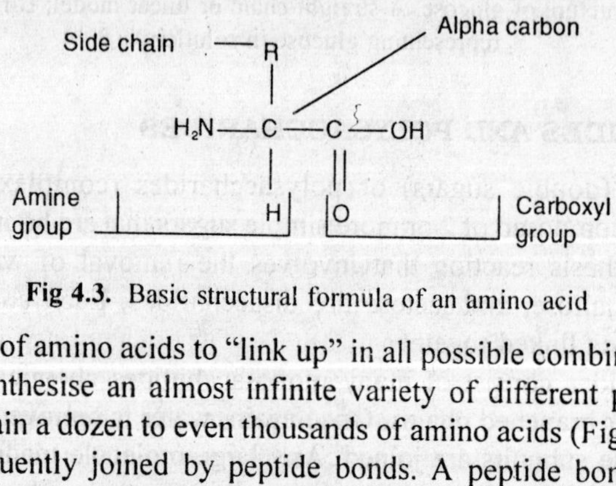

Fig 4.3. Basic structural formula of an amino acid

The ability of amino acids to "link up" in all possible combinations allows the cell to synthesise an almost infinite variety of different protein chains that may contain a dozen to even thousands of amino acids (Fig. 4.4). Amino acids are frequently joined by peptide bonds. A peptide bond is one that binds the carboxyl group of an amino acid to the amino group of another

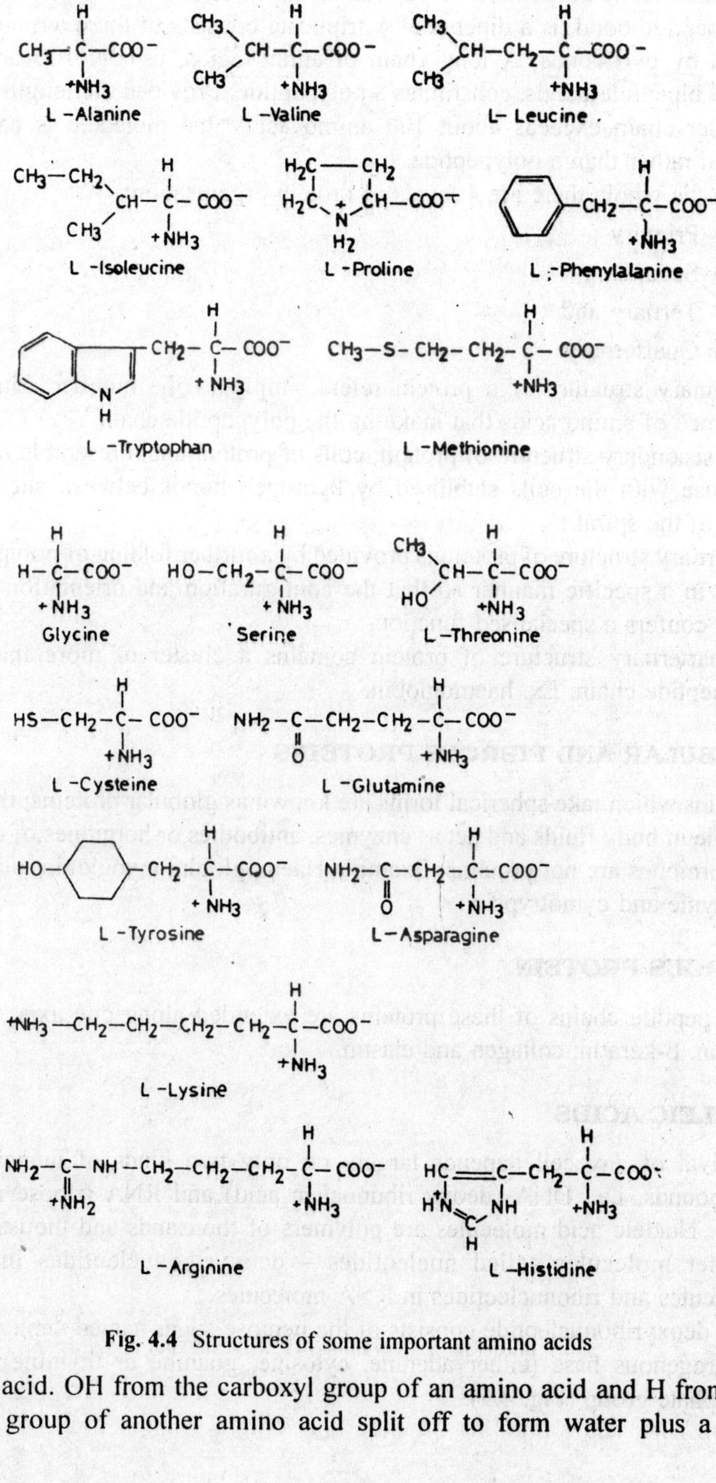

Fig. 4.4 Structures of some important amino acids

amino acid. OH from the carboxyl group of an amino acid and H from the amino group of another amino acid split off to form water plus a new compound called a peptide. A peptide made up of two amino acids, linked by a peptide bond, is a dipeptide. A triplicate consists of three amino acids linked by two bonds. A long chain of amino acids, usually 100 or more linked bipeptide bonds, constitutes a polypeptide. Provided the length of the polymer chain exceeds about 100 amino acids, the molecule is called a protein rather than a polypeptide.

On the whole there are 4 levels of protein organisation:

(i) Primary
(ii) Secondary
(iii) Tertiary and
(iv) Quarternary.

Primary structure of a protein refers simply to the number, kind and sequence of amino acids that make up the polypeptide chain.

In secondary structure of protein, coils of protein chain resemble a spiral staircase with the coils stabilised by hydrogen bonds between successive turns of the spiral.

Tertiary structure of protein is provided by a further folding of polypeptide chain in a specific manner so that the configuration and orientation of the chain confers a specialised function.

Quarternary structure of protein contains a cluster of more than one polypeptide chain. Ex. haemoglobin.

GLOBULAR AND FIBROUS PROTEINS

Proteins which take spherical forms are known as globular proteins; they are soluble in body fluids and act as enzymes, antibodies or hormones;of course, all hormones are not proteins. Example Haemoglobin, myoglobin, olbumin, lysozyme and cymotrypsin.

FIBROUS PROTEIN

Poly peptide chains of these proteins are extended along one axis. Ex. α-keratin, β-keratin, collagen and elastin.

NUCLEIC ACIDS

Survival of any cell depends largely on only two kinds of nucleic acid compounds, i.e., DNA (deoxy ribonucleic acid) and RNA (ribose nucleic acid). Nucleic acid molecules are polymers of thousands and thousands of smaller molecules called nucleotides – deoxyribonucleotides in DNA molecules and ribonucleotides in RNA molecules.

A deoxyribonucleotide consists of the pentose sugar named deoxyribose, a nitrogenous base (either adenine, cytosine, guanine or thymine) and a phosphate group (Fig. 4.5).

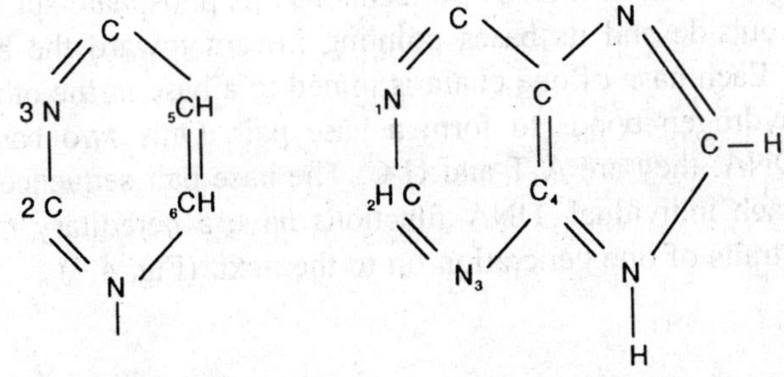

Fig. 4.5. DNA strands twined around a central axis

Ribonucleotides are similar but contain ribose instead of deoxyribose and uracil instead of thymine. Two of the bases in a deoxyribonucleotide, adenine and guanine, are called purine bases because they are derived from purine. Purines have double ring structure. Cytosine and thymine are derived from pyrimidine, so they are known as pyrimidines bases: Pyrimidines have a single ring structure (Fig. 4.5).

DNA molecules are the largest in a cell; they are very large polymers composed of many nucleotides. Two long polynucleotide chains compose a single DNA molecule. The chains coil around each other to form a double helix. A helix is spiral shaped, similar to the shape of a wire in a spring (Fig. 4.6).

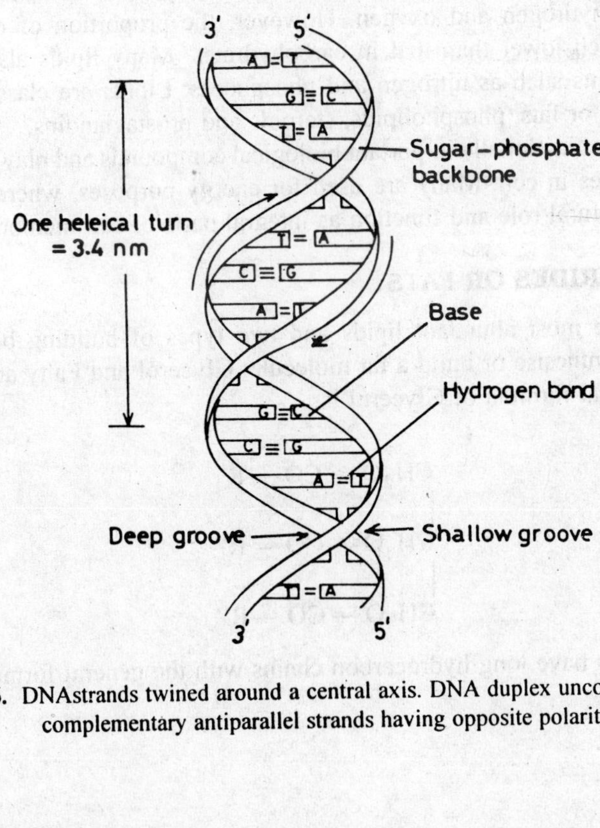

Fig. 4.6. DNAstrands twined around a central axis. DNA duplex uncoiled showing complementary antiparallel strands having opposite polarity

Each helical chain of a DNA molecule has its phosphate-sugar backbone toward the outside and its bases pointing inward toward the bases of the other chain. Each base of one chain is joined to a base in the other chain by means of hydrogen bonds to form a base pair. Only two base pairs are present in DNA; they are A-T and G-C. The base pair sequence in DNA is unique to each individual. DNA functions have a hereditary molecule by passing the traits of one generation on to the next. (Fig. 4.7)

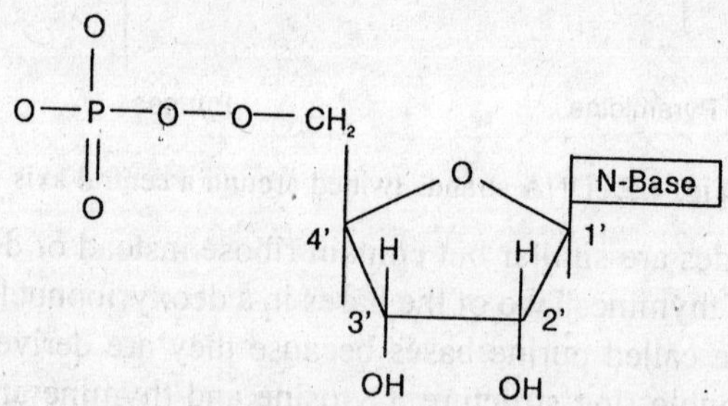

Fig. 4.7. Pentose in RNA is ribose while in DNA it is a 2' deoxyribose. The PO4 group is esterified usually at 5' position. The nitrogenous bases are linked to the 1' carbon.

LIPIDS

Lipids are water insoluble organic biomolecules. Lipids are composed largely of carbon, hydrogen and oxygen. However, the proportion of oxygen in lipids is much lower than that in carbohydrates. Many lipids also contain other elements such as nitrogen and phosphorus. Lipids are classified into triglycerides or fats, phospholipids, steroids and prostaglandins.

Lipids are functionally important biological compounds and play a number of major roles in cell. Many are used for energy purposes, whereas others serve a structural role and function as integral parts of cell membranes.

TRIGLYCERIDES OR FATS

They are the most abundant lipids and two types of building blocks are needed to synthesise or build a fat molecule: Glycerol and Fatty acids.

The general formula of Glycerol is:

$$\begin{array}{l} CH_2O-CO-R' \\ | \\ CH_2O-CO-R'' \\ | \\ CH_2O-CO-R'' \end{array}$$

Fatty acids have long hydrocarbon chains with the general formula—

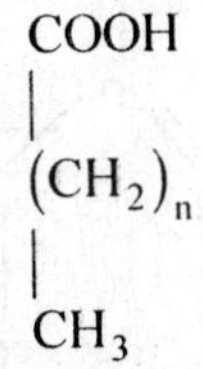

Most of the fatty acids which occur in living organisms an have even number of carbon atoms since they are synthesised by joining acetyl fragments. These are saturated fatty acids such as 16 carbon palmitic acid, and 18 carbon stearic acid.

All lipids do not act as energy reserves, though some are important structural constituents of cell membrane. This category of lipid is known as phospholipids: they are triglycerides in which phosphoric acid and nitrogenous compounds like choline, ethanolamine, inositol or serine replace one of the fatty acids. They are usually called derivatives of phosphotidic acid (Fig. 4.10). They occur both in plants and animals and some of them are specific to brain tissue (cephalins).

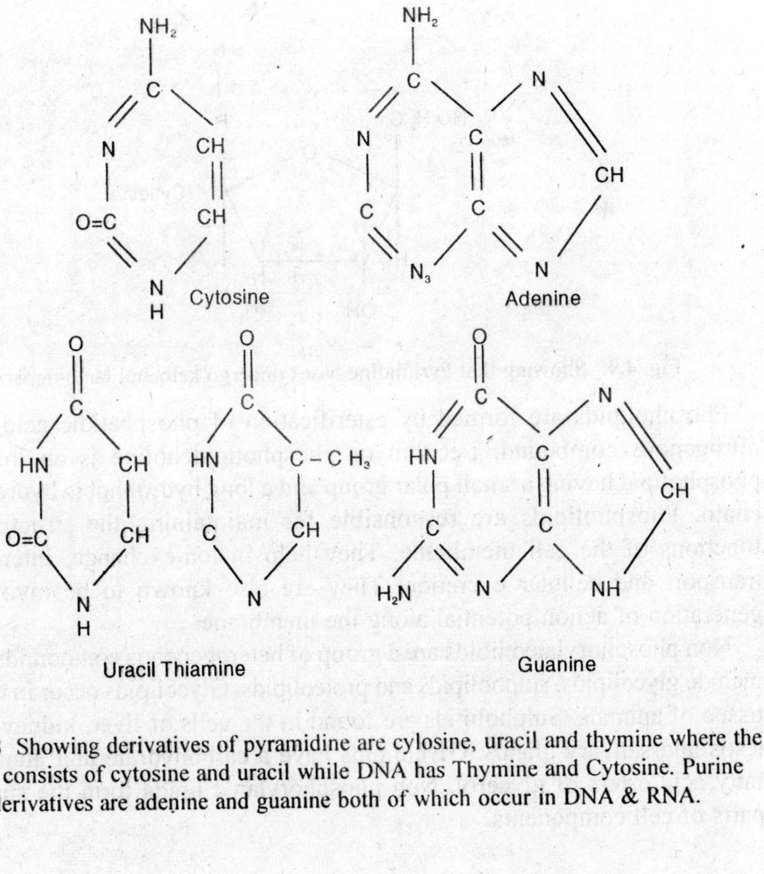

Fig. 4.8 Showing derivatives of pyramidine are cylosine, uracil and thymine where the RNA consists of cytosine and uracil while DNA has Thymine and Cytosine. Purine derivatives are adenine and guanine both of which occur in DNA & RNA.

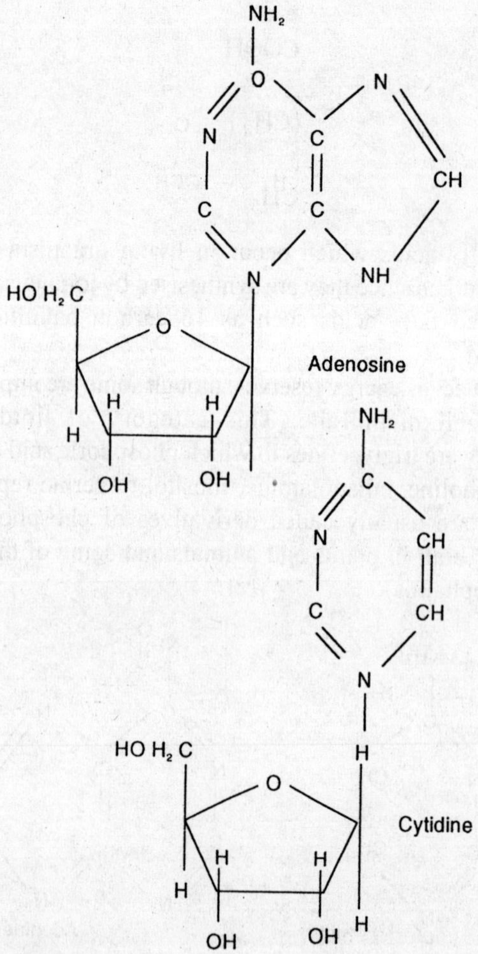

Fig. 4.9. Showing that Pyrimidine bases undergo ketoenol tantomerism

Phospholipids are formed by esterification of phosphatidic acid with a nitrogenous compound. Lecithin or phosphotidylcholine is an important phospholipid having a small polar group and a long hydrobhobic hydrocarbon chain. Phospholipids are responsible for maintaining the structure and functions of the cell membrane. They help in ion-exchange, intercellular transport and cellular excretion. They are also known to be involved in generation of action potential along the membrane.

Non phosphorylated lipids are a group of heterogeneous compounds which include glycolipids, sulpholipids and proteolipids. Glycolipids occur in nervous tissue of animals. Sulpholipids are found in the cells of liver, kidney, brain, testis and salivary glands. Glycolipids have a carbohydrate unit attached to fatty acid esters of glycerol. Non phosphorylated lipids form the structural parts of cell components.

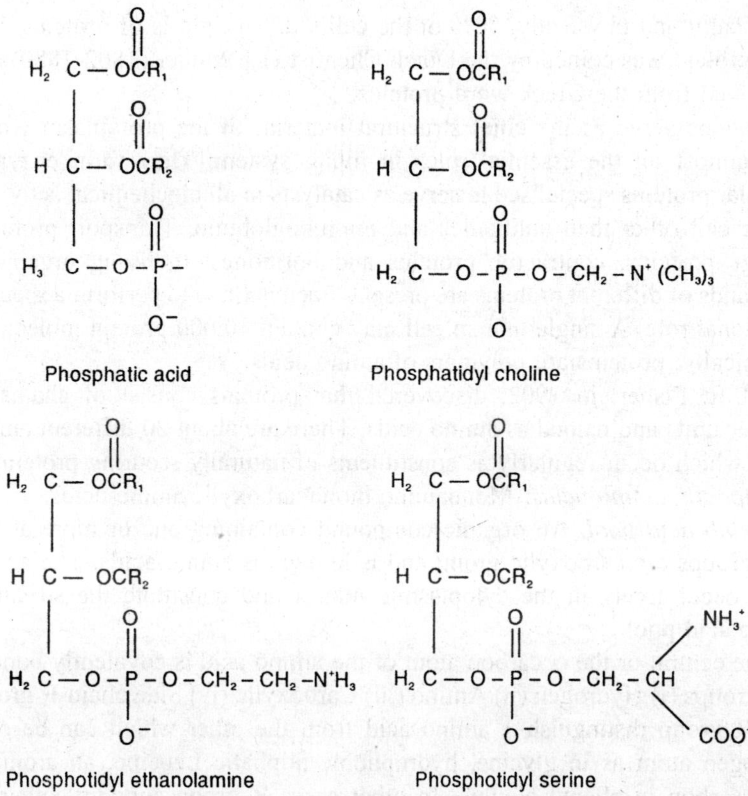

Fig. 4.10. Structures of some Phospholipids

Sphingolipids are associated with myelin sheath of nerve cells. Sphingolipids are absent in bacteria, but present in yeast, animal cell and plant tissues. Sphingolipids structural unit is either a ceramide or a sphingosine unit.

VITAMINS

Vitamins are organic compounds of diverse chemical nature required in small amounts for growth and reproduction. Vitamins A, B, C, E, F play an important role in cellular metabolism.

HORMONES

Hormones are organic complex compounds occurring in traces and regulate synthesis of mRNA, enzymes and physiological activities. Ex. Insulin, oestrogen, thyroxine and adrenaline.

Biochemical role of amino acid pool

Proteins are the most important macromolecules found in the cell both chemically and physically. 50% of the cell's dry weight is of proteins. The term protein was coined by the Dutch Chemist G.J. Muller (1802-1880) and is derived from the Greek word proteios.

Proteins serve as the chief structural material of the protoplasm which play almost all the essential roles in living system. They form enzymes globular proteins specialised to serve as catalysts in all biochemical activities of the cell other than antibodies and immunoglobulin, Transport proteins, storage proteins, contractile proteins and hormones. In living organisms thousands of different proteins are present. Each is fitted to perform a specific functional role. A single human cell may contain 10,000 protein molecules. Chemically, proteins are polymers of amino acids.

N.L.E. Fisher, in 1902, discovered that proteins consist of chains of smaller units and named as amino acids. There are about 20 different amino acids which occur regularly as constituents of naturally securing proteins.

Aliphatic amino acids: Monoamino monocarboxylic amino acids.

Amino acid pool: An organic compound containing one or more amino acid groups are carboxylic group and is known as amino acids. The amino acids occur freely in the cytoplasmic matrix and constitute the so called amino acid pool.

The central or the α carbon atom of the amino acid is covalently bonded to 4 groups (i) Hydrogen (ii) Amino (iii) Carboxylic (iv) Side chain-R group. The R group distinguish 1 amino acid from the other which can be only hydrogen atom as in glycine, hydrophobic aliphatic Leucine, an aromatic hydrocarbon in phenyl alanine. In other cases R group contains either an extra carboxylic group or extra amino group. Histidine contains N group Serine and Tyrosine have hydroxyl group in their side chain. Cystine possess thiol which is (SH) group.

Amino acids contain both basic and acidic groups and show amphoteric character. Such molecules unite with one another to form protein by losing H_2O. Such type of union is called Peptide linkage or Peptide bond. When two amino acids combine it is known as dipeptide bond. When three, it is known as tripeptide and when more than 3 it is known as poly or oligo. The various molecules unite to form peptones, proteases and proteins. The size, shape function of the protein is determine by the number of amino acids in the molecule or the type of amino acid or the distribution of amino acid in the molecule. The smallest molecular size is seen in ACT.H is 39 amino acid with a mol. weight of 4500 and the largest protein molecule size is seen in haemocyanin (invertebrate blood pigment) containing 8200 amino acids with 900,000 m.wt.

The ATP consists of purine base, adenine and pentose sugar ribose and three molecules of phosphoric acid (Fig. 4.11).

The adenine and ribose sugar collectively constitute the nucleotide which is having one or two or three phosphate groups. They form adenine, mono di or tri phosphate respectively. When ADP is linked with the last phosphate group, it is known as special energy rich bond because when the phosphate group of ATP is released, a large amount of energy 7300 cal is released. A common chemical bond releases only 300 cal of energy.

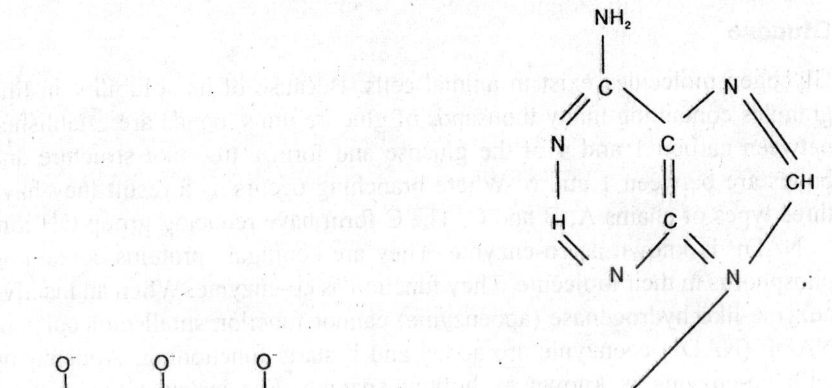

Fig. 4.11 Showing structure of adenosine triphosphate (ATP)

The chemical reaction which synthesises the energy rich bond also needs great amount of energy supplied by oxidation of food stuffs in the mitochondria and stored in the limited spaces of the cell but this is disposed of very quickly at the time of need in various cellular functions such as respiratory cycle, protein synthesis transmission, cell division and transportation.

NADP. Before proteins introduction to metabolism they split into amino acids which is accomplished by protease enzyme. Each peptide bond should be served with H_2O molecule, a hydrolytic reaction. Nitrogen is removed from the amino acid by oxydative deamination or trans amination and form NH_3. The remaining amino acids enter the metabolic stream in the form of ketoacids. Water is required for the process and two hydrogen are removed from Co enzyme NAD. NH_3 form is removed by converting into urea molecules by the liver cells before reaching to the kidneys through the blood. The energy derived from oxydative deamination depend upon the amino acid involved.

Glucose

Glycogen molecules exist in animal cells. Because of its solubility in tiny granules containing many thousands of glucose units, bonds are established between carbon 1 and 4 of the glucose and form a tree like structure and bonds are between 1 and 6. Where branching occurs as a result they have three types of chains A, B and C. The C form have reducing group OH ion.

NADP is known as co-enzyme. They are conjugate proteins containing phosphorus in their molecule. They function as co-enzyme. When an inactive enzyme like hydrogenase (apoenzyme) cannot function small molecules of NADP (NAD^+) coenzyme are added and it starts functioning. Apoenzyme with co-enzyme is known as hollow enzyme. For instance the enzyme hydrogenase is an apoenzyme which can function either with the coenzyme NADP+ or with NADP.

Lipids: are the organic compounds insoluble in H_2O and soluble in inorganic compounds like ether, acetal, benzene, etc. This character is because of the long chain of aliphatic hydrocarbon or the benzene ring in the molecules (non polar or hydrophobic), cooking oil, butter, ghee, wax, natural rubber, cholesterol, etc. The two major roles of lipids in cells (i) constitute the components of membranous organelles, plant pigment, keratins of carrots, lycopenes of tomatoes, Vit A, E and K, methol and Eucalyptus oil, and (ii) are stored as reserved energy source like fatty acids to be used when occasion occurs.

Lipids are made up of carbon, hydrogen and oxygen which is very small in number. Sometimes phosphorus, sulphur, nitrogen may also be present. Fats and oils are compounds of glycerol and fatty acids formed by the reaction of organic acids with alcohol. Glycerol is the only one kind $C_3H_8O_3$ which is same in all lipids. Fatty acids are amphypathic molecules having two regions (i) a long hydrocarbon chain which is hydrophobic and not very reactive chemically (b) is the carboxylic acid group which can be utilised in solution (COO^-) extremely hydrophobic and forms esters and amides.

In neutral solution, salts of the fatty acids form small spherical droplets or the mycelles in which the hydrocarbonyl group is at the surface and the hydrophobic carbon chain at the centre.

In the cells the fatty acids are esterified to other compounds are form saponifiable lipids. A fatty acid may either be saturated or unsaturated. Saturated fatty acids have long hydrocarbon chains ending in carboxyl group. The formula is CH_3-CH_2 (n times)-COOH. This type is commonly found in animals which is known as palmetic acid and in unsaturated fatty acids 2 to 6 carbon atoms of the hydrogen chain are linked by double bonds known as linolic acid.

Double bonds increase the flexibility of hydrocarbon and the chain giving the biological membrane more fluidity.

They are common in plants and animals living at low temperature. Ex. fish. Lipids in warm climatic atmosphere are saturated fatty acids.

Mammals are unable to synthesise certain fatty acids. Therefore they have to be supplied by diet known as essential fatty acids. Lipids are of 3 types: simple, compound and derived.

Simple lipids: They are alcohol esters of fatty acids. Ex. dry glyceroids in the presence of lypase and water break up again into glycerols and 3 fatty acids.

Compound lipids: Major constituents of the cell membranes contain fatty acids, alcohol, phosphorus, amino group carbohydrates. Ex. lecithine.

Derived lipids: They are without fatty acids. They are Vit A, E, K.

Glucose role in the cell: Glucose is stored in animal cells in the form of monopolysaccharides which is known as animal starch very much nutritive and scientifically it is known as glycogen found abundantly in liver and muscle cells of man and other animals. Glycogen is more soluble than starch existing in the cytoplasm as tiny granules. These glycogen molecules have different spectrum from largest molecules as glycosylic units. Each glycogen molecule is profusely branched (tree) chains of α glucose molecules. The glucosidic bonds are established between $(1 \rightarrow 6)$ carbon of glucose. A glycogen molecule contains 3 types of chains 'A', 'B' and 'C'. There is only 1 C chain which bears B and A chains and ends in the free reducing group which bears OH groups. B chains are attached directly to the C chain and bears one or more A chain. The A chain may also be linked to the C chain.

Energetics of glucose oxidation: Single molecule of glucose contains 6,86,000 calories, by *anerobic glycolysis.* Only 10% of it 6,86,00 calories can be released and cell can store only 45% of chemical energy by combustion of glucose in the form of ATP (36 ATP molecules). The rest of it has been used for the other cell functions.

But in aerobic conditions even by glycolysis or by Kreb's cycle can generate 2 molecules of ATP per molecule of glucose 38 ATP molecule are produced per glucose molecule on oxidation. Out of 38 two molecules are used for transportation of two NADH produces during glycolysis into mitochondria. Hence the net gain is 36 molecules. One high energy phosphate bond = 36000 cal.

$$C_6H_{12}O_6 + 36Pi + 36\ ADP + 6O_2 + 6H_2O \rightarrow 6\ CO_2 + 36\ ATP \rightarrow 12H_2O + 36000$$

When 2 ATP molecules are produced 72000 calories of energy are produced. The inorganic phosphorus Pi/O ratio in the electron transport chain since the formation occurs in 3 steps.

$$NADH + H^+ + 3ATP + Pi + \tfrac{1}{2}\ O_2 \rightarrow NAD^+ + H_2O + 3\ ADP$$

One way of indicating the ATP yield from oxidative phosphorylation is $\overline{P}/O$ ratio which is expressed as the moles of inorganic phosphate pi used per oxygen atom consumed. Thus in the above equation $\overline{P}/O$ is 3 because 3Pi and ½ O_2 is used.

Role of Na and K in the cell

Specific proteins in the active transport move to Na and K ions across the membranes to stimulate ATP.

It is ion pump or cation exchange pump which is driven by energy or ATP molecule to export $3Na^+$ outside the cell in exchange of the import of $2K^-$ ions inside the cell. Ex. electric organ of the Eel are very rich in this enzyme or pump. These ions are very imp for maintaining osmotic pressure and acid base balance.

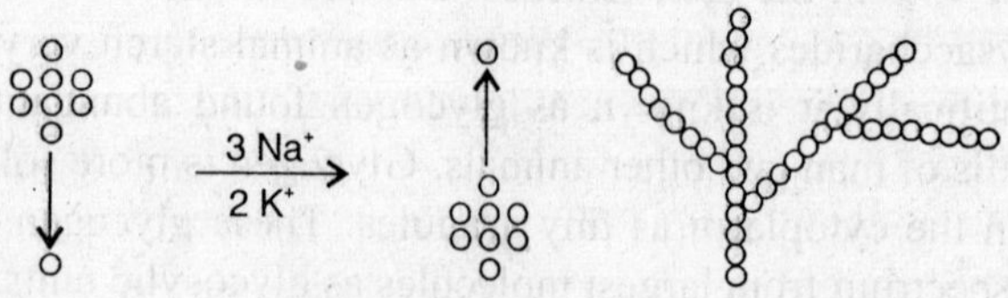

Fig. 4.12 Glucose tree

Biological role of carbonates and bicarbonates

In ion exchange permease protein catalyses an ion for 1 exchange of anions of bicarbonates across the membrane of erythrocytes. The rapid flux of an ion facilitates the transport in the blood of CO_2 from the tissue to the lungs. Waste CO_2 is released from the cell into the capillaries and diffuses across the membrane of the erythrocytes. In its gaseous form, CO_2 dissolves poorly in aqueous solution such as blood plasma, but inside the erythrocytes the potent enzyme carbonic anhydrase converts it into bicarbonate ions.

$$CO_2 + H_2O \xrightarrow[\text{anhydrase}]{\text{carbonic}} H^+ + HCO_3^-$$

SUMMARY

Chemical composition of a living cell explains that life is composed of small and large sized molecules, inorganic ions and water. Water of course, is most abundant in living organisms as the chemical reactions always proceed in aqueous medium. A large amount of water in cells exists in the form of individual molecules, and only about 5 per cent is found in bound form in association with other molecules. This small quantity of water contains chemical substances of different types in soluble or insoluble form, while some chemical substances may be present in colloidal suspension. Beside water living cells also contain large size molecules which can be classified according to their functions, such as production, utilisation, and storage of energy. There are other classes of compounds which participate in structure building repairs, catalytic functions, storage and transmission of hereditary information, and a host of other activities. Thus it is necessary to learn the structure and functions of these biochemical molecules.

EXERCISE

1. Write in detail structure and functions of carbohydrates in living organisms.
2. Give structures of the following:
   (a) a disaccharide commonly found in cane sugar.
   (b) a monosaccharide which is common in RNA.
   (c) an amino sugar present in peptidoglycans.
   (d) a disaccharide unit present in cellulose.
3. Distinguish between amilose and amilopectin.
4. Give the strtuctures and properties of a glycogen molecule.
5. Classify proteins on the basis of their conformations.
6. Protein configurations are associated with their functions. Comment.
7. Define a lipid. Why do we call it glyceride? When a triglyceride is hydrolised, what are the hydrolysis products?
8. Distinguish between satuarated and unsatuarated fats.
9. Write shortnotes on:
   (a) Fibrous protein
   (b) Globular protein
   (c) Purines
   (d) Pyrimidines
   (e) Glycolipids
   (f) Sphingolipids
10. Define a nucleotide and describe the structure of deoxyribose nucleotide.

5

Structure and Function of Cell Organelles

Like prokaryotic cells, eukaryotic cells are also bound by a plasma membrane and contain ribosomes. However, eukaryotic cells are quite complex and contain a nucleus, a variety of cytoplasmic organelles and a cytoskeleton.

The size of eukaryotic cell varies from 1u to 175000μ (175mm), The ostrich egg is the largest cell. Nerve cells are very long and some are as long as 3-3.5 feet The number of cells per organism also varies from a single cell in protozoans to millions of cells in higher plants and animals. Unlike prokaryotes, eukaryotic cells are organised into tissues and organs, all of which perform specific functions. Plant cells have a thick complex cell wall whereas the animal cell does not possess any such structure. The plant and animal cells also differ in some other properties. For example plant cells lack centrioles and lysosomes. (Fig. 5.1 & 5.2).

A typical eukaryote contains all the following organelles in its cell.

1. **Cell wall:** It is seen only in plant cells. This is made up of two layers i) Primary Wall ii) Secondary adjacent cells in a tissue have their primary hosts cemented by another layer called middle lamella which is due to pectin, magnesium, calcium which gives rigidity to the wall. The primary wall is thin and elastic and it is made up of pectin and cellulose. The secondary wall has three layers: Outer, Middle & Inner in a concentric manner. In case of sclerenchyma cells the thick secondary wall is made up of number of cellulose bundles called microfibrils.

Such structural details are possible only due to the electron microscope. The electron microscope has enabled a detailed study of tweenfibrils (Micelle) forms and fibrils. Each micelle measures about 250 Å. Each micelle is made up of hundred chains of cellular molecules and 3000 glucose molecules form a cellular chain. To connect the next continuing chain with the protoplasm there are minute perforations called plasmodesmata. Cell wall affords protection, with permeability characteristics, to allow the molecules to pass through metabolic activities of the cell.

2. **Plasma membrane:** It is the outermost covering of animal cells and is present beneath the cell wall in plant cells. It performs functions like transport, synthesis etc., besides giving the cell protection.

3. **Cytosol:** Plasma membrane is followed by the colloidal organic fluid called the cytosol; it responds to a variety of small molecules concerned with metabolism.

4. **Cytoplasm:** Though not an organelle, it is an integral part of any cell and it is in this watery fluid that all the organelles are present and a variety of metabolic reactions takes place.

5. **Golgi bodies:** They are cup shaped, located near the nucleus consisting of smooth, cisternae bounded vesicles which transport proteins.

6. **Endoplasmic reticulum:** It remains continuous with plasma membrane and the nuclear envelope. They are engaged in glycogenolysis and polypeptide synthesis.

7. **Ribosomes:** They act as the work benches of protein synthesis.

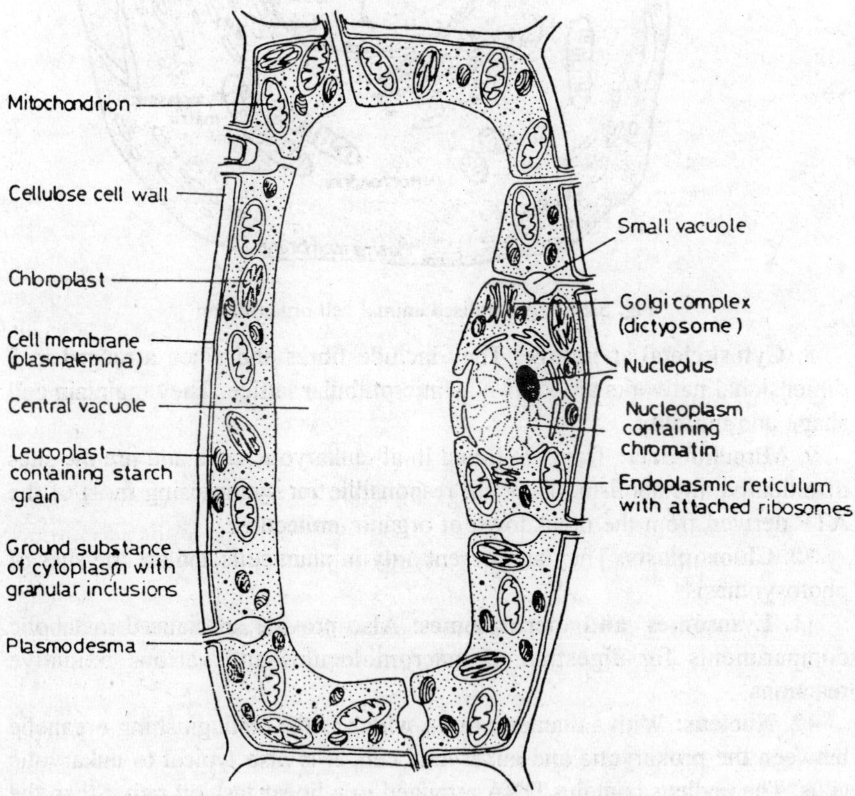

Fig. 5.1. Structure of a typical plant cell organisation

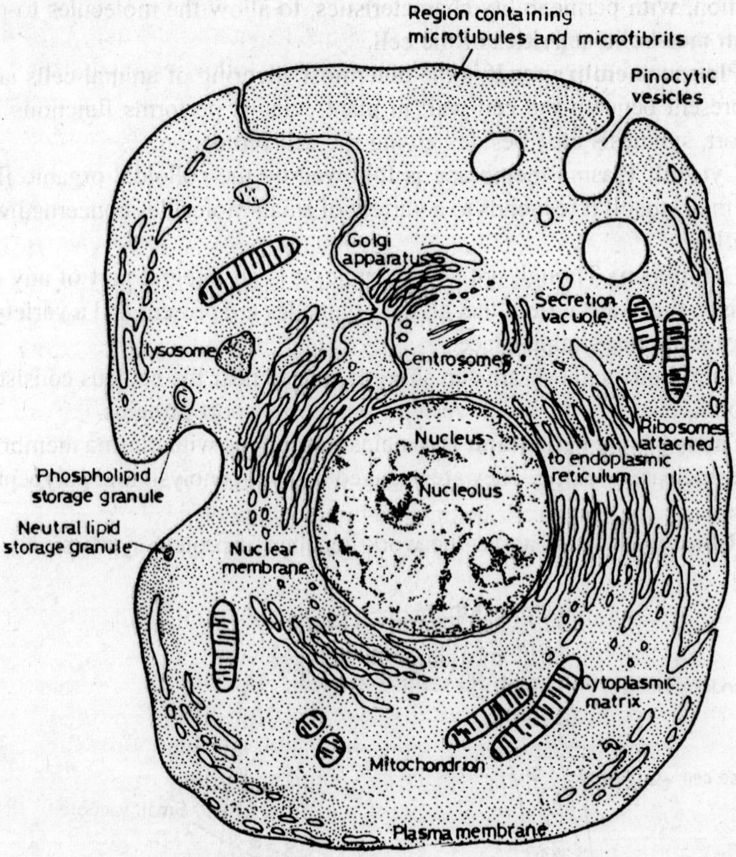

Fig. 5.2. A generalised animal cell organisation.

8. **Cytoskeletal structure:** They include fibres and when arranged in 3 dimensional networks are known as microtubular lattice. They maintain cell shape and mobility.

9. **Mitochondria:** They are found in all eukaryotic cells and are the sites of oxidative metabolism. They are responsible for synthesising most of the ATP derived from the breakdown of organic molecules.

10. **Chloroplasts:** They are present only in plant cells and are the sites of photosynthesis.

11. **Lysosomes and peroxisomes:** Also provide specialised metabolic compartments for digestion of macromolecules and various oxidative reactions.

12. **Nucleus:** With a diameter of 5μ which is the distinguishing organelle between the prokaryotic and eukaryotic cell, it is also typical to eukaryotic cells. The nucleus contains DNA arranged in a linear fashion rather than the circular fashion, as in prokaryotes. The nucleus is also the site of DNA replication and RNA synthesis.

All these organelles provide compartments in which different metabolic activities are localised, thus helping to perform a variety of complex reactions without interfering with each other. The structure and the functions performed at various sites in the cell will be detailed in the following sections.

PLASMA MEMBRANE

The structure that separates the cell content from the external environment is the plasma membrane. It is a thin film (6 to 10 nm thick) constituting continuous lipid bilayer with proteins intercalated in or adherent to both surfaces. The plasma membrane can be resolved only with the electron microscope, which reveals its numerous infoldings and differentiations as well as the different functions that establish connections with neighbouring cells. The main function of plasma membrane is to control selectively the entrance and exit of the materials. The plasma membrane is covered and reinforced by the cell wall in plant cells and by the cell coat in animal cells.

The cell wall is responsible for the rigidity of most plant tissues. This structure mainly consists of cellulose fibres into which other substances like lignin, which is the main component of timber may be incorporated. There are funnels ranging through the cell wall that is plasmodesmata which allow communication with other cells in a tissue. The cell coat covered in most animal cells is made of glycoproteins, glycolipids and polysaccharides that may extend the thickness of the cell membrane and continue far beyond it. Besides protection, the cell coat is involved in molecular recognition between cells; it contains enzymes and antigens and is fundamental in the association of cell in a tissue.

Cell membrane term was coined by Nageli and Cramer in 1865 and Plowe coined the term plasmalemma to plasma membrane.

In 1902 Overton postulated that the plasma membrane is composed of a thin layer of lipid. In 1926 Gorter and Grendel found that the lipid content of hemolyzed erythrocytes was sufficient to form a double layer of lipid molecules over the entire cell surface.

Other information comes from the study of interfacial tension of different cells. Tension at a water-oil interface is about 10-15 dyn/cm, whereas surface tension of cells is almost nil. The low tension is due to the presence of protein among the lipid components. In fact, when a very small amount of protein is added to a model lipid water system, the surface tension is lowered.

To explain all these, many molecular models of cell membrane were proposed by many cell biologists. Some of them are:

a) *Lamellar model:* Danielle and Dawson proposed that the plasma membrane is sheath like and composed of two layers of lipid which are sandwiched between two continuous layers of proteins. This is known as sandwich model or Trilamellar model (1935).

b) *Unit membrane model:* Robertson (1959) proposed his famous unit membrane model. According to him, the unit membrane consists of bimolecular lipid leaflet sandwiched between outer and inner layers of proteins organised into the pleated sheet configuration (protein-phospholipid-phospholipid-protein). Such an arrangement was presumed to be basically the same in all cell membranes. The observations of thin section by electron microscopy has led to the concept of trilayered plasma membrane which is also called the unit membrane model.

c) *Micellar model:* According to this model a central core (micelle) of hydrophobic phospholipids is surrounded by the hydrophilic protein molecules.

d) *Protein crystal model:* According to this model the phospholipids are intercalated in the gaps or cavities of crystal like protein structure.

All these models and unit membrane model, however, are certainly an oversimplification.

Gorter and Grendel in 1934 performed an experiment to study the arrangement of the components of the plasma membrane.

A known number of RBCs were taken and lysed by treating the RBC with hypotonic solutions. This was done in beakers containing water. After the lysis of the cells the air-water interphase shows a layer of phospholipid molecules spread over the surface of water. As they already knew the number of RBC they could approximately measure the area occupied by the membranes of the RBC on the air-water interphase which now had a single layer of phospholipid molecules. However, they found that the area occupied by the single layer of phospholipid molecules on the air-Water interphase was double the area occupied by the membranes in RBC as per calculation. This led to the conclusion that the phospholipids were not present in a single layer but a double layer.

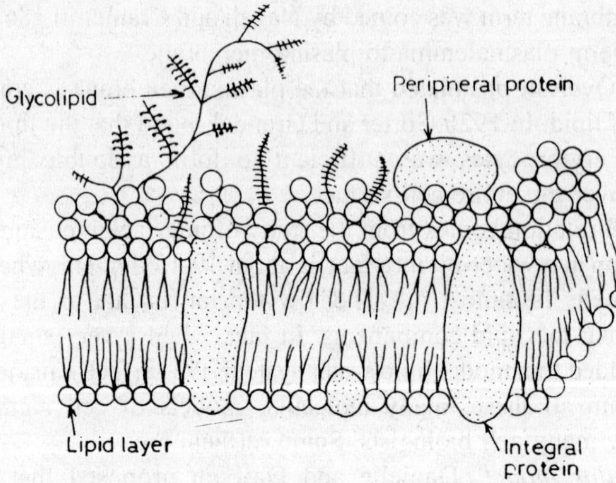

Fig. 5.3. Fluid-mosaic model of plasma membrane proposed by Singer and Nicholson (globular proteins are seen floating in the bimolecular lipid matrix)

The fluid mosaic model:The most popularly accepted model of cell membrane is the fluid mosaic model proposed by Singer and Nicholson

(1972). This postulates that lipid and proteins are dispersed in a kind of mosaic arrangement and that biological membranes are quasi-fluid structures in which both lipids and the proteins are able to move within the bilayer, i.e, the proteins are embedded in a lipid bilayer in such a way that proteins float in lipid sea. (Fig. 5.3). The lipid molecules mainly consist of phospholipids which are arranged in two layers. Each phospholipid molecule is made up of a glycerol head connected to a tail of long chain fatty acid-hydrocarbons, which always contain 1 or 2 double bonds and give the tail rather a kinked structure.

Many types of proteins, namely extrinsic, intrinsic and transmembrane proteins are present interpersed with the lipid molecules. Extrinsic proteins are the proteins which are superficially attached to the lipid molecules. The integral proteins of the membrane are intercalated to a greater extent into a rather continuous lipid bilayer. These integral proteins are amphipathic with polar regions (hydrophilic) protruding from the surface and nonpolar (hydrophobic) regions embedded in the hydrophobic interior of membrane (Fig. 5.4).

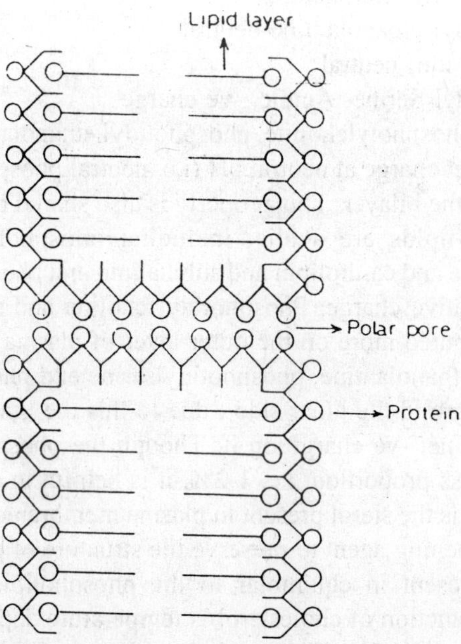

Fig. 5.4 Danielli-Davson model of plasma membrane showing pores in the membrane

Proteins and Phospholipids are not tightly associated with each other and have freedom to move along the surface of plasma membrane. Phospholipid molecules may also move along the cell membrane thus making the plasma membrane a viscous fluid rather than a rigid structure. In fact, the cohesive forces between lipids and proteins are relatively weak interactions, i.e. ionic hydrogen bonds and mainly hydrophobic in character. Some of the lipids and proteins on the outer surface are frequently associated with sugars to form glycolipids and glycoproteins.

COMPOSITION OF PLASMA MEMBRANE

In general, the plasma membrane is composed of

(i) lipids (Phospholipids, cholesterol)
(ii) proteins
(iii) oligosaccharides bound to lipids (glycolipids) and proteins (glycoproteins).

Lipids: The main lipid components of the plasma membrane are phospholipid, cholesterol and glycolipids. Their proportions vary in different cell membranes.

Phospholipids: They are loosely aranged and form 20-79% of the cell membrane. The major proportion (50%) of membrane phospholipid is represented by

(a) Phosphotidylcholine-neutral
(b) Phosphotidyl ethanolamine-neutral
(c) Sphingomyelin-neutral
(d) Phosphatidyl Serine- Acidic, -ve charge

Among these phosphotylcholine, phosphotidyl ethanolamine and sphingomyelin have no net charge at neutral pH (i.e, neutral phospholipids) and tend to pack tightly in the bilayer. This property is also shared by cholesterol. 5 to 20% of phospholipids are acidic, including phosphotidyl inositol and phosphotidylserine and cardiolipin and sulphalipin and phosphotidyl glycerol. They have a negative charge. Phosphotidyl choline and sphingomyelin are neutral and distributed more on the outer layer of plasma membrane.

Phosphotidyl ethanolamine, phosphotidylserine and phosphotidyl inositol are distributed more in the inner side ; due to this the inner layer of plasma membrane gets a net -ve charge on it. Though the phosphotidyl inositol is present in very less proportion, i.e. 1-2%, it is helpful in cell signalling.

Cholesterol: It is the sterol present in plasma membrane and gives stability or acts as strengthening agent to preserve the structure of plasma membrane. Cholesterol is present in equimolar to the phospholipids in cytoplasmic membrane. The function of cholesterol is temperature dependent . At higher temperature cholesterol interferes with movement of fatty acid chain and prevent permeability of solutes. At low temperature cholesterol interferes with hydrophobic interactions avoiding freezing of the membrane.

Basic functional ability of plasma membrane depends on the structure of phospolipid molecules:

(i) Phospholipids have hydrophilic heads on either surface and hydrophilic tails are present in the centre. Due to this arrangement the water soluble ions or molecules cannot enter through plasma membrane.

(ii) The hydrophobic tails are not straight but are kinked. These long chains of fatty acids can move freely due to which tight packing of these chains is not possible. Phospholipids are able to rotate on their own axis.

MEMBRANE PROTEINS

Proteins represent the main component of most biological membranes. Numerous enzymes, antigens and various kinds of receptor molecules are present in plasma membrane. These proteins were identified by Singer and Nicholson. The membrane may consist of 30-50% of proteins. For every 50-100 molecules of lipid one protein molecule is present. Membrane proteins have been divided into peripheral and integral proteins according to the degree of their association with the membrane, and the methods by which they can be solubilized.

PERIPHERAL PROTEINS

They are present only on the surface of the plasma membrane, due to their hydrophilic nature, and are bound to the plasma membrane by ionic bonds and protein-protein interactions. They are easily dissociated by mild treatments like high pH, hypotonic solutions, high ionic strength, metal ion chelating agents, etc., The dissociated proteins are soluble in water and are usually free of lipids and molecularly dispersed in neutral aqueous buffers. Their dissociation will not cause actual disruption of the plasma membrane; for example, i) spectrin may be removed from red cell ghosts by chelating agents. ii) cytochrome-C found in mitochondria is easily removed in high salt solutions.

INTEGRAL PROTEINS

They represent more than 70% of the membrane proteins. They pass through the membrane and are associated with hydrophobic regions of the lipid bilayer also. They are exposed to either side of the membrane, and when present throughout the membrane they are called transmembrane proteins and consist of both hydrophilic and hydrophobic regions, i.e. they are amphipathic. They are dissociated from the plasma membrane by surface acting agents like detergents, hydrophobic bond breaking organic solvents and chatropic agents. Their dissociation is accompanied by the dissociation of the whole plasma membrane.

OTHER PROTEINS

The oligosaccharide group of glycoprotein formed from the ribosomes attached to endoplasmic reticulum integrates with phosphotidyl inositol present in the inner leaflet membrane by GPI anchor (Glycosyl Phosphotidyl Inositol Link) to form a GPI molecule. These are known as GPI linked proteins. Other than these, some more proteins are seen in plasma membrane formed or synthesized by free ribosomes. In erythrocyte or leucocyte membrane, the leucoproteins consists of Hexose, Hexosamine, Fucose, NAG and NANA-N-acetyl-neuraminic acid (sialic acid). The proteins which contain sialic acid in their glyco group are called sialo proteins and these proteins are recognised by endothelial receptors. Because of the amphipathic nature of the transmembrane proteins, they cannot be easily crystallised and hence their 3 dimensional structure cannot be identified. The transmembrane protein has 3 regions: The middle portion is α-helical structure; above and below it, peripheral proteins are present which accept light and send to L,M,N portions of α-helical structure. It consists of single polypeptide chain of 247 amino acids, which traverse 7 times the lipid bilayer. Each of the traversing segments is an α-helix with some 25 amino acid residues. The oligosaccharides of the proteins and lipids on the outer side of the membrane form a layer around the membrane called Glycocalyx in eukaryotes, and its basic function is to protect and also work as binding sites or receptors.

FUNCTIONS OF THE PLASMA MEMBRANE

Main functions of plasma membrane are:

(i) Cell Permeability

(ii) Proteins present on cell membrane may act as carriers for transport of materials and may help in cell recognition or may act as receptors for informational molecules.

CELL PERMEABILITY

Permeability is fundamental to the functioning of the living cell and to the maintenance of satisfactory intracellular physiological conditions. This function determines which substances can enter the cell, many of which may be necessary to maintain its vital processes and the synthesis of living substances. It also regulates the outflow of excretory substances and wastes from the cell.

The presence of membrane establishes a net difference between the intracellular fluid and extracellular fluid. This may be fresh water or salt water in unicellular organisms growing in ponds or seas, but in multicellular organism, the internal fluid i.e. blood, lymph and especially the interstitial fluid, is in contact with the outer surface of the cell membrane. One of the functions of the cell membrane is to maintain a balance between the osmotic pressure of the intracellular fluid and that of the interstitial fluid.

Passive transport: Passive transport occurs along the lipid membrane depending upon the concentration gradient and does not involve any proteins of energy. The molecules move across the lipid membrane from higher to lower concentration. Molecules like water can pass through passive transport. Small polar molecules and hydrophobic molecules are permeable by passive transport whereas the charged molecules, larger molecules and certain hydrophilic molecules cannot pass through passive transport.

Molecules like oxygen, carbon-di-oxide and polar molecules like water, nonpolar solvents like Benzene and alcohol are permitted. Amino acids are not permitted as they are large and charged. Glucose is also not allowed due to its larger size. Small ions like H^+, Ca^{+2}, Cl^-, Na^+, K^+ are also not allowed by passive transport due to their charge (Fig. 5.5).

FACILITATED TRANSPORT (DIFFUSION)

Occurs by two different ways—

(a) occurs along the concentration gradient.

(b) occurs against the concentration gradient.

Proteins are involved in this transport. Two types of protein transporters are seen.

(i) carrier proteins

(ii) channel proteins

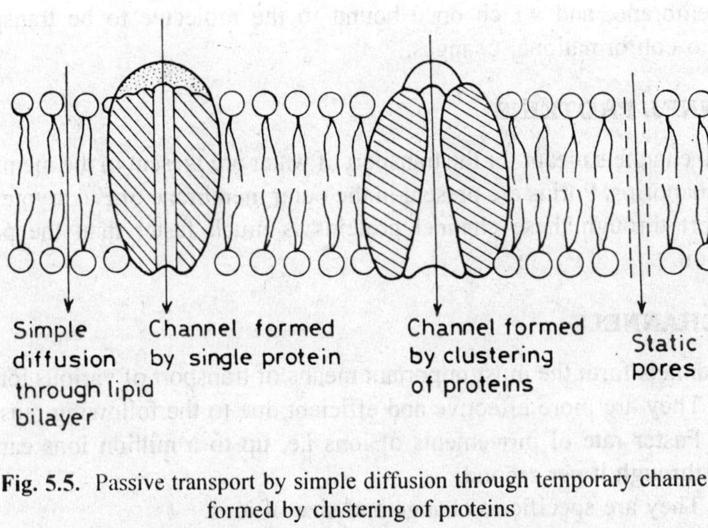

Fig. 5.5. Passive transport by simple diffusion through temporary channels formed by clustering of proteins

CARRIER PROTEINS

The transport of many different molecules across the membrane shows a high degree of specificity, i.e, the permeability of a molecule is related to its chemical structure. While one molecule may readily enter the cell, another of the same size but slightly different molecular structure may be completely avoided. A clear example of the selectivity is provided by the transport of two isomers (Glucose and Galactose) into a bacterial cell. Although the only difference between them is in the position of -OH group at carbon 4, these two sugars penetrate the membrane by different transport mechanisms. This type of selectivity is attributed to transport proteins or carrier proteins or permeases. It is thought that a permease functions in a manner somewhat similar to an enzyme or a receptor, in having a binding site able to recognize the molecule to be transported. Some permeases can transport only if there is a favourable concentration gradient. Glucose concentration is always less inside the cell, as it is actively metabolised inside the cell, except in liver where active outward flow of glucose molecules is seen.

The carrier mechanism has also been postulated to mediate the entrance of the glucose into intestinal cells. This mechanism is activated by Na^+. The carrier ion in this model has polar amino acids or binding site for Glucose, and should exist in two conformational states, one with low affinity for glucose and another with high affinity. A carrier model can be postulated in which the glucose site is initially with a high affinity state and binds the glucose molecules at the surface of the cell. In the second step the carrier is translocated and the glucose site changes to a low affinity releasing the molecule into the cell. The two affinities are modulated by Na^+ concentration inside and outside the cell. This carrier mechanism follows fixed pole mechanism in which the carrier is represented by integral proteins that traverse the membrane, and which once bound to the molecule to be transported undergo conformational changes.

CHANNEL PROTEINS

Special channels meant for the transport of water are present in the membrane. Ex.Aquaporins. Porins are present in the outer membrane of Eukaryotes. The transport through these channel proteins is much faster than the passive diffusion.

ION CHANNELS

Ion channels form the most important means of transport of various ions and water. They are more effective and efficient due to the following reasons:

(1) Faster rate of movements of ions i.e, up to a million ions can pass through it per second.
(2) They are specific to change and size of ions.
(3) They are not always open.

Based on the type of signals they require to close and open, they are of two types:

(i) Ligand gated channels
(ii) Voltage gated channels

LIGAND GATED CHANNELS

They open and close in response to the binding of neurotransmitters or some other signaling molecules.

Ex. The binding of Acetylcholine to the acetylcholine transport molecule determines the opening or closing of the negatively charged hydrophobic gates. Hence acetylcholine acts as signalling molecule and when it binds to the molecule, the gates open due to conformational change and allow only +vely charged ions like Ca^{+2}, K^+, etc. to pass through them. But -vely charged ions are not allowed to pass through them because of -vely charged hydrophobic gates.

VOLTAGE GATED CHANNEL

The opening and closing of the voltage gated channel depends upon the electron potential or charge of the cell Ex. Na^+ pump, K^+ pump.

NA$^+$ PUMP

Generally Na^+ are pumped outside the cell and the Na^+ pump aids in this.

Generally Na^+ size is small and appropriate to the size of the Na^+ pump and so passes through it along with water. But as K^+ is larger than Na^+, K^+ cannot pass through a Na^+ pump.

K$^+$ PUMP:

Generally K^+ are pumped inside the cell and the channel of K^+ pump has a binding site and so it can pass through it along with water. But Na^+ size is very small to bind with the active site in K^+ Pump, and hence only K^+ions can pass through them.

GROUPS TRANSLOCATION:

Translocation of a molecule inside the cell, which involves the chemical alteration of that molecule,is called Group translocation. Group translocation always occurs against the concentration gradient but without true expenditure of energy. Ex. Glucose is transported by group translocation into cell as glucose-6 phosphate by addition of a phosphate molecule outside the cell. Here the ATP is used up in the phosphorylation of Glucose, i.e. for biochemical reaction but not actually for transport. This glucose 6 phosphate may follow Glycolysis in the cell. Sugars are generally transported by this method.

ACTIVE TRANSPORT

Active transport involves transport of molecules via the cell membrane with expenditure of energy; energy procured from ATP hydrolysis or electron potential of the cell, without the chemical alteration of the molecule against the concentration gradient. Ex. Na^+, K^+ pump. It includes protein molecules with 3 binding sites for Na^+ and 2 binding sites for K^+ and can pump 3 Na^+ outside and $2K^+$ inside the cell by utilisation of only one ATP molecule.

In normal configuration the Na^+ binding sites are exposed to the cytoplasm and the Na^+ which are always pumped outside the cell bind to these binding sites when the protein molecule is phosphorylated by the hydrolysis of ATP.

As soon as the Na^+ binds to the active sites, conformational change takes place and the Na^+ binding sites are exposed to the cytoplasm and so K^+ from outside the cell binds to these active sites which causes dephosphorylation of protein molecule and thus leading to conformational change. The K^+ are released inside the cell. So only the ATP molecule is utilised to pump out 3 Na^+ and to pump in 2 K^+ ions by active transport.

CA2⁺ PUMP

$Ca2^+$ Pump is also similar in structure to the Na^+, K^+ pump and helps in active transport of $Ca2^+$ ions outside the cell by ATP hydrolysis. The concentration of Ca^{+2} inside the cell is 0.1 mm. The concentration of Ca^{+2}outside the cell = 1mm. But the Ca^{2+} are transported outside the cell. During this transport some other ions may be transported inside the cell simultaneously.

CYSTIC FIBROSIS

It is a disease caused by malfunctioning of ion channels namely CFTR i.e. Cystic Fibrosis Transport Regulator Channel.

The cells are very sensitive to the Ca^{+2} concentration. Ca^{+2} acts as a

– cell signalling molecule.
– H Pump is more frequent in plant cells and bacteria.
– Na^+, K^+ pumps are frequent in animal cells.
– The entry and exit of the water molecules in the cell is governed by the turgor pressure which develops due to the presence of a rigid cell wall.

ION GRADIENT ACTIVE TRANSPORT

The concentration of ions inside or outside the cell is called ionic gradient. The cell always maintains the ionic gradient by pumping of ions like Cl^-, Ca^{+2}, Na^+, H^+, K^+. If +ve charge is formed outside of the cell, OH^- ions are always present inside the cell having -ve charge. This is the stable gradient always present around the cell. Any cell maintains an ion gradient across its membrane by active transport. Cells contain a net +ve ion gradient on the outer surface, where large amount of H+ and +ve charged ions are accumulated. The inner surface of the cell membrane maintain a -ve gradient by the accumulation of OH- ions. The ion gradient keeps fluctuating across the membrane due to the movements of ions into or outside the cell against the concentration gradient by active transport. When Na+ & Ca^{+2} is transported out of the cell, against the concentration gradient, there is a reduction in net ionic gradient, across the cell membrane. The movement of cation across the cell membrane is thus compensated by the inward flow of equal number of cations. This mechanism of maintaining ionic equilibrium in the cell membrane is called “Donnan’s equilibrium.

TRANSPORTERS

Three types of transporters are seen in the cell membrane namely:

(a) Uniporter
(b) Symporter and
(c) Antiporter

Uniporter: This type of transporter transfers only one specific molecule either outside or inside the cell through it. Ex. Na^+ Pump.

Symporter: This type of transporter transports one type of ions and some other ion or molecule can also pass along these ions in the same direction. Ex. Glucose enters into the cell along with Na+ ions through Na+ Channel along the concentration gradient.

Antiporter: This type of transporter allows one type of ions inside and other type of ions outside the cell simultaneously. Ex. N^{a+}, K^+ Pump - Na^+ outside K^+ insideK+ - Ca^{+2} Pump-K+ inside Ca^{+2} outside the cell.

OTHER FUNCTIONS OF PLASMA MEMBRANE

Endocytosis: Endocytosis is related to the activity of plasma membrane. It may not take place in all the cells. It includes phagocytosis and pinocytosis.

Phagocytosis: The ingestion of the food particles inside the cell by the formation of a phagosome of vesicle by specialised cells called phagocytes is called phagocytosis, whereas Macrophages and neutrophils are called pinocytosis.

Process: Once the phagosome or pinosome are formed, they are associated with lysosome to form phogolysosome and pinolysosome into which the hydrolytic enzymes are released. Active digestion takes place and simpler compounds or waste materials are formed in the phagolysosome. The useful material is secreted in the cell and finally the phagosome with debris or waste material is sent or vomited out of the cell by a process known as exocytosis, emiocytosis (or) cell vomiting.

Vesicles: Vesicles are globular structures that bud off from the cell membrane during endocytosis. They are of two types: 1) Nonspecific Vesicles 2) Specific Vesicles.

1) *Nonspecific vesicles:* These are more or less equal to phagosomes or pinosomes, which may later combine with lysosomes, endoplasmic reticulum or golgi bodies depending upon the type of molecules or on the nature of the residue in vesicles.

- If it contains complex materials – it associates with lysosomes for digestion into simpler compounds.
- If it contains proteins – they are passed to endoplasmic reticulum for sorting, marking or for Glycosylation. If the proteins present are useful to the cell, then they are targeted to specific spaces; otherwise, they are thrown out if they are not required.
- chances of union of vesicles with golgi bodies is rare.

– Sometimes these vesicles may also combine with endosome (which is a vesicle that is present in the cell for a long time before forming secretory vesicles).

2) **Specific vesicles:** Specific vesicles contain receptors which facilitate them to capture particular material. Ex. Intake of cholesterol. There are certain pits in the mammalian cells called "clatherine coated pits", coated by a substance called clatherine, which contains LDL receptors. Cholesterol is ingested in the form of LDL - Low Density Lipoprotein. The LDL gets bound to the LDL receptors and slowly the pit will be filled with LDL. Then it forms clatherine coated vesicles which will be transferred specifically to the lumen of endoplasmic reticulum to get degraded to form cholesterol. This cholesterol is then passed to a specific site via endoplasmic reticulum, Golgi apparatus. Sometimes LDL may be sent outside the cell or may be loaded on the inner walls of blood vessels, thus reducing the width of the blood vessels and causing high blood pressure (B.P).

SPECIALISED STRUCTURES OF THE CELL MEMBRANE

The differentiations of the cell membrane corresponds to regions specially adapted to different functions such as absorption, secretion, fluid transport, mechanical attachment or interactions with neighbouring cells.

(i) **Microvilli:** In the intestinal epithelium microvilli are prominent and form a compact structure that appears under the light microscope as a striated border. They represent the cytoplasmic processes covered by plasma membrane or the projections of the plasma membrane. They are mainly involved in the absorption of nutrients and are supported by cytoskeletal elements like microtubules and microfilaments. The actin filaments are attached to the tip of the microvilli by -actinin and their function is to produce contraction in the microvilli. The outer surface of microvilli is covered by a coat of filamentous material (fuzzy coat) composed of glyco-protein molecules.

(ii) **Interdigitation:** These are structures formed by folding of the plasma membrane into finger like or digit like structures which are predestined to form other specialised structures.

(iii) **Special invaginations:** Special invaginations of plasma membrane are seen forming the chambers between them, and mitochondria are present in these chambers helping in the production of energy for transport.

(iv) **Zonula occludens** (tight junctions): Tight junctions are specially differentiated regions that seal the intercellular space thus preventing the passage of fluid to and from the lumen. These junctions are present just below the apical border of the cell. In a polarised epithelium, one of the functions of the tight or occluding junctions is the maintenance of different intercellular environments in the apex and basolateral regions. The tight junction could act as a barrier to the diffusions of macromolecules or lipids in the bilayers, so as to differentiate the apical from basolateral portion, of the cell membrane, each having their own particular composition and physiological characteristics.

Mechanical adhesion between cells is mainly done by the desmosomes, of which there are two types: belt and spot desmosomes. The difference between them resides in their localisation, extension and relationship with the microfilaments.

Belt desmosomes (Zonula adherence or Terminal bars): Intermediary junctions are generally found at the interphase between columnar cells, just below the region of tight junctions. They form a band that girdles the inner surface of the cell membrane. This band contains a web of (6nm) actin filaments and another group of interwoven, intermediate filaments of 10nm. Actin microfilaments are contractile and intermediate filaments play a structural role. The plasma membranes of the adjacent cells come closer to each other and cytoplasm gets thickened into a gel in this region.

Macula adherence (Spot desmosome): They appear as darkly stained bodies on the cell surface under the light microscope. They represent localised circular areas of contact, about 0.5 nm in diameter, in which the plasma membranes of the two adjacent cells are separated by a distance of 30 to 50 nm. The electron microscope reveals a midline structure or intercellular core and two parallel desmosomal/plasma membranes. Under each plasma membrane there is a discoidal intracellular plaque, towards which numerous filaments called tonofibrils or tonofilaments converge. These tonofibrils are not contractile and belong to common type of intermediate keratin filaments. They are rigid structures.

Hemidesmosomes: Along the basal surface of some epithelial cells separate desmosomes called hemidesmosomes may be observed. These are similar to desmosomes in fine structure, but represent only half of them, the outer surface frequently being substituted with collagen fibrils. Some tonofibrils may be present.

Gap junction: The Gap junction or nexus or communication junction appears as a plaque-like contact in which the plasma membrane of adjacent cells are in close-apposition or tightly connected, separated by a space of only 2 to 4 nm. The cytoplasm is absent between plasma membranes in this region. It mainly binds the cells together and helps in cell communication.

Inter membrane canaliculus or intercellular canaliculus: They are hexagonal structures formed by the infolding of the plasma membrane cytoplasm which is thickened at this region. It basically helps in maintaining communication between cells in the form of ions.

Cell wall: The cell wall constitutes a kind of exoskeleton that provides protection and mechanical support for the plant cell. This includes maintenance of a balance between the osmotic pressure of the intracellular fluid and the tendency of water to penetrate the cell. The cell wall of fungi is made up of chitin, a polymer of N-acetyl glucosamine. The exoskeleton of crustaceans (Ex. scorpion) and Insecta (Ex. cockroach) is also made up of chitin.

Cell wall of plants: The wall consists of a microfibrillar network lying in a gel like matrix of interlinked molecules. The microfibrils are mostly cellulose, consisting of straight polysaccharide. These are the glucose chains, which by intra, and intermolecular hydrogen bonding produce the structural unit known as microfibril. Microfibrils in turn bind among themselves with non cellulose polysaccharides and proteins to form the cell wall. In addition to protein, the gel matrix contains some polysaccharides and lignin. The major polysaccharide fractions are pectin substances (containing Galactose, Galacturomic acid, arabinose, sometimes Rhamnose may be present) which are soluble in water and Hemicellulose (consisting of glucoronic acid, xylose, fucose, mannose, glucose, etc.) which are extracted with alkalis. Lignin is found only in mature cells and is made up of an insoluble aromatic polymer resulting from the polymerisation of phenolic alcohols. Suberin consists of Gums and Resins alongwith some aromatic residues and suberic acid.

CYTOSOL

Cell cytoplasmic organelles can be separated by ultracentrifugation and what is left behind is soluble fraction known as cytosol.

Physical nature of cytoplasm: It appears jelly like, transparent, and colourless fluid. Granier coined a name ergatoplasm for this material. Its basophilic nature is due to the presence of ribonuclease.

Cytoplasm is a heterogeneous mixture of opaque granules and free organic compounds imparting it a colloidal naturé. Its ground material has a unique property of changing its consistency, sometimes behaving as a viscous material. The peripheral zone of cytoplasm is thick and jelly like, called plasmagel. Around the nucleus, it is thin-liquefied, called plasmasol or endoplasm. A large number of diverse chemical nature substances are also dissolved in cytoplasm, which has a high percentage of water. Proteins are most abundant, of which 20-25% are in the form of soluble proteins including enzymes. Carbohydrates, inorganic salts, lipids and lipoidal substances are also found in cytoplasm.

Cytoplasm exhibits the phenomenon of phase reversal in which colloidal solution undergoes sol-gel transformation. The sol state is the liquid phase containing more water and less solids whereas the gel state is semi-solid and jelly like, with less water and more solids. Sol gel transformation depends on temperature and occurs in response to metabolic activities of the cell. Cytoplasm exhibits viscosity changes intracellular motion and amoeboid movements.

Cytoplasm suspended particles follow Brownian motion, in which motion of the particles is incoherent. Particles also possess electric charge either negative or positive. As identical charged particles repel, movement occurs.

Cytoplasm also shows amoeboid movement and intracellular motion.- Cyclosis. Amoeboid movement of protozoans, leucocytes of blood is due to frequent changes of cytoplasm from sol to gel state and vice versa. Stomatal guard cells in the leaves are capable of absorbing water and removing it as per the need.

Chemical nature of cytoplasm: Cytoplasm contains 90% of water and the remaining 10% includes organic and inorganic compounds in various proportions.

The most important compound present in cytoplasm is water as it forms dispersion medium of living matter. Living cells contain 60 to 90% water, whereas dormant cell tissues contain 10 to 20% only. As water has several unique properties such as high heat capacity, heat of vaporisation, latent heat of fusion, surface tension and dielectric constant compared with other solvents, these properties influence the behaviour and functions of the cell.

Other properties of water make cytoplasm a compound of fundamental significance like

(i) Water being an excellent solvent, it dissolves many organic and inorganic substances.

(ii) As water acts as a solvent, for charged substances, it dissociates itself in protons and negatively charged OH^- ions.

(iii) Water increases the dissociation of electrolytes.

(iv) Water serves as a means of dispersal of energy, food and waste matter.

(v) Due to its weak conductivity to heat, it loses or gains heat slowly. As a result, water helps in maintaining constant body temperature in homeothermic animals.

Among the 20 elements that constitute cytoplasm, carbon, hydrogen, nitrogen and oxygen are most abundant and the rest are in traces. Elements C, H, N, O, P and S are responsible for the formation of building blocks of all biomolecules. Inorganic compounds are found as ions, such as Na^+, Mg^{2+}, Po^4 2^-. cl^-, K^+ and Ca^{2+} in traces. Some major element ions -$Co3^+$, No_3^- are also present. All these ions are important for carrying out physiological functions. $Ca2^+$ is important in causing gelation of cytoplasm, muscle contraction and coagulation of blood. K^+ helps in reducing the viscosity of cytoplasm and muscle relaxation.

GOLGI BODIES

Golgi bodies were identified by Camillo Golgi in 1989. In plants and lower invertebrates it is also called as dictyosome. It is also considered as Golgi complex. The Golgi complex contains parallelly arranged,flattened sac like structures, which lack ribosomes.

Golgi complex or Golgi bodies occur in all cells except the prokaryotic cells and eukaryotic cells of certain fungi, sperm cells of bryophytes and pteridophytes, cells of mature sieve tubes of plants and mature sperm, and red blood cells of animals. Their number in plant cells may vary from several hundred to a single organelle in some algae. In animal cells usually a single golgi complex is present. In oocytes there may be many such bodies. The Golgi body may be scattered throughout the cell or may be generally localised at the place near the nucleus, in the shape of a cup. It gives out several vesicles which later on fuse with the plasma membrane, or form distinct organelles themselves.

Golgi bodies of plant cells are about 1-3μin length and 0.5μ height. It is composed of flattened membrane bound sacs (cisternae) and associated vesicles. Each Golgi complex usually contains about 3-7 flat tubular cisternae. Insects have up to 30 cisternae. The Golgi body has a distinct polarity in both structure and function. The face of Golgi body faces the Endoplasmicreticulum and is usually convex and is called the cis-face or the forming phase. The proteins from the Endoplasmic reticulum enter the Golgi body from this side. The opposite side which is concave is called the transface or the maturing face from which the proteins exit. The Golgi complex consists of three functioning distinct compartments. The cis-Golgi network, the golgi stack and the trans-golgi network. Transport vesicles carry the proteins between these networks. In addition,membrane tubules may also connect the Golgi compartments.

A Golgi body is surrounded by a differentiated region of cytoplasm, where ribosomes, glycogen and other organelles are scarce or absent. This zone is called the zone of exclusion. In conjunction with the endoplasmic reticulum, lysosomes etc. the Golgi body forms an extensive inter communicating membrane system called the endomembrane system. This forms an exclusive area inside the cell, where various processing mechanisms and transport continuously occur, separated from the cytoplasm. The Golgi body is majorly involved in transport, processing of proteins, other than synthesizing polysaccharides and cell wall components in plants (Fig. 5.6).

Functions: Proteins enter the Golgi from the cis-golgi network which mainly serves to receive the transport vesicles from the Endoplasmic reticulum. Proteins marked to be present within the Endoplasmic reticulum are recognised and returned to the Endoplasmic reticulum by recycling pathway with the help of receptors and signals. Other proteins are transported to the Golgi stack, which is the site of most metabolic activities of the Golgi apparatus. The proteins are processed here and then transported to the trans-golgi network where the final stages of processing are completed. The trans-golgi network acts as a sorting centre, directing the proteins and other molecules to the lysosomes, plasma membrane or the cell exterior. Another important function is to make the N-linked oligosaccharide that has to be added in endoplasmic reticulum. The proteins targeted for lysosomes are added with mannose-6-phosphate residue as the last step before which several sugars are removed. In the Golgi body carbohydrates are added to the serine or threonine residues (Fig. 5.6a).

Golgi bodies also participate in lipid metabolism, i.e. synthesis of glycolipids and sphingomyelin.

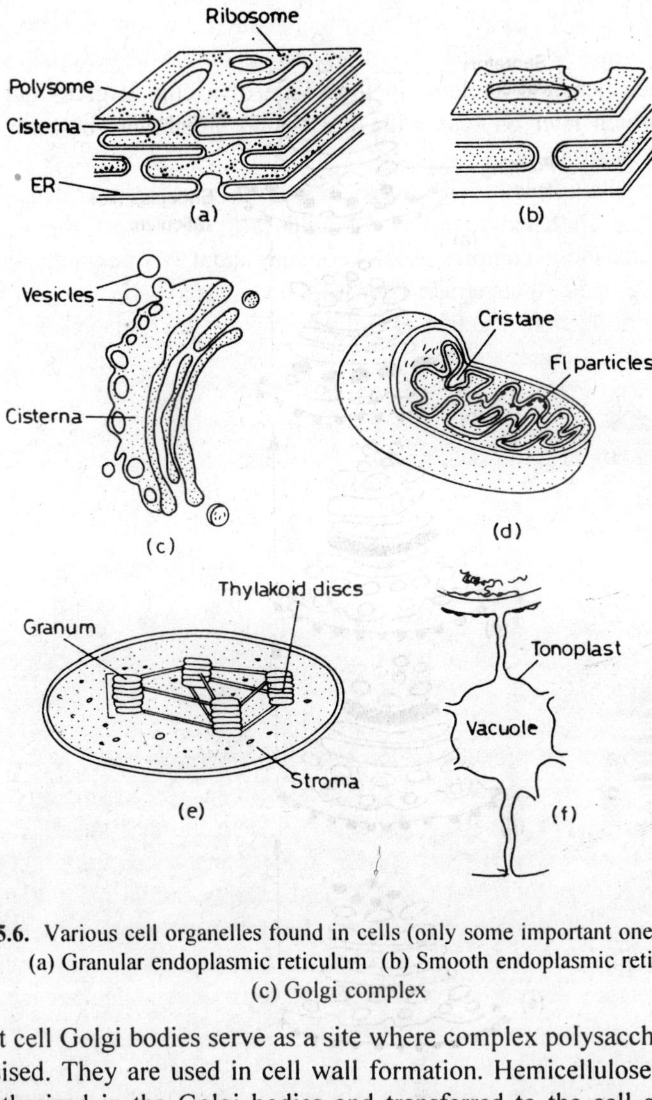

Fig. 5.6. Various cell organelles found in cells (only some important ones given) (a) Granular endoplasmic reticulum (b) Smooth endoplasmic reticulum (c) Golgi complex

Plant cell Golgi bodies serve as a site where complex polysaccharides are synthesised. They are used in cell wall formation. Hemicellulose of pectin are synthesised in the Golgi bodies and transferred to the cell surface by transport vesicles.

Proteins, lipids and polysaccharides are transported from Golgi body to their final destinations, through the secretory pathway. Proteins are carried to the plasma membrane by bulk flow.

In yeasts and plant cell, proteins are transported from the Golgi body to the vacuole as they lack lysosomes. Golgi body contains various enzymes viz., phospholipase ADPase, ATPase, CTPase, Thyamine pyrophosphatase, galactose transferase, Glucose-6-Phosphotase, etc.

Holocrine, apocrine and merocrine like cellular secretions are caused due to Golgi bodies.

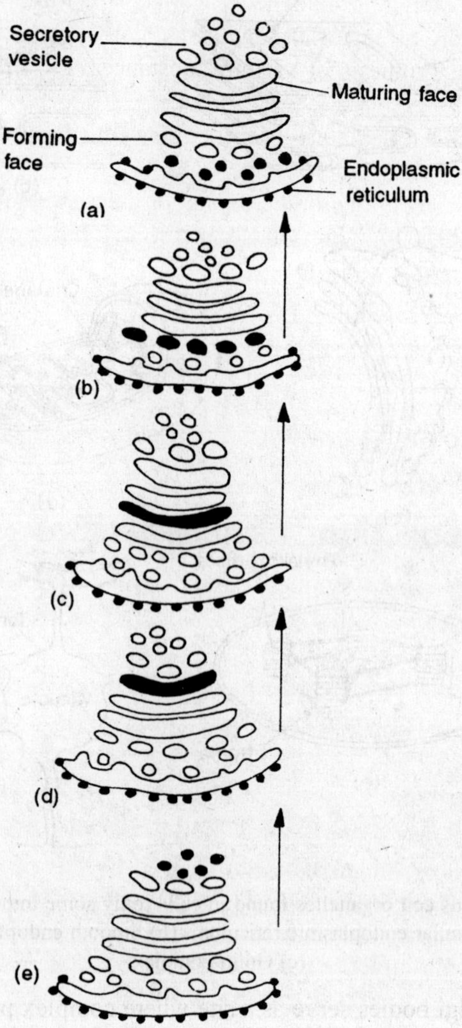

Fig. 5.6a. Membrane flow through Golgi complex: (a) vesicles bud off the endoplasmic membrane, (b) vesicles fuse with the Golgi membranes, (c) and (d) secretions move through the Golgi cisternae, and finally (e) released

ENDOPLASMIC RETICULUM (ER)

The name Endoplasmic Reticulum was coined by Porter (1953) after observing in liver cells. It is a network of tubules and sacs which extend from nuclear membrane throughout the cytoplasm. Cells involved in protein synthesis and storage contain large amount of ER, whereas erythrocytes, (RBC) eggs and embryonic cells lack ER.

ER consists of cisternae, vesicles and tubules.

a) **Cisternae** are long flattened sac like structures with 40-50 mμ in diameter. They are arranged in bundles paralelly. They occur commonly in cells involved in synthesis of various substances. Ex. brain, notochord and pancreas.

b) **Tubules** are reticulately branched system. They occur almost in all cells with a diameter of 50-500 mμ.

c) **Vesicles** are oval membrane bound vascular structures with a diameter of 25-500 mμ they remain isolated in the cytoplasm and are abundantly found in pancreas.

Structurally, it is trilaminar 50-60 Å thick membrane having continuation with the nuclear membrane Golgi body and plasma membrane. Depending upon their surface morphology, E.R. is of two types, viz., Rough E.R. and Smooth E.R.

The rough E.R is coated with ribosomes whereas the smooth E.R. is devoid of ribosomes.

SMOOTH ENDOPLASMIC RETICULUM (AGRANULAR) (SER)

Cells involved in lipid synthesis are rich in S.E.R. Also it is found in glycogen storing liver cells, adipose cells, interstitial cells, leucocytes and spermatocytes. Muscle cells also possess SER; here it is called the sarcoplasmic recticulum. Retinal cells possess them as tightly packed vesicles and hence are called the myeloid bodies (Fig. 5.6).

FUNCTIONS

1) **Synthesis of fats:** Besides providing mechanical support to the cell the SER is actively involved in lipid synthesis. As the fats are highly hydrophobic, they are synthesised in the cytosol. Although some lipids are synthesised in association with other membranes, most of them are synthesised at SER and are transported through the E.R. to their destination.

Eukaryotic membranes are basically composed of phospholipids, glycolipids and cholesterol. Phospholipids are mainly derived from glycerol. They are synthesised on the cytosolic side of the E.R. membrane with the involvement of the enzyme glycerol-3-phosphate and coenzyme A.

2) **Glycogenolysis:** The smooth E.R. is involved in the breakdown of glycogen, through the action of glucose-6-phosphotase. Glycogenolysis by smooth E.R. takes place only in the cells of fasted animals.

3) **Detoxification:** Smooth E.R. performs an important role in detoxification in liver cells. The detoxification reactions also handle the various metabolic waste products like fatty acids, bile salts, steroids and haeme recovered from haemoglobin breakdown.

4) **Cholesterol metabolism:** S.E.R. also plays a role in cholesterol and lipid metabolism in the liver and secretion of CL^- ions in the HCL secreting

cells of the stomach, secretion of triglycerides in the intestinal absorptive cells, synthesis and secretions of steroid hormones by mammalian gonads. The pigments of the retinal cells are also synthesised by S.E.R. from vitamin A.

GRANULAR OR ROUGH ENDOPLASMIC RETICULUM (RER)

Endoplasmic reticulum coated with ribosomes is called the rough E.R. In the cells which are involved in active protein synthesis, such as plasma cells, pancreatic cells, endocrine glands cells and liver cells, E.R. is coated with ribosomes extensively, as these ribosomes take up basic dyes and hence are termed as basiphilic bodies, Nissls bodies etc. Ribosomes are work benches of protein synthesis (Fig. 5.6).

FUNCTIONS

(1) **Sorting of protein & secretion:** Lysosomes or Plasma membrane are synthesised on the membrane bound ribosomes which enter the E.R. for being sorted and secreted.

The m-RNA'S to be translated on membrane bound ribosomes contain a set of unique codons just 3' of the initiation site and translation of these codons yields a unique sequence at the amino terminal of the growing polypeptide called the signal sequence, which triggers attachment of the ribosome to the membrane.

Protein processing is related to the E.R. and the 20 amino acid sequence is not necessary for the functioning of the protein and is involved in tagging the polypeptide to the E.R.

Transmembrane proteins are translocated partially into the lumen of E.R. Further translocation is stopped by a stop-transfer sequence on the polypeptide. The lumen of E.R. contains various enzymes and proteins which are involved in proper folding and functioning of proteins.

(2) **Glycosylation of proteins:** Proteins which enter E.R. are glycosylated on specific aspargine residues on the protein, while their translation is still in process. Glycosylation involves the addition of an oligosaccharide to the aspargine residues. A membrane bound enzyme called oligosaccharyl transferase transfers the oligosaccharide to the Aspargine. Further synthesis of glycerides, fatty acids and fatty acids biosynthesis takes place in E.R. membrane due to the involvement of various enzymes.

COMMON FUNCTIONS OF BOTH S.E.R. AND R.E.R.

(1) An ultra structural framework to the cell is provided.
(2) Transport by permeases and carriers like in plasma membrane.
(3) Increases the area of enzymatic reactions.
(4) Different proteins are secreted from the E.R. to Golgi apparatus and then to secretory vesicles. The proteins destined to E.R. are initially sent to Golgi appratus and then again brought back to the E.R. with

the help of specific receptors and signals added to it.

(5) They organise intracellular signals.

(6) E.R. forms new nuclear envelope after cell division.

(7) Sarcoplasmic reticulum in muscle cells is involved in Ca++ ions releasing during muscle contraction.

RIBOSOMES

Ribosomes are small globular organelles of 250 Å diameter. Robinson and Brown in 1953 have reported, firstly, ribosomes in bean roots. In 1956 Palade, identified ribosomes from Amphibian oocytes and discovered the presence of RNA in ribosome. Chloroplasts mitochondria and nucleus also possess ribosomes. Ribosomes or ribonucleo protein particles are situated either on the wall of endoplasmic reticulum or are present in cytoplasm independently.

It is estimated that mammalian cells contain 10 million ribosomes per cell and in E.coli 10,000 - 20,000 per set. Ribosomes are many in actively metabolising cells. They have 2 subunits and are of 2 types.

a) *70s Ribosomes:* are smaller with a sedimentation coefficient of 70s. They are found in Bacteria, Mitochondria and Chloroplast.

b) *80s Ribosomes:* are larger in size with a sedimentation coefficient of 80s found in cells of higher plants & animals.

Each of these is consisting of two subunits, one larger & other smaller subunit. 70s one is coupled with a 50s subunit and a 30s subunit. 80s subunit is coupled with 60s & 40s.

S = Sved berg units (Sedimentation coefficient)

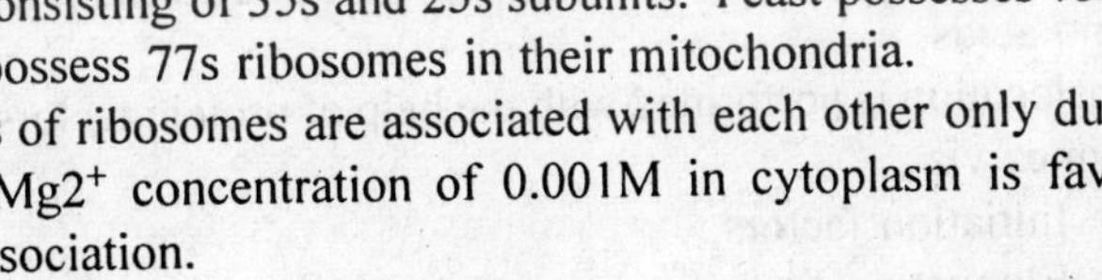

However, certain exceptional ribosomes are seen in mammalian mitochondria, consisting of 35s and 25s subunits. Yeast possesses 73s ribosomes and fungi possess 77s ribosomes in their mitochondria.

Subunits of ribosomes are associated with each other only during protein synthesis. $Mg2^{+}$ concentration of 0.001M in cytoplasm is favourable for subunits association.

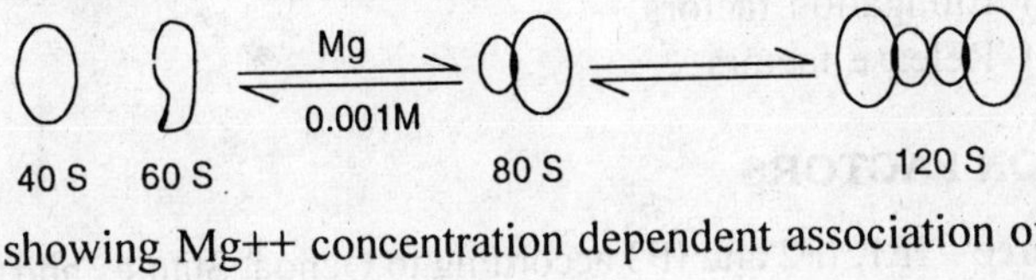

Figure showing Mg++ concentration dependent association of ribosomes or subunits.

Chemical composition: Proteins and RNA are the composition of ribosomes without any lipid content.

Polysomes: Ribosomes are often found in groups; such conditions are considered as polysomes or polyribosomes.

Structure: The 40s and 30s subunits are rather curved and consist of a head, a base, cleft and a platform. The larger subunit is either triangular or rounded in shape. It has two concave sides and one flattened surface.

The cleft of smaller subunit lodges m-RNA and a small gap between the two subunits accommodates t-RNA to bind to it. The concavity of larger subunit accommodates polypeptide.

Ribosomal RNA and proteins: RNA and Proteins are major constituents of Ribosomes. Ribosomal RNA (rRNA) are present in more than 80% of the RNA present in cells.

Prokaryotic ribosomes possess 3RNA molecules whereas in Eukaryotes, there are 4 rRNAs.

The rRNA contains several nucleotides as follows:

23sr-RNA = 3300 Nucleotides.
16sr-RNA = 1650 "
5sr-RNA = 120 "

Protein ratio differs from Prokaryotic ribosome to Eukaryotic ribosome.
E.Coli ribosome possess 40-60% rRNA. and 37% Protein
Eukaryotic ribosome possess 40-44% rRNA and 60% Protein.

In 1960, the role of ribosomes in protein synthesis was enlightened. It is considered that the rRNA provides a structural background for the proteins to catalyse different reactions; further the comparisons with primitive self catalytic RNA (Ex.RNA-ase-P& Ribozymes) led to prove the catalytic role of RNA.

FUNCTIONS

(A) Ribosomes are involved in protein synthesis; specific function being codon and anticodon recognition between t-RNA and m-RNA. This is brought about by r-RNA. r-RNA also participates in protein synthesis due to its base-pairing quality.

(B) To form the polypeptide chain, peptide bond formation between the amino acids.

(C) Translocation is performed with the help of protein factors and certain enzymes viz.,
   (i) Initiation factors,
   (ii) Elongation factors,
   (iii) Release factors.

INITIATION FACTORS

They are three – IF1, IF2 and IF3 according to Ochoa, Stanley and Iwanowski (1968).

IF1: It weighs 92K daltons and helps in the binding of formyl methionine tRNA to 30s subunits.

IF2: It weighs 80K daltons. It has an SH group associated with it, which regulates the binding of GTP.

IF3: It weighs 30K daltons and helps in the binding of m-RNA to the 30s subunit. It also acts as a dissociation factor for 70s ribosomes. Eukaryotes have 6 IFs.

Elongation factors: Two elongation factors are identified. They are EFT and EFG.

(1) EFT is temperature sensitive factor and is of two type i.e. EFTu AND EFTs.

EFTu: Elongation factor temperature unstable; it helps in the transfer of amino-acetyl tRNA to the A-site (GTP is converted to GDP)

EFTs: Elongation factor temperature stable: Its function is to activate the inactivated EFTu - GDP to EFTu - GTP (GDP is converted to GTP)

(2) EFG is also called as G-factor and also as translocase as EFG along with GTP helps in the movement of ribosome on the mRNA (Translocation) during protein synthesis.

Release factors: 2 release factors RF1 and RF2 are available in prokaryotes. They identify the termination codons UAA, UAG and UGA and help in the termination of polypeptide chain.

Enzymes: An enzyme known as peptidyl transferase is the 50s ribosomal subunit, which helps in the formation of peptide bonds. This enzymes nature is similar to that of ribozymes, i.e. catalytic RNA along with some proteins.

Cytoskeletal structure: The term cytoskeleton is applied to the microtubules, microfilaments and also intermediate filaments spread through the cytosol. Tubulin (in Microtubules) Actin, Myosin, Tropomyosin (in Microfilaments) are the main proteins present in the cytoskeleton.

Microtubules: De Robertis and Franchi in 1953 have identified microtubules in nerve cells, and were called as neurotubules or nerve tubules. Sabatini, Bansch, Barnette in 1963 have explained actual structure of microtubules. Ledbetter and Porter have found microtubules in plant cells.

Microtubules are found in Cilia, and flagella, basal granules, Centrioles, Kinetochore, polarcaps in plants, spindle fibres; axoneme in protozoa, cortex of meristematic cells in plants and elongated cells like lens cells and sperms.

Structure: Mitrotubules are long, unbranched and hollow, completely made up of proteins. In length they range several micrometers & in dia 210-250 Å. Microtubule is composed of protein sub units that are globular and are called "tubulin protein".

Two different monomers, Tubulin-a and Tubulin-b are present. The a & b subunits are associated as dimer pH to form the microtubule depends upon pH, temperature, cation concentration and GTP. It was found that microtubules assemble at 37°C. Mg++, Ca++ are also necessary for the association of these units to form a microtubule. Besides Tubulin protein, microtubules also contain MAP - Microtubule Associated protein, which is involved in microtubular assembly.

FUNCTIONS

Microtubules help in movement of Cilia, and Flagella, Cyclosis or Cytoplasmic streaming, maintaining the cell structure movement of chromosomes during cell division, the division of the cell during cytokinesis. Microtubules involve in shaping the cell during cell differentiation. Further they help protozoans to procure their food, transport of particulate matter, including determination of intrinsic polarity in certain cells.

Microfilaments: Microfilaments consist of Actin Filaments which are smaller in structure than microtubules and very active. They were first observed in muscle cells in 1947.

Actin filaments consist of monomers. They are proteinaceous in nature and each actin monomer is known as G-Actin (Globular Actin) which polymerise to form F-actin (Fillamentous actin). Each G-actin is made up of 375 amino acids. Association and Dissociation is dependent upon Cation concentration. Polymerisation of monomers is also helped by Actin binding proteins. The actin monomers attached to ATP residues mainly attack at +ve end. While the actin monomers with ADP residues are no residue remain at the -ve end. There is a rotation of 160° in between one molecule and other when an actin filament is formed which results in the formation of double stranded helix like appearance.

Intermediate filaments: They are unbranched structures having a diameter of about 100Å, which are intermediate between microtubules and microfilaments. These are not directly involved in the cell mobility, but give mechanical strength to the cell. More than 50 types of intermediate filaments have been identified. They have been divided into 6 types.

(1) Acidic Keratin (epithelial cells)
(2) Neutral Keratin & Basic Keratin (epithelial cells)
(3) Vimentin in (WBC & Fibroplast)
Desmin (Muscle cells)
Peripherin (Peripheral nervous system)
(4) Neurofilament proteins (In nerves)
(5) Nuclear lamins (Nuclear lamina)
(6) Nestin (Central Nervous System)

MITOCHONDRIA:

Mitochondria were first described by Altmann in 1890 as Bioblasts. The term Mitochondrion was coined in 1897 by Benda. They are commonly referred to as power houses of the cell and specialize in the synthesis of adenosine triphosphate (ATP), the useful chemical form of energy.

They are present in all aerobic eukaryotic cells except in RBC. They are uniformly distributed in the cytoplasm. In some cases they are located near the nucleus. In general, mitochondria are located near such structures where energy requirements are heavier.

Structure: Mitochondria are surrounded by a double membrane system consisting of an inner and outer mitochondrial membrane separated by an inter membrane space. The inner membrane forms numerous folds called cristae which extend into the interior of the matrix.

Outer mitochondrial membrane is freely permeable to small molecules. It contains proteins called Porins, which form channels that allow the free diffusion of molecules smaller than about 6000 daltons. Composition of the intermembrane space is similar to the cytosol with respect to ions and small molecules (Fig. 5.7).

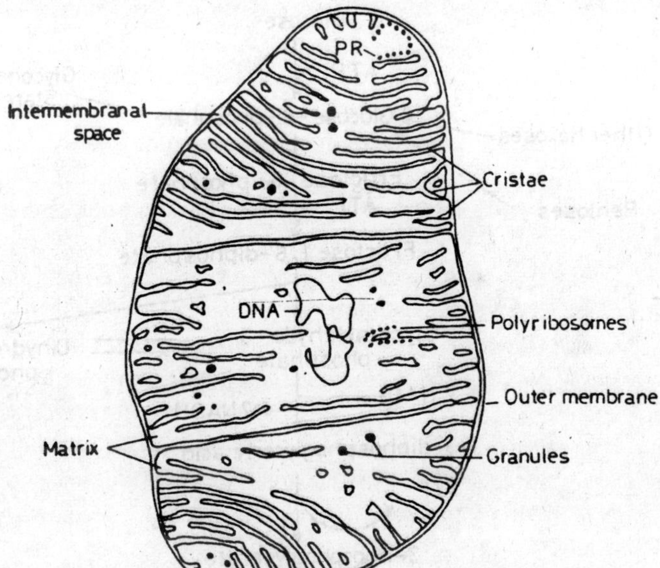

Fig. 5.7. Diagrammatic presentation of a mitochondrium showing typical internal structure

Mitochondria contain their genetic system, which is separate and distinct from the nuclear genome of the cell. Mitochondrial genomes are usually circular.

Functions: Oxidative phosphorylation (Fig. 5.7a):

Majority of the usable energy obtained from the breakdown of carbohydrates or fats is derived by oxidative phosphorylation which occurs within the mitochondria. The breakdown of glucose by glycolysis and citric acid cycle yields a total of four molecules of ATP and ten molecules of NADH and two molecules of FADH2. They are then transferred to molecular

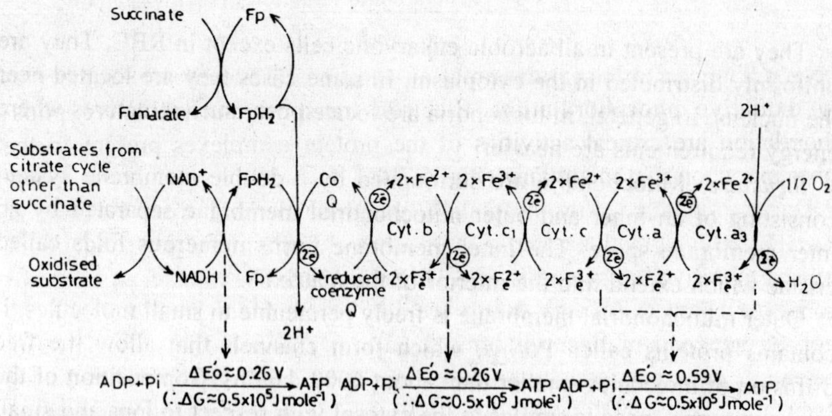

Fig. 5.7a. Electron transport chain and the locations where ATP is synthesised

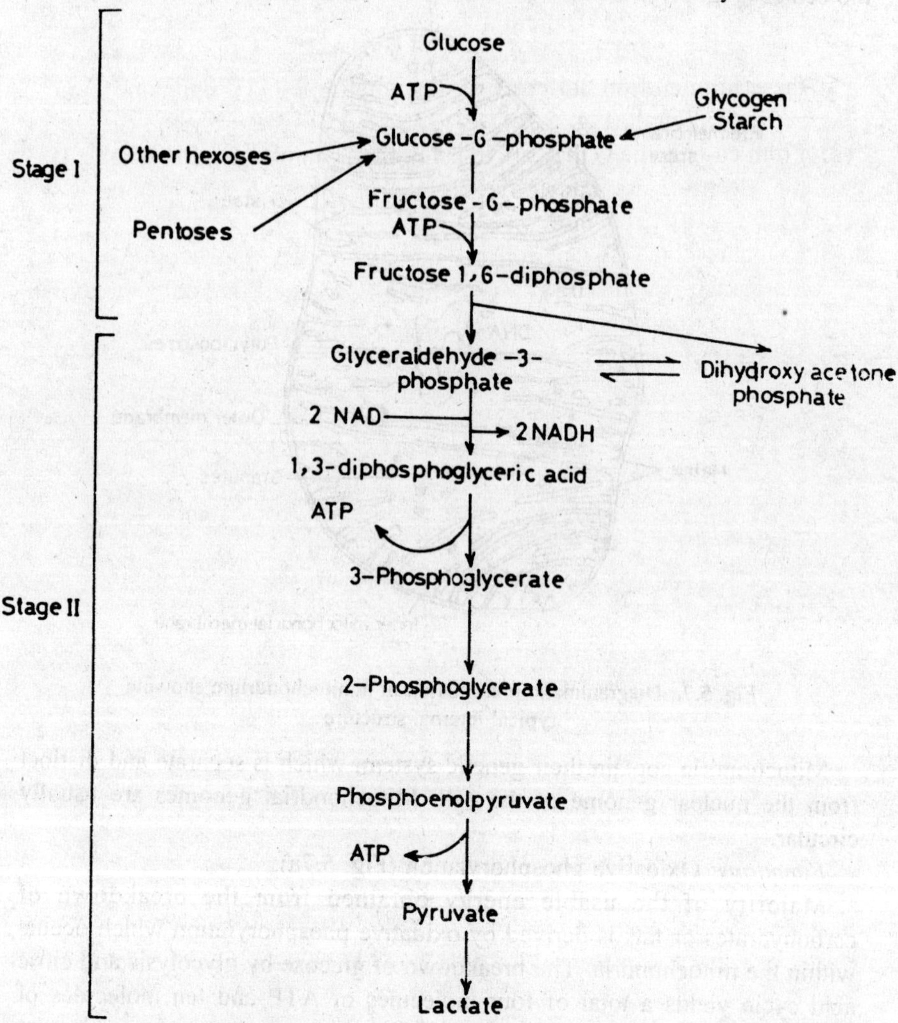

Fig. 5.7b. The two stages of glycolysis

oxygen, coupled to the formation of an additional 32 to 34 ATP molecules by oxidative phosphorylation. Electron transport and oxidative phosphorylation are critical activities of the protein complexes present in the inner mitochondrial membrane (Fig. 5.7b).

During oxidative phosphorylation, electrons derived from FADH2 & NADH combined with oxygen and the energy released from these reaction is utilised to derive the synthesis of ATP from ADP. The transfer of electrons from NADH to O2 is a very energy yielding reaction with Δ G° = 52.5 Kcal/mol for each pair of electrons transferred to be harvested in a usable form; the energy must be graudally produced by the passage of electrons through a series of carriers which constitute the electron transport chain (Fig. 5.8).

The steps of electron transport are as follows:

(1) Electrons from NADH enter the electron transport chain in complex I. These are first transferred to flavin mononucleoxide (FMN) and these form an Iron Sulfer carries to co-enzyme-Q (Ubiquinone) Δ G° = 52. 5Kal/mol. (Fig. 5.7c).

(2) From co-enzyme Q they are transferred to complex III. In complex III the electrons are transferred from cytochrome b to cytochrome c (Δ G° = 10.1K cal/mol).

(3) Cytochrome-C is a peripheral protein of the mitochondrial inner membrane. The electrons are then transferred to complex IV which is also called cytochrome oxidase, where they are finally transfered to O2 (Δ G° = –52.5kcal/mol.

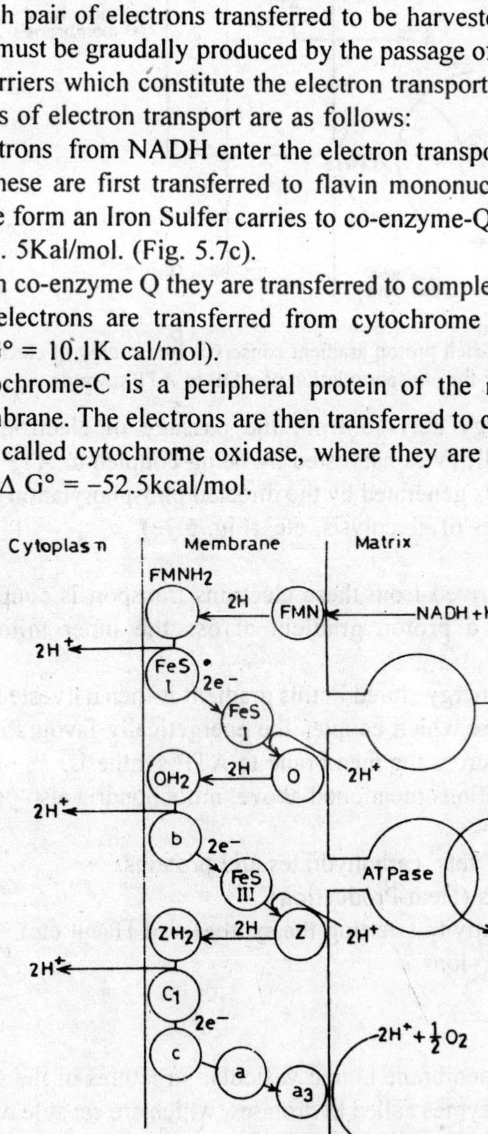

Fig. 5.7c Showing development of electrochemical proton gradient across the inner mitochondrial membrane, explaining chemiosmotic hypothesis

(4) When electrons are transferred from succinate to FADH2 Complex II receives them unlike complex I in the other process when electrons are accepted from NADH (Fig. 5.7d).

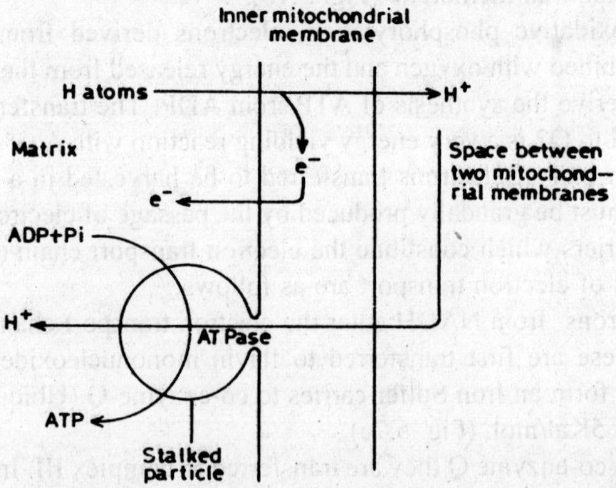

Fig. 5.7d. The energy-rich proton gradient conserves free energy of electrons and ultimately accounts for phosphorylation of ADP to ATP towards the matrix

(5) The free energy derived from the passage of electrons through complexes I, III, IV is harvested by being coupled to ATP synthesis.

(6) Here the ATP is generated by the directed phosphorylation of ADP to ATP as in cases of glycolysis, etc. (Fig. 5.7e).

(7) The energy derived from these electrons transport is coupled to the generation of a proton gradient across the inner mitochondrial membrane.

(8) The potential energy stored in this gradient is then harvested by a fifth protein complex which couples the energetically favourable flow of protons back across the membrane to ATP synthesis.

Other than the functions mentioned above, mitochondria also perform the following functions:

(1) Degradation of fats, carbohydrates and proteins.

(2) Thermiogenesis (Heat Production).

(3) Biosynthetic activity (starting the synthesis of Haem etc).

(4) Storage of Ca^{++} ions.

LYSOSOMES:

Lysosomes are tiny, membrane bound vesicular structures of the cell which possess destructive enzymes called hydrolases, which are capable of breaking down macromolecules. Deduve in 1955 reported about lysosomes.

Lysosomes occur in all mammalian cells except in mature RBC. Many plant cells also are devoid of it. Pancreatic cells, liver cells, leucocytes, spleen cells have large number of lysosomes. Macrophages, too, have large number of lysosomes.

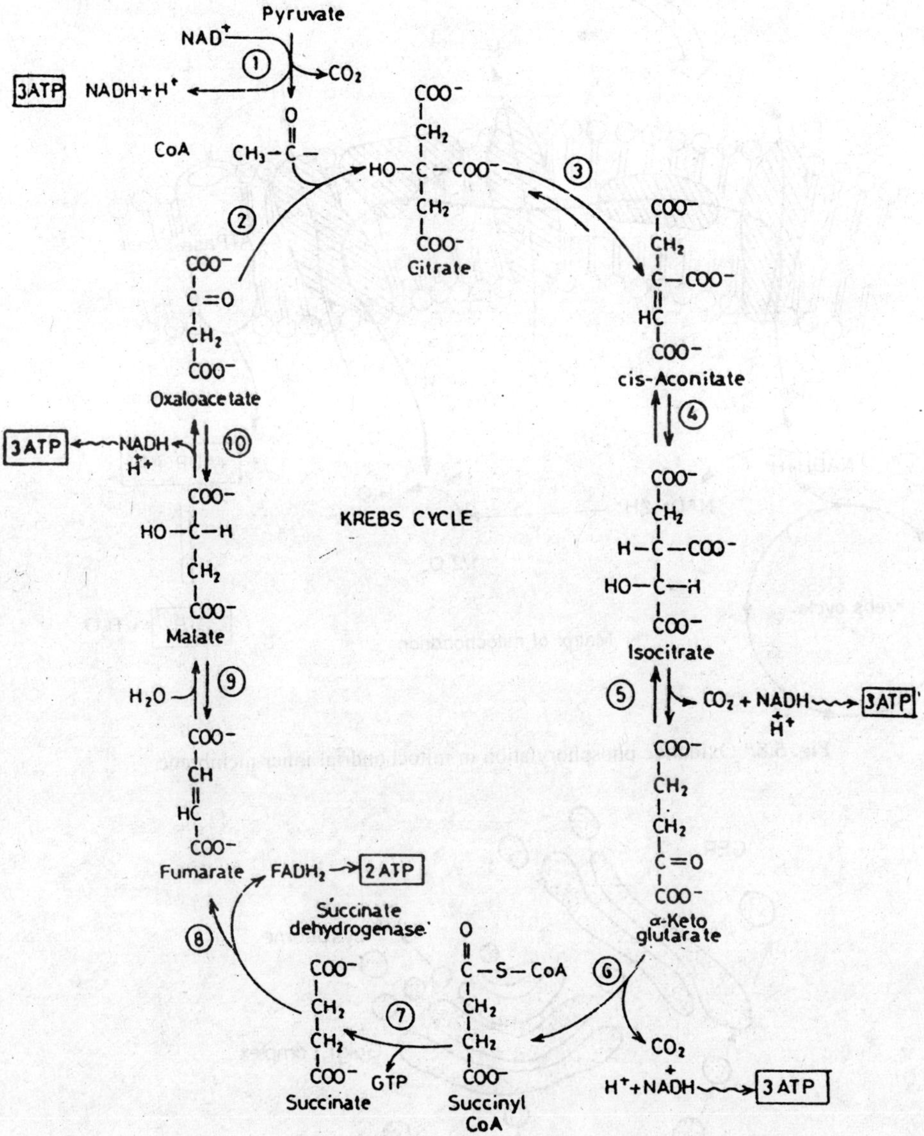

Fig. 5.7e. The tricarboxylic acid cycle (Krebs cycle)

Lysosomes are spherical in shape. Size varies from 0.2 to 0.8μm; in kidney cells their size is 5μ; they are larger in phagocytes and leucocytes. They are round vacuolar structures filled with dense material including acid hydrolases and phosphatases. They are bounded by a unit membrane (Fig. 5.9).

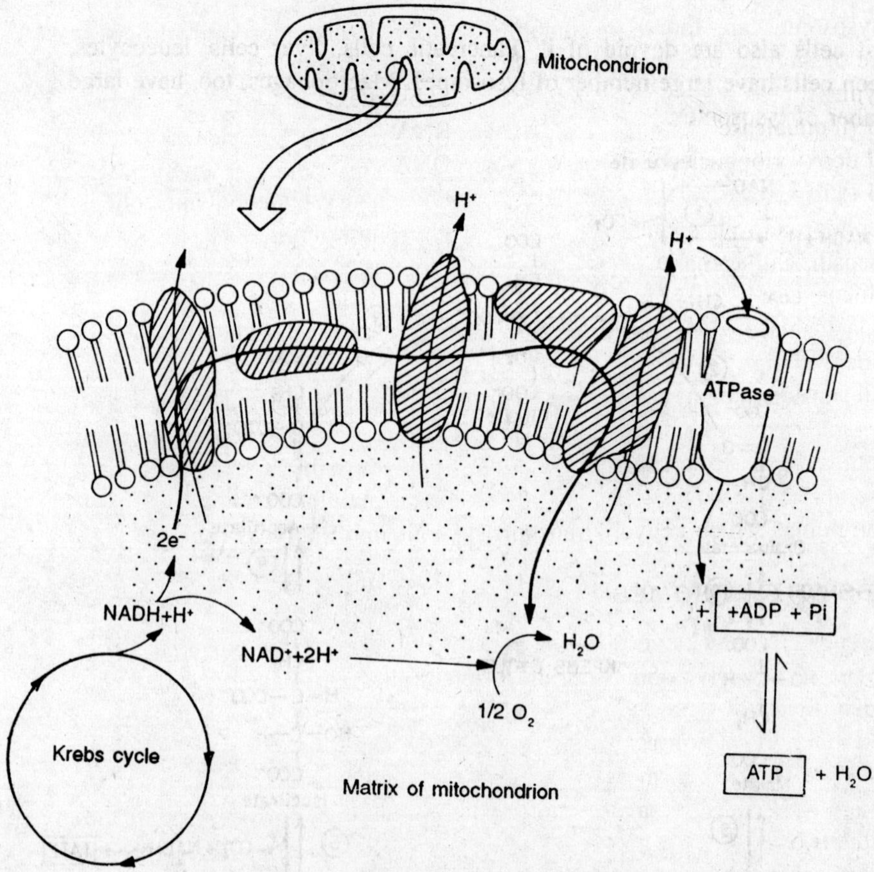

Fig. 5.8. Oxidative phosphorylation in mitochondrial inner membrane

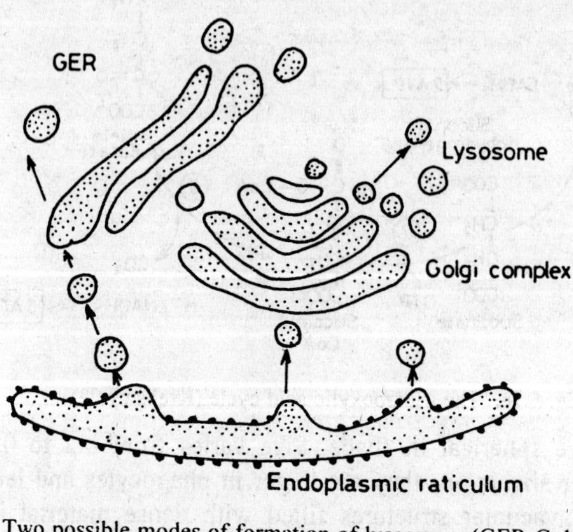

Fig. 5.9. Two possible modes of formation of lysosomes (GER Golgi-associated endoplasmic reticulum)

Lysosomes are found to contain 40 different enzymes or hydrolases. Some of the enzymes and their substrates are listed below:

Enzymes	Substrates
Acid ribonuclease	RNA
Acid deoxy ribonuclease	DNA
Acid Phosphatase	Phosphate esters mononucleotides
Phosphoprotein phosphatase	Phosphoproteins
Cathepsin, Collagenase	Proteins
α-Glucosidase	α-Glucosides ← Glycogen
β-Galactosidase α-Mannosidase }	Mucopolysaccharides
β-Glucoronidase }	

All lysosomal enzymes are acid hydrolases and proteases that are active at pH5 (acidic), but not at neutral - pH of the cytoplasm. This helps as a precaution against the damage of cellular organelles and components due to lysosomal damage. The acidic pH of lysosome is maintained with the help of proton pumps which actively transport H+ ions into the lysosome.

KINDS OF LYSOSOMES

Different cells may have different lysosomes or the same cell may have different forms of lysosomes during different phases of development. They are identified as 4 types.

1) **Primary lysosomes:** Newly formed lysosomes are considered primary lysosomes. They are formed as vesicles which bud off from the Golgi body. They are rather inactive and do not involve in the digestive function of the lysosome.

2) **Secondary lysosome:** Secondary lysosomes are also heterophagosomes or digestive vacuoles. When cells feed on foreign or exogenous substances by phagocytosis, the plasma membrane forms membrane bound vesicles called phagosome or pinosome. Primary lysosomes fuse with phagosome and form secondary lysosome or phagolysosome. In neutrophils, the secondary lysosome is formed after the fusion of specific granules with the phagosome. Secondary lysosome functions to mix hydrolases with the ingested particles in the phagosome thus promoting digestion of these substances.

Once the complex particles in the phagosome are digested to simple compounds, they diffuse out of the lysosome leading to diminishing size of lysosome. The reduction in surface area occurs by inward budding of lysosomal membrane. Many such vesicles are progressively formed leading to the accumulation of vesicles inside lysosome. Such a body is called multivesicular body.

3) **Autophagic bodies:** Lysosomes are also involved in the formation of autophagic vacuoles. These vacuoles are formed as a result of phagocytosis of the intracellular organelles which are non functional. These vesicles are called autophagosomes. The primary lysosomes fuse with autophagosomes and form autophagolysosomes which are actively involved in the digestion

of phagocytosed components in the vesicles by the action of lysosomal hydrolases.

4) **Residual bodies:** Though most of the materials are digested inside the lysosome some cannot be digested; such materials remain in lysosome to form resudual bodies. Contents of residual bodies are released from the cell by exocytosis.

FUNCTIONS

(1) Digestion of large extracellular particles by the formation of phagolysosomes. They are also responsible for killing of bacteria and viruses that are phagocytosed, by fusion with phagosome (Fig. 5.10).

(2) During starvation lysosomes are seen actively involved in the digestion of intracellular organelles for energy.

(3) Autolysis also occurs with the help of lysosome. In insects and amphibian larvae, during metamorphosis loss of various body parts occurs by lysosomal action. This process is called autophagy and lysosomes are commonly considered suicide bags of the cell.

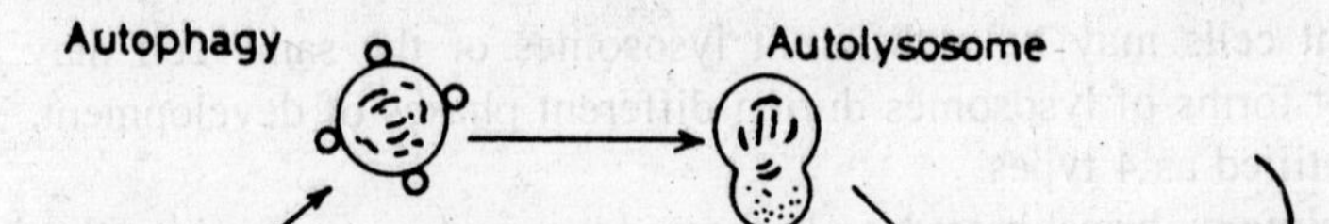

Fig. 5.10. Cellular digestion in lysosomes showing crinophagy and formation of residual bodies

(4) Extracellular digestion by the secretion of lysosomal enzymes in some gametes like sperm also done by lysosomes.

(5) Lysosomes are also supposed to initiate mitosis in cells.

LYSOSOMES AND DISEASES

As a result of defects in normal lysosomal function various diseases occur.

Hydrogen peroxide is an important chemical in phagocytic killing and the disease is called chronic granulomatic disease. This character is explained by an X-linked recessive condition. In this disease due to absence of H2O2 in lysosomes, killing of bacteria and fungi also do not occur. Subsequently many macrophages just phagocytose bacteria and retain them within themselves and accumulate in an area. This condition is called Granuloma and the disease is termed "Chronic granulomatous diseases".

Gaucher's disease occurs where lysosome lacks an enzyme "gluco-cerebrosidase"which cleaves glucocerebroside to glucose and ceramide. Due to the deficiency the substrate i.e. glucocerebroside, accumulates inside the lysosome causing disease.

Diseases also occur when lysosome fails to fuse with phagosome. At this time bacteria present in phagosome divide and multiply and cause disease. The diseases like Gaucher's disease which result from the deficiency of lysosomal hydrolasers are called lysosomal storage diseases as they result in accumulation of a particular product in lysosome.

NUCLEUS

The nucleus was discovered by Robert Brown in 1831. The nucleus is considered to be the core of the cell. It commands all metabolic events. Miescher (1869) discovered nucleoproteins and nucleic acids and suggested that it is an important component of the nucleus. In 1888, Waldeyer observed and designated chromosomes.

The number of nuclei per cell may vary from one to many. Generally, cells contain a single nucleus, and such cells are called uninucleate cells. Paramaecium protozoan contains 2 nuclei and is called as Binucleate. Some cells possess 3 or more nuclei and are called as Polynucleate. Polynucleate cells of plants are called coenocytic cells while those of animals are called syncytial cells.

The nucleus may be elliptical or discoid in shape; it is covered with a nuclear envelope which contains nuclear pores involved in transport of molecules across the nuclear membrane. Nuclear membrane separates the nucleus from cytoplasma and peroxides the structural framework of nucleus.

Nuclear membrane: is derived from the membrane system of the cell. It is a double layered structure; outer layer is continuous with endoplasmic reticulum. Several nuclear pore complexes exist on this membrane (Fig. 5.11).

Nucleus contains a number of proteins, such as RNA polymerase, DNA polymerase, several types of cytochromes and histones; all are synthesised in cytoplasm of the cell and transported into the nucleus.

NUCLEAR MATRIX

Within the nuclear envelope the nucleus body possesses a dense jelly like mass composed of karyolymphs and chromatin. Karyolymph is rich in proteins with a small percentage of DNA, RNA and Phospholipids. In this matrix DNA replication, transcription and transport of substances take place. Nuclear metrix has a network of interchromatinic matrix.

There is a nucleolus in the nucleus; its number may vary in different cells. Cells lacking nucleolus show little or no protein synthesis. Nucleolus or nucleoli are situated along certain chromosomes. The morphology of nucleolus undergoes change during cell cycle. During prophase it disappears and during Metaphase it reappears.

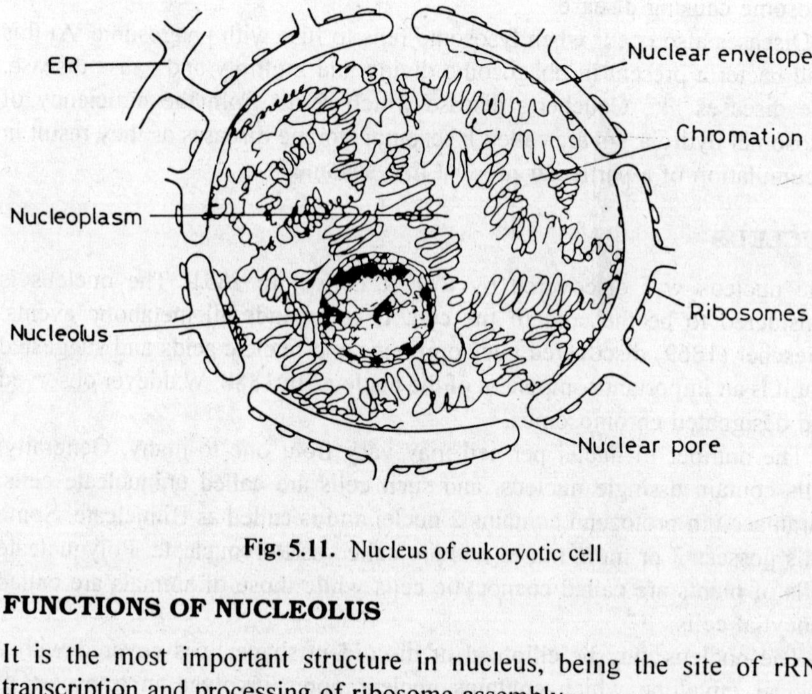

Fig. 5.11. Nucleus of eukoryotic cell

FUNCTIONS OF NUCLEOLUS

It is the most important structure in nucleus, being the site of rRNA transcription and processing of ribosome assembly.

It possesses genes for the 5s, 58s, 18s and 28s tRNA. The 58S, 18S and 28s, rRNA are transcribed as a set and the 5srRNA is transcribed out of the nucleolus. There are around 2000 'copies of the genes that code for 5.8s, 18s and 28srRNA and around 2000 copies of the genes that code for 5s rRNA.The 5.8s,18s and 28s rRNA. genes are distributed on five different chromosomes in humans (13, 14, 15, 21 and 22) and 5s genes are present on chromosome No.1. Repeated ribosomal genes serve to produce a large number of ribosomes required for protein synthesis.

Nucleolus does 3 major functions i.e. (i) Transcription of the genes that code for ribosomal RNA. ii) Processing of periribosomal molecule and iii) Assembly of ribosomal subunits.

The nucleolus organiser region of the chromosome consists of multiple copies of the genes that code for rRNA, some of which are redundant. The nucleolar genes produce rRNA, hence they are called ribosomal genes which are rich in nucleotides containing guanine and cytosine.

The 45s rRNA is methylated, cleaved and reduced to 18s and 28s units, and released in the cytoplasm where they remain either free or bound to the endoplasmic reticulum.

The circular DNA molecules present in the extra nucleolus like bodiesproduce large amounts of rRNA. These extra nucleoli and the rDNA disappear gradually with the progress of embryogenesis.

Chromatin: is considered as the most important part of the nucleus as it forms the basis of heredity. It contains DNA, RNA and protein in a compact form, in which most of the DNA sequences are functionally inactive and inaccessible due to binding of proteins.

Chromatin is constructed of several subunits of each 200 base pairs of DNA organised by basic proteins called histones, into a bead like structure. The DNA cannot be directly packaged into chromatin. Therefore, it has 3 levels of organisations:

(i) winding of the DNA into bead-like particle, which are invariant competant of euchromatin, heterochromatin and chromosomes.

(ii) coiling of series of beads into a helical chain to form the chromatin fibre.

(iii) finally, the level or organisation determined by the packaging ratio of the fibre itself.

CHEMISTRY OF CHROMATIN

Chromatin is a viscous substance containing DNA, RNA and PROTEINS. It consists of more proteins than DNA. The proteins are of histones and non-histones. RNA is less than 10% of the mass of DNA and is mostly in the form of nascent chains that are still associated with the DNA template.

FUNCTIONS OF THE NUCLEUS

(1) Nucleus carries the genetic material, i.e DNA.

(2) DNA also helps in regulation of gene expression by selective import of proteins into the nucleus. Proteins responsible for genome structure and organisation are all imported into the nucleus selectively. They include histones, DNA polymerases, RNA Polymerases, transcription factors and splicing factors. These proteins are targeted to the nucleus by specific amino acid sequences called nuclear localisation signals. These signals direct their transport through the nuclear pore complex.

In the absence of ATP proteins containing nuclear localisation, signals bind to nuclear pore complex but do not pass through the pore.

Transport of proteins through nuclear pore is an energy dependent process. It requires ATP and GTP.

Some proteins move within the nucleus and some keep moving in and out of the nucleus. These proteins are shuttling proteins. Import of proteins is regulated by either phosphorylation or by binding of proteins which cover the nuclear localization signals for importing, binding and subsequent protein transport.

RNAs are transported out of the nucleus; they are transported as ribonucleoprotein complexes. One of the proteins associated with the RNA acts as a signal for transport.

(3) Nuclear matrix provides ultraskeletal structure for the nucleus and acts as a support for various reactions and domains.

Nuclear region contains a cluster of rRNA genes. They are very actively transcribed by RNA Polymerose I.

Primary transcript of the rRNA genes is a large 45s pre-RNA which contains 18s, 28s and 5-8s rRNA. External transcribed regions are present on both 5' and 3' ends of the pre-rRNAs and 2 internal transcribed spacers lie between the 5, 8s, 18s and 28s rRNA. Removal of ETs and ITs give rise to independent rRNA fragments. This process is called splicing and is helped by an array of mRNAs. After splicing the rRNAs along with proteins, and 5sRNA are packed in the nucleus to form Ribosomal subunits, and they are transported out of the nucleus into cytoplasm.

PEROXISOMES

Protozoans, cells of fungi, plants, liver and kidney of vertebrates possess certain membrane bound spherical bodies of 0.2 to 1.5 μm diameter in close association with endoplasmic reticulum, mitochondria or chloroplast. They are called microbodies or peroxisomes.

Peroxisomes are small, membrane enclosed organelles that contain enzymes involved in a variety of metabolic reactions, including several aspects of energy metabolism. Though peroxisomes are morphologically similar to lysosomes, they are assembled like mitochondria and Chloroplasts from

proteins, which are synthesized on free ribosomes and imported into peroxisomes as completed polypeptide chains. Though peroxisomes do not contain own genomes, they are similar to mitochondria and chloroplasts in that they replicate.

Peroxisomes occur in many animal cells and in a wide range of plants. They occur in all photosynthetic plants, brown algae, fungi and protozoans.

Peroxisomes in which glyoxylate cycle occurs are often called Glyoxysomes. They are found in yeasts, neurospora, oil rich seeds etc.

FUNCTIONS

Peroxisomes contain at least 50 enzymes involved in biochemical pathways in different types of cells.

(1) Peroxisomes are originally defined as organelles which carry out oxidation reactions leading to the production of hydrogen peroxide.

(2) As H_2O_2 is harmful to the cell they also contain catalase and enzymes which decompose hydrogen peroxide either by converting it to water or by using it to oxidise another organic compound.
(3) They are thus involved in oxidation of a variety of compounds,viz., uric acids, amino acids, and fatty acids.
(4) Oxidation of fatty acids is of great improtance, as it provides a major source of metabolic energy.
(5) In animal cells, fatty acids are oxidised only in peroxisomes.
(6) Peroxisomes are also involved in lipid biosynthesis. In animal cells cholesterol and dolichol are synthesised in peroxisomes.
(7) In liver, peroxisomes are involved in the synthesis of bile acids, which are derived from cholesterol.
(8) Peroxisomes contain enzymes required for the synthesis of plasmalogens, a family of phospholipids. Plasmalogens are important tissue components in heart and brain.
(9) Peroxisomes play a two way important role in plant cells. In seeds they are responsible for conversion of stored fatty acids to carbohydrates, which provide energy and raw materials for growth of the germinating plant. It occurs through a series of reactions forming the glyoxylate cycle. Thus peroxisomes are also called Glyoxysomes.
(10) In leaves they are involved in photorespiration which serves to metabolise a side product formed during photosynthesis. During photosynthesis Co_2 is converted into carbohydrates by calvin cycle. Through successive steps Glycine is transported to mitochondria, where it is converted to serine. Serine is sent back to peroxisome, where it is converted to glycerate which again enters the calvin cycle in the chloroplast. Thus it helps in utilising most of the carbon in the glycolate.

PEROXISOME ASSEMBLY

Peroxysomal proteins are synthesised on free ribosomes in the cytosol and are imported to the peroxisomes as complete peptide chains. Two signals are involved in targeting proteins into peroxisomes. One is the simple sequence of 3 amino acids ser-lys-leu, and the second one is a nine amino acid sequence. However, the import mechanism is not well understood.

CHLOROPLAST

Chloroplasts were first described by Nehemiah Grew and Antony Van Leeuwenhock in the 17th century. Dutrochet in 1837 observed that chlorophyll is essential for O_2 evolution from plants. Meyer first described the structure of chloroplast in 1883. Chloroplasts belong to to the group of plastids, a term coined by Schimper in 1883.

Chloroplasts are characterised by the presence of green chlorophyll pigments. There are several types of plastids, which are classified as per presence or absence of pigments, viz., proplastid, amyloplast, leucoplast, elioplast, chloro-amyloplast, chromoplast, chloroplast, pheoplast, rhodoplast, etc.

Structure: Chloroplast is in many ways similar to mitochondria. Both of these are involved in generation of metabolic energy, contain their own genome and perform various critical functions.

Plant chloroplasts are large organelles (5-10 μm long) bounded by a double membrane-chloroplast envelope. Besides these two, chloroplast has a third inner membrane system-thylakoids membrane; this forms a network of

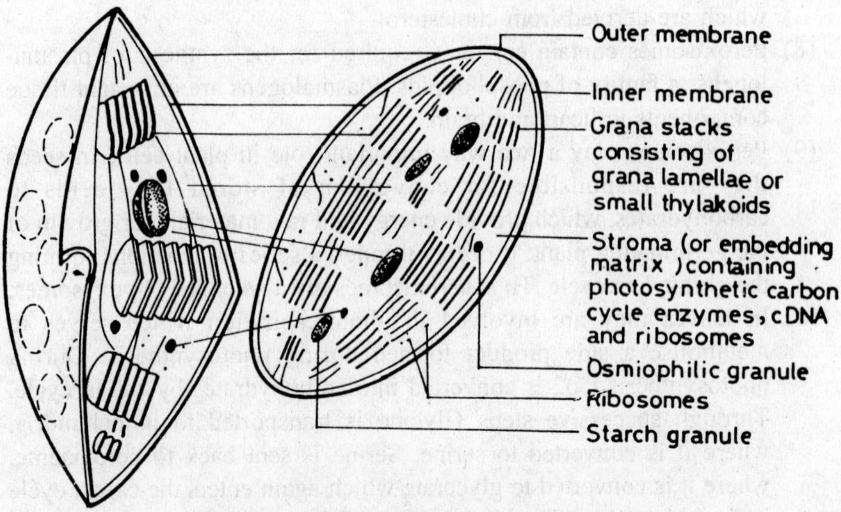

Fig. 5.12. Internal structure of the chloroplast

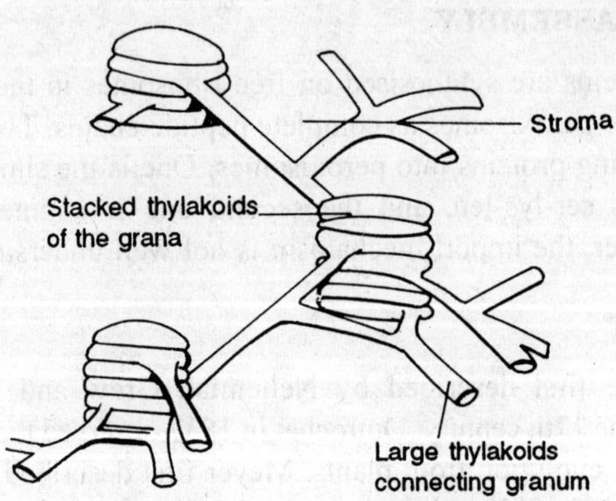

Fig. 5.12a. Internal structure of chloroplast showing arrangement of thylakoids in the stroma. The grana are interconnected

flattened discs called thylakoids. Thylakoids are arranged in stacks called grana. These three membranes divide chloroplasts into three different compartments (Fig. 5.12 and Fig. 5.12a).

1) Intermembrane space between two membranes of chloroplasts envelope.
2) Stroma which lies inside the envelope but outside the grana.
3) Thylakoid lumen (Fig. 5.13).

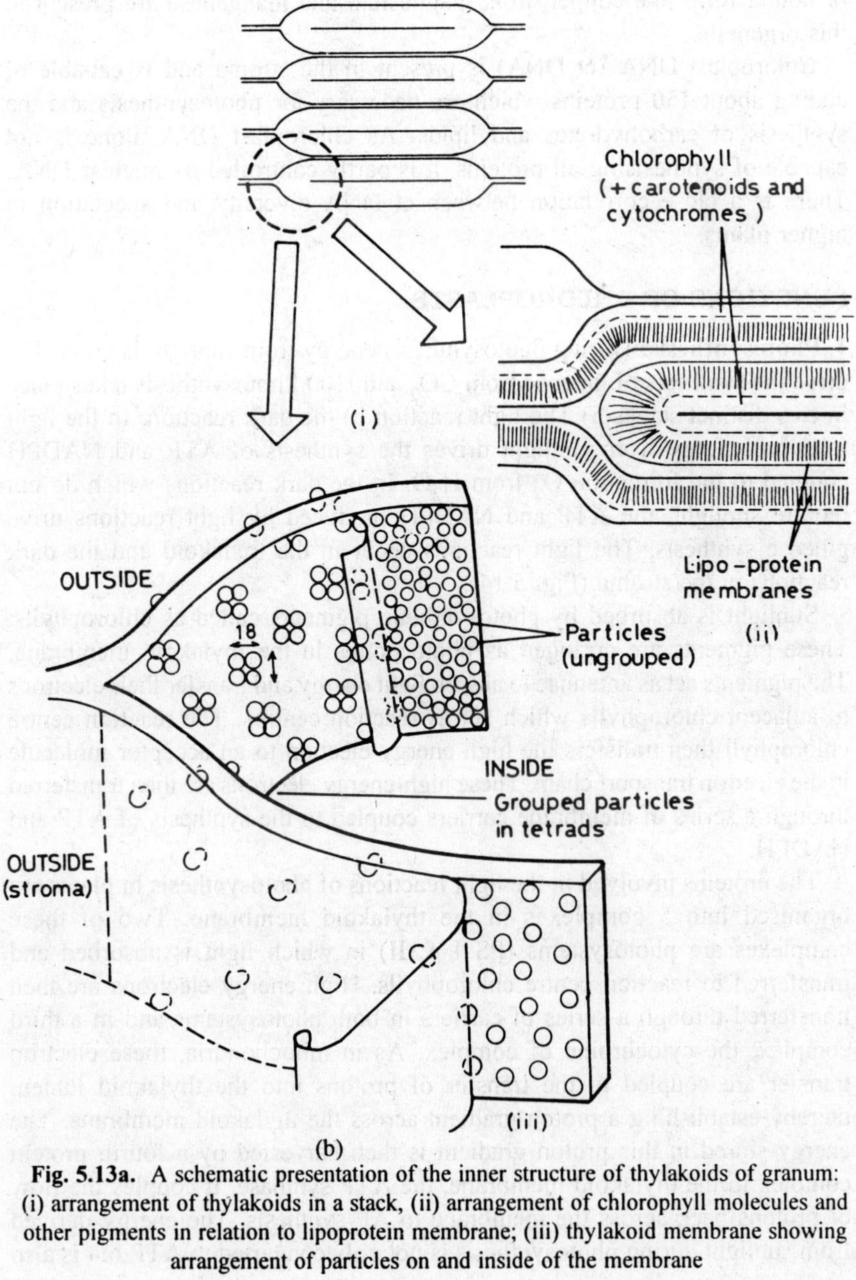

Fig. 5.13a. A schematic presentation of the inner structure of thylakoids of granum: (i) arrangement of thylakoids in a stack, (ii) arrangement of chlorophyll molecules and other pigments in relation to lipoprotein membrane; (iii) thylakoid membrane showing arrangement of particles on and inside of the membrane

Chloroplasts are rich in proteins, lipids, carbohydrates, nucleic acids, minerals and pigments. They also contain phospholipids such as phosphotidyl glycerol, phosphatidic acid, inositol and ethanol amine. DNA & RNA are also associated with chloroplasts. As chloroplasts synthesise starch, they store carbohydrates. Haeme and Chlorophyll are the major pigments in chloroplats, but chlorophyll-A is more prominent. Several elements in free or bound form like copper, iron, magnesium and manganese are present in this organelle.

Chloroplast DNA (ct DNA) is present in the stroma and is capable of coding about 150 proteins which are necessary for photosynthesis and the synthesis of carbohydrates and lipids. As chloroplast DNA alone is not capable of synthesising all proteins, it is partly controlled by nuclear DNA. There is a close correlation between ct DNA diversity and speciation in higher plants.

FUNCTIONS OF CHLOROPLASTS

1. **Photosynthesis:** During photosynthesis energy from sunlight is utilised to drive the synthesis of glucose from CO_2 and H_2O.Photosynthesis takes place in two distinct stages a) The light reaction b) the dark reaction. In the light reaction, energy from sunlight drives the synthesis of ATP and NADPH coupled to the formation O_2 from H_2O. In the dark reactions which do not require sunlight, the ATP and NADPH produced by light reactions drive glucose synthesis. The light reactions occur in the thylakoid and the dark reactions in the stroma (Fig. 5.14).

Sunlight is absorbed by photosynthetic pigments called as chlorophylls. These pigments are arranged as photocentres in the thylakoid membrane. The pigments act as antennae to absorb light energy and transfer their electrons to adjacent chlorophylls which act as reaction centres. The reaction centre chlorophyll then transfers the high energy-election to an acceptor molecule in the electron transport chain. These high-energy electrons are then transferred through a series of membrane carriers coupled to the synthesis of ATP and NADPH.

The proteins involved in the light reactions of photosynthesis in plants are organised into 5 complexes in the thylakoid membrane. Two of these complexes are photosystems (PS I & II) in which light is absorbed and transferred to reaction centre chlorophylls. High energy electrons are then transferred through a series of carriers in both photosystems and in a third complex, the cytochrome of complex. As in mitochondria, these electron transfer are coupled to the transfer of protons into the thylakoid lumen, thereby establishing a proton gradient across the thylakoid membrane. The energy stored in this proton gradient is then harvested by a fourth protein complex in the thylakoid membrane, the ATP synthase. It couples the flow of protons back across the membrane to AP synthesis. The energy derived from sunlight during photosynthesis is not only converted to ATP but is also used to synthesise NADPH required for the subsequent conversion of CO_2 to carbohydrate the subsequent conversion of CO_2 to carbohydrate. One photosystem is used to synthesise ATP and the others for NADPH synthesis. (PSI for NADPH and PS II for ATP).

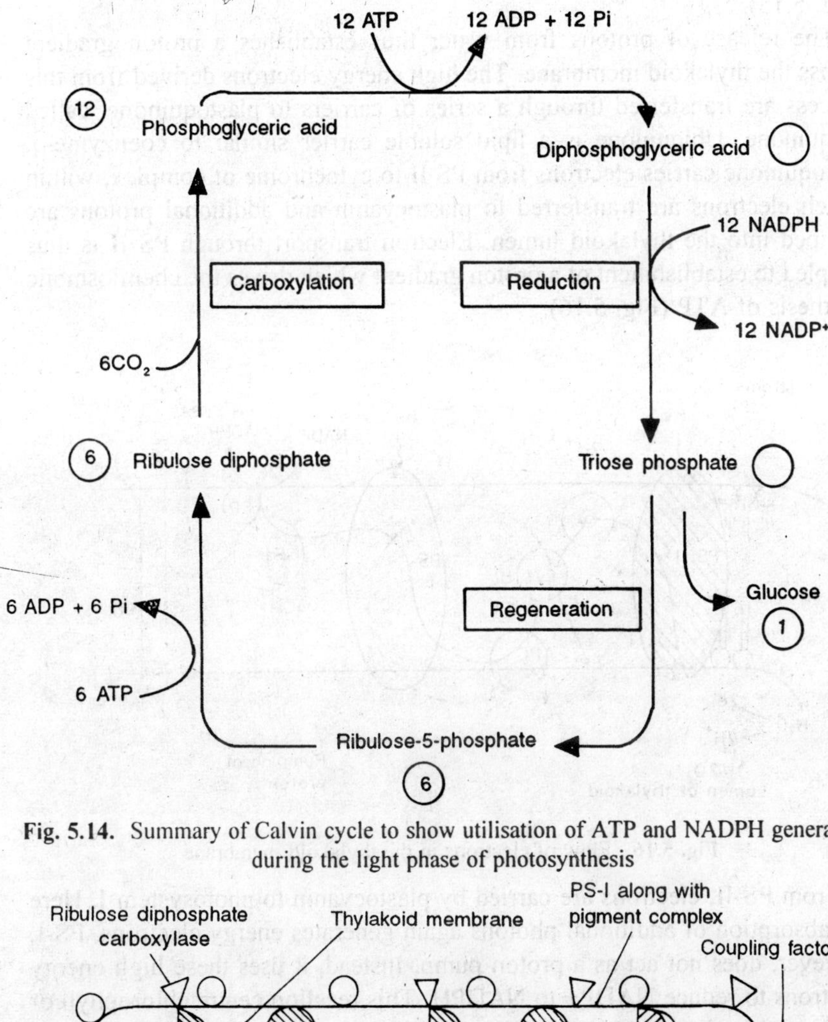

Fig. 5.14. Summary of Calvin cycle to show utilisation of ATP and NADPH generated during the light phase of photosynthesis

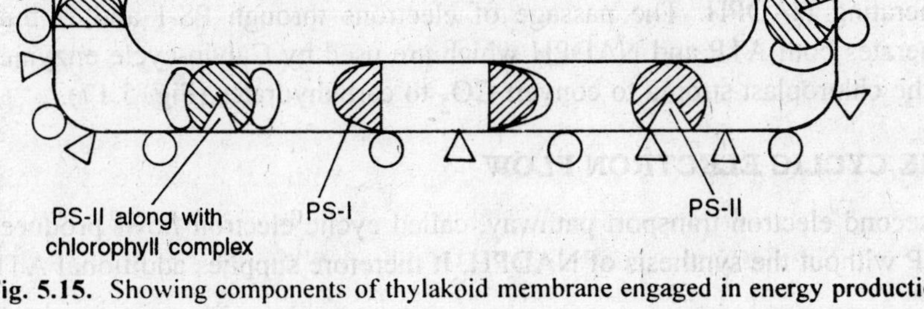

Fig. 5.15. Showing components of thylakoid membrane engaged in energy production

The pathway of electron flow starts as PS II which is homologous to the reaction centre described in phodospirillum. At PS II the energy derived from absorption of photons is used to split water molecules to molecule O_2 and protons. This reaction takes place within the lumen of the thylakoid (Fig. 5.15).

The release of protons from water thus establishes a proton gradient across the thylakoid membrane. The high energy electrons derived from this process are transferred through a series of carriers to plastoquinone, called ubiquinone. Ubiquinone is a lipid soluble carrier similar to coenzyme-q. Platoquinone carries electrons from PS II to cytochrome of complex, within which electrons are transferred to plastocyanin and additional protons are pumped into the thylakoid lumen. Electron transport through PS II is thus coupled to establishment of a proton gradient which drives the chemiosmotic synthesis of ATP (Fig. 5.16).

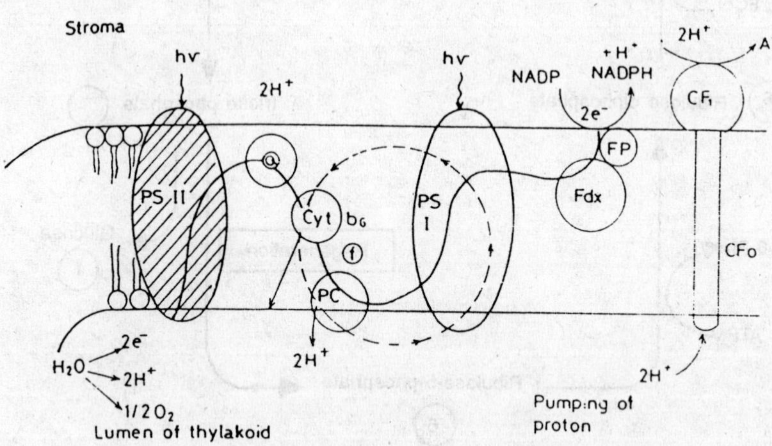

Fig. 5.16. Flow of electrons in the thylakoid membrane

From PS-II, electrons are carried by plastocyanin to photosystem I. Here the absorption of additional photons again generates energy electrons. PS-I, however, does not act as a proton pump. Instead, it uses these high energy electrons to reduce NADP+ to NADPH. This reaction centre chlorophyll of PS I transfer its electrons through a series of carriers of ferredoxin. The enzyme NADP reductase then transfers electrons from ferredoxin to NADP+ generating NADPH. The passage of electrons through PS-I and II thus generates both ATP and NADPH which are used by Calvin-cycle enzymes in the chloroplast stroma to convert CO_2 to carbohydrates (Fig. 5.17).

THE CYCLIC ELECTRON FLOW

A second electron transport pathway, called cyclic electron flow, produces ATP without the synthesis of NADPH. It therefore supplies additional ATP for other metabolic processes. In cyclic electron flow, light energy harvested at PS I is used for ATP synthesis rather than NADPH synthesis. Instead of being transferred to NADP, high energy electrons from PS I are transferred to cytochrome of complex. Electron transfer through cytochrome of complex

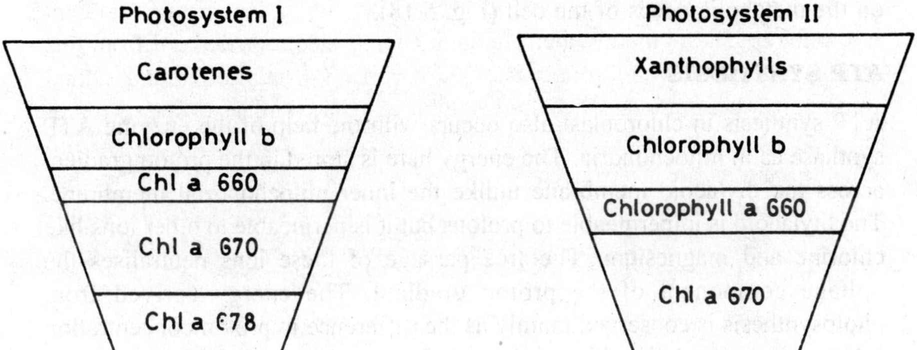

Fig. 5.17 Light gathering pigments involved in photosynthesis

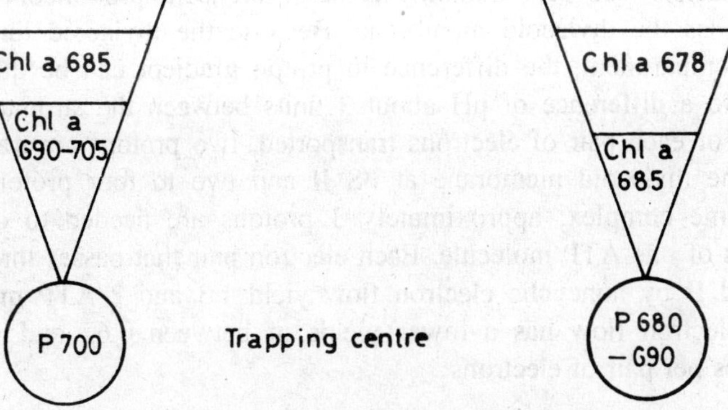

Fig. 5.18 Cyclic phosphorylation

is then complex as PS II, to the establishment of proton gradient across the thylacoid membrane. Plastocyanin then return these electrons to PS II in a lower energy state, completing a cycle of electron transport in which light harvested at PS I is used to pump protons at the cytochrome of complex. Electron transfer from PSI can thus generate either ATP or NADPH depending on the metabolic needs of the cell (Fig. 5.18).

ATP SYNTHESIS

ATP synthesis in chloroplast also occurs with the help of the enzyme ATP synthase as in mitochondria. The energy here is stored in the proton gradient across the thylacoid membrane unlike the inner mitochondrial membrane. The thylacoid is impermeable to protons but it is permeable to other ions like chlorine and magnesium. The free passage of these ions neutralises the voltage component of the proton gradient. The energy derived from photosynthesis is conserved mainly as the difference in proton concentration (pH) across the thylacoid membrane. Because the thylacoid lumen is a closed compartment, the difference in proton gradient can be quite large leading to a difference of pH about 3 units between the stroma and the lumen. For each pair of electrons transported, two protons are transported across the thylacoid membrane at PS II and two to four protons at the cytochrome complex; approximately 3 protons are needed to drive the synthesis of one ATP molecule. Each electron pair that passes through the PS I and II by noncyclic electron flow yield 1.3 and 2 ATP molecules. Cyclic electron flow has a lower yields of between 0.67 and 1.3 ATP molecules per pair of electrons.

THE CHLOROPLAST GENOME

Chloroplast contain their own genome reflecting their evolutionary origins from photosynthetic bacteria. They have multiple copies of genomes per genome. The chloroplast genome is more complex than the mitochondrial genome. It ranges from 120 to 160 kb and contains approximately 120 genes. The genome of the chloroplasts of some plants have been completely sequenced. The chloroplast genome encodes both RNA and proteins involved in the gene expression. Four rRNAs and 30tRNA species have been encoded by the genome, the rRNA include 23s, 16s, 5s and 4.5rRNA. The chloroplast genome unlike mitochondria codes for all the tRNA needed for the translation of the proteins, follow the universal genetic code.

Genes encoded by the chloroplast DNA

Function	Number of Genes
Genes for genetic appratus	4
rRNA (23S, 16S, 5S, 4.5S)	
rRNAs	30

Ribosomal proteins	21
RNA Pol subunits	4
Genes for photosynthesis	
PS I	5
PS II 12	
Cyt of complex	4
ATP Synthase	6
Ribulose biphosphate carboxylase	1

IMPORT AND SORTING OF CHLOROPLAST PROTEINS

Nuclear genes code for many of the chloroplast proteins. These proteins are synthesised on cytosolic ribosomes and then imported into the chloroplast as completed polypeptides; sorting of proteins to enter the chloroplastis quite complex as the chloroplast has 3 internal separate compartments, the lumen, the inter membrane space and the stroma. Proteins are imported into the chloroplast by target sequences of N-terminal amino acids of 30-100 sequences, called transit peptides. The transit peptides direct the proteins translocation across the membranes. These are then cleaved by proteolytic cleavage. The transport of proteins across the envelope requires ATP. The proteins that are to be incorporated into the thylacoid lumen are generally targeted by two signal peptides. One is cleaved as it enters the stroma and the protein now has a second signal which helps in entry into the lumen.

OTHER PLASTIDS

Chloroplasts are only one of the members of a large family of plasbids called plastids. All plastids contain the same genome as the chloroplast. They differ in structure and function. They lack the inner membrane system (thylacoid) and the photosynthetic apparatus.

Plastids that contain pigments are generally called chromoplasts. They lack chlorophyll but contain carotenoids. They are responsible for yellow, orange and red colors of some flowers and fruits. Leucoplasts are non-pigmented plastids which store a variety of energy sources in nonphotosynthetic tissues. Amyloplasts store starch and eleioplasts store lipids.

All plastids including chloroplasts are derived from proplastids, which are small 0.5 to 1μ diameter; they are present in organelles of rapidly dividing roots and shoot cells. They mature into plastids. Mature plastids are able to shift from one type to the other. Chlorophyll containing chloroplasts change into chloroplasts containing other pigments during ripening of fruits. The development of plastids is signalled by environmental conditions. They develop into matured ones with an intermediate vesicles, called Etioplasts, which exists in dark; when in light they develop into chloroplasts with the fusion of vesicles into thylacoid membrane systems.

SUMMARY

Cytoplasm, the term is applied to the ground substance of a living cell, that fills the interior of the cell, containing many types of organelles, each of which carries out a well-defined set of functions. This is the internal content of the cell matrix, which is the seat of most of the metabolic activities. It may, however, undergo differentiation in different types of cells, a characteristic attributed to the genes located in the chromosomes of the nucleus. The cell matrix contains highly organised array of cytoskeletal elements, the organelles and their functions are described.

EXERCISE

1. Explain in detail the physical properties and chemical organisation of plasma membrane.
2. What is the unit membrane hypothesis?
3. Explain in detail the chemical composition of plasma membrane.
4. Explain in detail about Active Transport and Passive Transport mechanisms.
5. Explain the differences between Active and Passive Transport mechanisms.
6. What are the types of transporters of cell membranes? Write in detail.
7. Write in detail physico chemical nature of cytoplasm.
8. Explain the structure and functions of Golgi bodies.
9. Explain the structure and functions of endoplasmic reticulum on the basis of enzymes associated with them.
10. Explain in detail about ribosomes and their functions.
11. Write in detail about cytoskeletal structures?
12. List the organelles derived from endoplasmic reticulum and discuss briefly.
13. Mitochondria are energy converters; explain the mechanism of energy transformation by mitochondria.
14. Describe lysosomes and narrate polymorphism in lysosomes.
15. Explain the structure and functions of nucleus.
16. Explain the structure and functions of peroxisomes.
17. Describe the structure of chloroplast.
18. Narrate the functions of chloroplast.
19. What do you know about chloroplast genome?
20. Write shortnotes on:
    - (a) Diffusion
    - (b) Cholestrol
    - (c) Integral proteins
    - (d) Ion channel
    - (e) Ligandgated channel
    - (f) Ca++ pump

(g) Vesicles
(h) Microvilli
(i) Tight junctions
(j) Ribosomal RNA
(k) Microtubules
(l) Microfilaments
(m) Secondary lysosomes
(n) Lysosomes and Diseases
(o) Nucleolus
(p) Plastids

6

Cell Cycle and Cell Division

THE CELL CYCLE

Growth of an organism means an increase in the number of its cells, with individual cell remaining about same size. Self reproduction of the cell is the most fundamental characteristic. All cells reproduce by dividing into two, with each parental cell giving rise to two daughter cells. These newly formed daughter cells can themselves grow and divide, giving rise to a new cell population formed by the growth and division of a single parental cell and its progeny.

Division of cells must be carefully regulated and co-ordinated with both cell growth and DNA replication, in order to ensure the formation of progeny cells containing intact genomes. It is important that growth, synthesis of DNA, chromosome segregation and cytokinesis should all take place in an orderly manner. As it is known, all eukaryotic organisms have a regulated pattern of cell division and follow the pattern strictly. Somatic cells divide by a process known as Mitosis, which involves chromosome condensation, spindle formation, alignment of chromosomes, and distribution into daughter cells in equal numbers, maintaining the ploidy of the cell from generation to generation. Genetic cells divide by a process known as Meiosis which is similar to mitosis but results in the formation of daughter cells which have half the number of chromosomes as that of the parent (haploid) However, all eukaryotic cells including yeasts replicate their DNA once before entering mitosis/meiosis. Controls exist within the cell cycles of all eukaryotes to ensure that once cells have replicated their DNA they do not do so again, before proceeding to cell division.

In eukaryotic cells, progression through the cell cycle is controlled by a series of protein kinases that have been conserved from yeasts to mammals. In higher eukaryotes, this cell cycle machinery is regulated by growth factors that control cell proliferation. It is the disturbance or the defects in the cell cycle regulation that brings about abnormal proliferation of cells as in cancer cells.

EUKARYOTIC CELL CYCLE

The division cycle of most cells consists of four co-ordinated processes:

a) Cell Growth,
b) DNA Replication,
c) Distribution of duplicated chromosomes to daughter cells, and
d) Cell Division.

In bacteria, cell growth and DNA replication occurs throughout most of the cell cycle, and duplicated chromosomes are distributed to daughter cells in association with the plasma membrane. In eukaryotes, the cell cycle is more complex and consists of four discrete phases. Cell growth is a continuous process, but DNA synthesis occurs only in one phase of the cell cycle and the replicated chromosomes are then distributed to daughter nuclei by a complex series of events before cytokinesis. Progression between these stages of cell cycle is controlled by a conserved regulatory cellular signal that controls cell proliferation.

PHASES OF CELL CYCLE

Human cells in culture are typical examples of eukaryotic cells which divide every 24 hours. The cell cycle is divided hereinto two basic parts, Mitosis and interphase. Mitosis (Nuclear division) is the most dynamic stage of the cell cycle, which involves the separation of chromosomes and ends in cell division. However, both these events last only for an hour and 95% of the cell cycle is thus spent in interphase.

During interphase the chromosomes are condensed and are distributed throughout the nucleus so that the nucleus appears to be uniform. However, at the molecular level interphase is a very dynamic phase which involves growth and DNA replication both preparing the cell for the subsequent cell division in an orderly manner. The cell grows at a steady rate throughout interphase with most of the dividing cells doubling the size between subsequent cell division. The timing of DNA synthesis thus divides the cell cycle into 4 discrete phases, the 'M' Phase, the 'G' Phase, the 'S' Phase and the 'G2 Phase'.

THE 'M' PHASE

This phase of the cell cycle corresponds to mitosis, followed by cytokinesis. This phase lasts for about one hour in the human cell which takes 24 hours to complete the cell cycle. In other cell types like yeast the whole cell cycle takes only 90 minutes to pass through all the stages. Even shorter cell cycles occur in early embryonic cells, shortly after fertilisation of the egg. In this case, cell growth does not occur. Instead, they rapidly divide the egg cytoplasm into smaller cells. DNA replication occurs rapidly here and 'M' phase alternates with 'S' phase, i.e. DNA synthesis phase.

The M Phase is the phase in which mitosis occurs. This phase includes various orderly steps which end up in cell division giving rise to two daughter cells. Mitosis starts with the condensation of chromosomes followed by movement of centrioles apart from each other towards the oposite sides of the nucleus. Then microtubules grow from the two centrioles forming spindles. Some of the spindles attach themselves to kinetochores of the chromosomes followed by the alignment of these chromosomes in the metaphase plate or the equatorial plate. Then the spindles start to pull apart the chromatids, characteristic of the anophase and slowly the chromosomes move apart towards the poles. This event is followed by cytokinesis in which the cell divides into two and results in the formation of 2 daughter cells which enter the next phase of the cell cycle, the 'G' phase (Fig. 6.1).

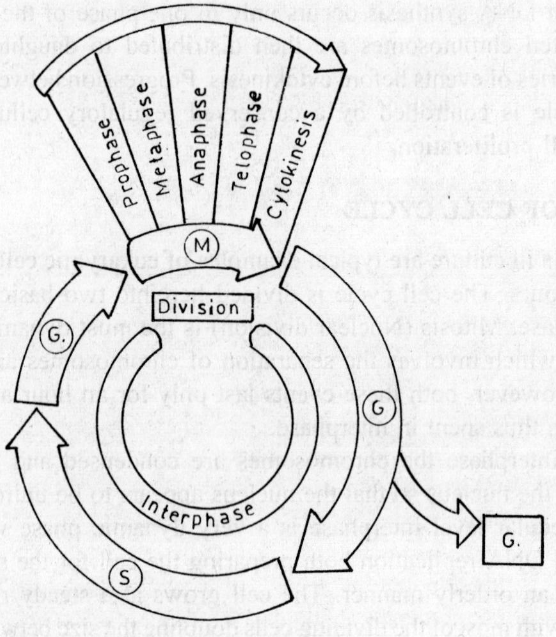

Fig. 6.1. Cell cycle

The G1 Phase: : The M Phase is followed by a G1 phase (Gap 1) which corresponds to the interval between mitosis and initiation of DNA Synthesis. During this phase the cell is metabolically active and grows continuously but does not replicate its DNA. In a human cell the G1 phase may last about 11 hours. In early embryonic cells, as already stated, the M phase is followed by 'S' phase, there being no 'G'1 phase. This indicates that these cells do not grow, but just divide and replicate their DNA continuously till a certain point. In contrast to embryonic cells some cells in adult animals cease to divide altogether. Ex. nerve cells. Other cells divide occasionally when needed, so as to replace cells that have been lost because of injury or cell death. Ex. skin fibroblasts, liver cells, kidney cells and cells of the lung. These cells, after entering the G1 phase, come out of it after sufficient growth and enter a quiescent stage of the cycle called the G o phase, where they remain metabolically active but no longer proliferate unless called on to do so by appropriate extra cellular signals.

Chromosome condensation

↓

Centrioles move apart to opposite sides of the nucleus

↓

Microtubules grow from centrioles forming spindles

↓

Spindle fibres attach to kinetochores of chromosomes

↓

Chromosomes align on the metaphase plate

↓

Anaphase movement of chromatids to opposite poles

↓

Cytokinesis - cell divides

↙ ↘

Daughter cell in G1 phase | daughter cell in G1 phase

Flow chart of events occurring in M-phase.

The 'G' 2 Phase: The completion of DNA synthesis in the S phase is followed by G2 phase (Gap2) during which cell growth continues and proteins are synthesised in preparation for mitosis. The significant characters of cells in this phase is that they contain double amount of DNA i.e. (4x). This phase lasts about 4 hours in a typical human cell. The G2 phase cells are involved in active protein synthesis, especially those proteins which are required for the mitotic division. The DNA content of the cells can be determined by analysis of fluorescence intensity of individual cells in a flow of cytometer of fluorescence activated cell sorter.

REGULATION OF CELL CYCLE

Progression of cells through the division cycle is regulated by extra cellular signals from the environment, as well as by internal signals that monitor and co-ordinate the various processes that take place during different cell cycle phases. Different cellular processes such as cell growth, DNA replication and mitosis must be coordinated during cell cycle progression.

A major cell cycle regulatory point in many type of cells occurs in the late G1 and controls progression from G1 to S. This check point was defined in budding yeast (Saccharomyces cerevisiae). It was named as START (Fig. 6.2).

Once cells have passed START, they are committed to enter 'S' phase and undergo one cell division cycle. Passage through START is a high regulated event in yeast cell cycle. It is controlled by external signals, such as the availability of nutrients as well as by cell size. If nutrients are not properly available the cells arrest their cell cycle at START and enter into the Go phase rather than the S phase. Thus START represents a decision point at which the cell determines whether sufficient nutrients are available to support progression through the rest of the division cycle. Polypeptide yeast mating factors that signal yeast mating also arrest cell cycle at START. In addition to serving as a decision point for monitoring extracellular signals, START is the point to which cell growth is co-ordinated with DNA synthesis and cell division. It functions analogously to START in yeasts. However, the passage of animal cells through the cell cycle is regulated primarily by the extra cellular growth factors that signal cell proliferation, unlike yeast cells where the regulation is by avaialability of nutrients. In the presence of appropriate growth factors, cells pass the restriction point and enter S phase. Once it has passed through the restriction point, the cell is committed to proceed through 'S' phase and the rest of the cell cycle, even in the absence of further growth factor stimulation. In the absence of growth factors, cells arrest at the cycle at the restriction point and enter S phase. Once it has passed through the restriction point, the cell is committed to proceed through S phase and the rest of the cell cycle, even in the absence of further growth factor stimulation. In the absence of growth factors cells arrest at the cycle at the restriction point and enter into Go phase where the cell remains inactive for long periods Ex. skin fibroplast.

Some cells are also regulated in the G2 phase. This is seen in yeasts, where it is regulated primarily by control of transition from G2 to M. In animals, the primary example of cell cycle control in G2 is provided by oocytes. Vertebrate Oocytes remain arrested in the G2 Phase for long period of time until their progression to 'M'Phase is triggered by hormonal stimulation. Extracellular signals can thus control cell proliferation by regulating progression from the G2 to M as well as G1 to S phases of the cell cycle.

CHECK POINTS OF CELL CYCLE

Other than the controls described earlier, events that take place during different stages of the cell cycle must be coordinated with one another so that they occur in appropriate order. This coordination between different phases of the cell cycle is dependent on a system of check points and feedback controls that prevent entry into the next phase of the cell cycle until the events of the preceding phase have been completed.

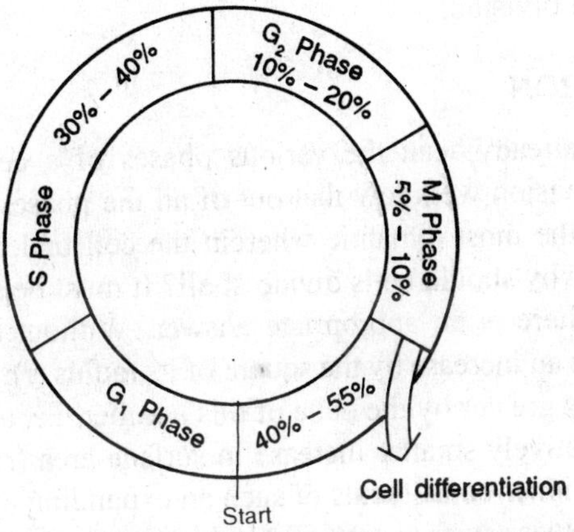

Fig. 6.2. Cell cycle: G_1 is pre-DNA synthetic phase; S phase is synthetic phase representing period of DNA synthesis; G_2 phase is post-DNA synthetic period; M phase is the mitotic phase from which some cells leave the cycle to undergo differentiation. Duration of each phase is given in %

Several cell cycle check points function to ensure that incomplete or damaged chromosomes are not replicated and passed on to daughter cells. One of the most clearly defined of these check points occurs in G2 and prevents the initiation of mitosis until DNA replication is completed. This G2 checkpoint senses unreplicated DNA and generates a signal that leads to cell cycle arrest. Operation of G2 check point prevents the initiation of 'M' phase before completion of 'S' phase, so cells remaining G2 until the Genome have been completely replicated. The cell cycle can also be arrested at the G2 check point in response to DNA damage. DNA damage also arrests the cell cycle at the G1 phase and the end of 'M' phase, which ensure damaged DNA repair, replication and also separation of chromosomes only after complete complementation has occured. Further the cell cycle is regulated by regulators like MPF (Maturation promoting factor) which is a dimer of Cd c and cyclin, a set of protein kinases. MPF is a conserved regulator of the cell cycle and is composed of two key subunits Cd c, and cyclin B. MPF activity is controlled by periodic accumulation and degradation of cyclin B during cell cycle progression. Phosphorylation of Cd c has further been provided to be the other regulatory mechanism.

In mammalian cells Cyclin B synthesis begins in the S phase. Cyclin B then accumulates and forms complexes with Cd c protein kinase. Once activated the Cd c protein kinase phosphorylates a variety of target proteins that initiate the events of M Phase. In addition, Cd c activity triggers the degradation of cyclin B , which occurs as a result of ubiquitin mediated proteolysis. The proteolytic degradation of cyclin B then inactivates Cd c leading the cell to exist mitosis, undergo cytokinesis and return to interphase. Thus the cell cycle is well regulated and plays a significant role in the life of a cell and cell division.

CELL DIVISION

As we have already seen the various phases of a cell's life, i.e., during growth and division we know that out of all the phases in the cell cycle the 'M' Phase is the most dynamic wherein the cell undergoes a programmed division. But why should cells divide at all? It must be surprising to put this question but there is an appropriate answer. Without cell division growth occurs through an increase by the square of its radius (r); its volume increases proportionately greater by the cube of this number, i.e. an increase in size (r) produces a relatively smaller increase in surface area (r) than in volume (r). Therefore, the inner constituents of such an expanding organism would have proportionately less surface area from which to obtain food, oxygen and the various metabolic necessities as well as less surface from which to secrete their metabolic products and wastes. In the absence of division, death would quickly ensure both as a result of these causes and because of the many physical stresses and accidents that could rupture as a membrane. Therefore, some form of cell division is necessary for the maintenance of a cell.

Eukaryotic cell division, as mentioned earlier, is quite complex and involves a unique mechanism. Two basic types of cell division have been identified according to the pattern of division and the behaviour of chromosomes. The division which results in the distribution of daughter chromosomes in equal numbers to each of the daughter cells and exactly the same number as in the parent cell is called Mitosis. This type of division takes place in all somatic cells. In the other type of cell division the daughter cells have only half the number of chromosomes as in the parent cell. This division is called Meiosis. It takes place in the formation of gametes or spores which are required for sexual reproduction. The first account of Mitosis was provided by A. Schneider in 1873. Walter Flemming who observed cell division in salamander and showed that nuclear division involves longitudinal splitting of the chromosomes. In 1882, he coined the term Mitosis. Cell division in prokaryotes and eukaryotes can generally be divided into three types i.e., direct cell division-amitosis, indirectcell division-mitosis and reduction division-meiosis.

MITOSIS

The cell division occurring in all somatic cells wherein the two daughter cells formed contain an equal number of chromosomes and the same number as that of the parent cell is called Mitosis. In this division the ploidy of the cell is maintained. Mitosis can be divided into four different phases. However, before entering the M phase (Mitotic), the cell is said to be in a resting stage, i.e. the interphase.

The interphase

Interphase is the period between successive cell divisions where the cell grows in size and also prepares for Mitosis. This phase includes the G1 phase, S phase and the G2 phase of the cell cycle. In all the phases the cell is not resting or inert, but is actively involved in protein synthesis and DNA replication preparing itself for Mitosis. Hence, this can be called the preparatory phase. In this phase the cytoplasm is rich with ribosomes and protein constituents. The nucleus is distinct with a nucleolus. The chromatin is rather decondensed. The time taken for mitotic division however varies from species to species (Fig. 6.3).

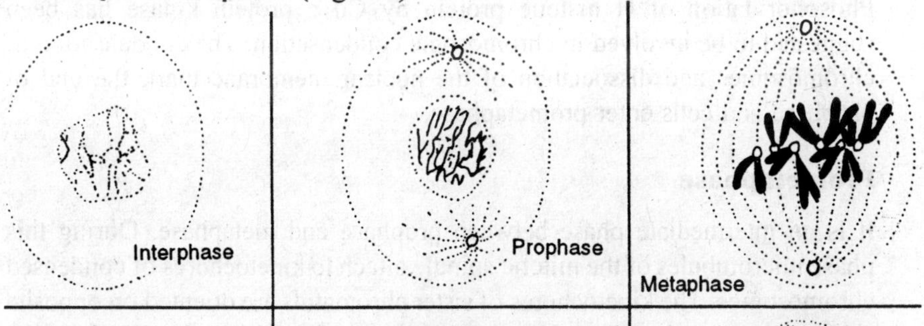

Fig. 6.3. Stages in mitosis. Two daughter cells are produced at the end of telophase

Prophase

It is the first stage in mitotic division. Beginning of the prophase is marked by the appearance of condensed chromosomes each of which consists of two sister chromatids. These newly replicated DNA molecules remain intertwined throughout the S and G2 phase becoming untangled during the process of chromatin condensation. The condensed sister chromatids are then held together at the centromere, which is a DNA sequence to which proteins bind to form the kineto chore-the site of spindle microtubule attachment. In addition to chromosome condensation, cytoplasmic changes leading to the development of mitotic spindle initiate during prophase. The centrosomes separate and move to opposite sides of the nucleus, serving as two poles of the mitotic spindle, which begins to form during late prophase. In the end of the prophase the breakdown of nuclear envelope is seen. In most cells breakage of the nuclear envelopes marks the end of prophase of mitosis. The dissociation of nuclear lamina results from the phosphorylation of the lamines. This causes the filaments to break down into individual lamine dimerse, which is catalysed by the Cd c protein kinase. Simultaneously, the nuclear membrane also breaks up into fragments and then into vesicles.

The other major change that occurs in prophase is the chromosome condensation. The interphase chromatin, i.e. packaged into nucleosomes condenses approximately 1,000 fold to form the chromosomes. This condensation is needed to allow the chromosomes to move along the mitotic spindle without becoming tangled or broken during their distribution to daughter cells. DNA in this highly condensed stage can no longer be transcribed and so RNA synthesis stops and the nucleolus also disappears. Phosphorylation of H histone protein by Cd c protein kinase has been suggested to be involved in chromosome condensation. The condensation of chromosomes and dissociation of the nuclear membrane mark the end of prophase and cells enter prometaphase.

Prometaphase

It is an intermediate phase between prophase and metaphase. During this phase microtubules of the mitotic spindle attach to kinetochores of condensed chromosomes. The kinetochores of sister chromatids are oriented on opposite sides of the chromosome, so they attach to microtubules aminating from the opposite poles of the spindle. The chromosomes shuffle back and forth until they eventually align on the metaphase plate in the centre of the spindle. At this stage the cell has reached metaphase.

Metaphase

By the time the cell reaches metaphase, the chromosomes are completely condensed, most probably by MPF (Cd c -Cyclin B) mediated phosphorylation of H Histone protein. The nuclear membrane disappears allowing the spindle microtubules to attach to the kinetochores of the chromosomes. The proteins assembled in the kinetochore include motors that drive the movement of chromosomes towards the minus ends of spindle microtubules, which are anchored to the centrosome. Other than these motors, the centrosome also has some motor proteins bound to the centrosome which pull the chromosomes towards the centrosome. This attraction is opposed by a force called polar wind that repels chromosomes from spindle poles. Consequently, the chromosomes in the prometaphase move back and forth. Microtubules from opposite poles of the spindles eventually attach to the two kinetochores of sister chromatids and balance the forces acting upon them. Thus finally, chromosomes are aligned on the equatorial plate, the identification of metaphase. The spindle formed during the process generally contains three types of tubules.

(a) Kinetochore microtubule that attach to kinetochores
(b) Polar microtubules which overlap with one another in the centre of the cell
(c) Astral microtubules that radiate outward from the centrosome toward the cell periphery.

Once chromosomes are aligned the cells proceed into the next phase of cell division, i.e. the anaphase.

Anaphase

The progression from metaphase to anaphase results from the activation of a ubiquitin mediated protiolysis system that degrades cyclin B and thereby inactivate MPF. However, the transition from metaphase has been identified to be dependent upon a target protein X, i.e suggested to have a role in holding the two sister chromatids together at the centromere. Once this protein is degraded the chromatids are free and the movement of these chromatids to opposite poles occurs by the motor aided depolymerisation of the microtubules.

Anaphase is characterised by the movement of sister chromatids to opposite poles. Chromosome movement has been identified to proceed via two mechanisms called Anaphase A and Anaphase B. Anaphase A consists of the movement of chromosomes towards spindle poles along the kinetochore microtubules. This type of chromosome movement is thought to be driven by a kinetochore associated motor protein that translocates chromosomes along spindle microtubules in the minus end direction, towards the centrosomes. The action of these proteins is coupled to disassemble the shortening of the kinetochore microtubules. It is thought that motor activity and microtubule disassembly both contribute to chromosome movement. Anaphase B refers to the separation of spindle poles themselves. Spindle poles separation can occur by overlapping polar microtubules sliding against one another, pushing the spindle poles apart. This movement is aided by positive end directed motor proteins. Thus, separation of sister chromatids to opposite poles mediated by various proteins marks the end of the shortest phase of mitosis. Anaphase is followed by the final phase of mitosis before before cytokinesis – the Telophase.

The telophase

Telophase is marked by the reassembly of nuclei in each of the daughter cells, still to be separated, and the decondensation of chromosomes also is seen. In the telophase two nuclei form around the separated sets of daughter chromosomes. Chromosome decondensation, reassembly of nuclear envelope appear to be signalled by inactivation of Cd c Kinase, which leads to dephosphorylation of the proteins that were phosphorylated and the initiation of mitosis resulting in the exit from mitosis and the reformation of interphase nucleus.

The initial step in reformation of nuclear envelope is the binding of vesicles formed during nuclear membrane breakdown to the surface of chromosomes. This is followed by reassembly of the nuclear pore complexes, reformation of nuclear lamina and chromosome condensation to form a complete single nucleus. The nucleolus too reforms and transcription of rRNA genes begins, completing the return from mitosis to an interphase nucleus.

Cytokinesis

The completion of mitosis is usually accompanied by Cytokinesis, giving rise to two daughter cells. Cytokinesis in animal cells is mediated by a contractile ring of actin and myosin II filaments that form beneath the plasma membrane. The cell is cleaved in a plane that passes through the metaphase plate perpendicular to the spindle. Cleavage proceeds as construction of actin - miosin filaments pulls the plasma membrane inward, eventually pinching the cell in half.

The mechanism of cytokinesis is different in higher plants, which are surrounded by rigid cell walls. Rather than being pinched in half by a contractile ring these cells divide by forming new cell walls and plasma membranes inside the cell.

Thus, at the end of mitotic division, a cell gives rise to two daughter cells which contain the same number of chromosomes as that of the parent cell. These cells again grow in size and undergo mitotic division, the net result being the growth of the organism.

Meiosis

The term Meiosis was coined by J.B.Farmer in 1905. Meiosis is a reductional division that results in the formation of four daughter cells. Each daughter cell contains half the number of chromosomes as that of the parent cell. These daughter cells are thus haploid and also called a reduction division. Mitosis generally takes place in somatic cells but meiosis always occurs in cells of the reproductive organs resulting in the formation of haploid gametes. These gametes again fuse and form diploid zygotes (Fig. 6.4).

Meiosis can be classified into 3 types:

(a) Terminal
(b) Intermediate
(c) Zygotic meiosis.

(a) *Terminal meiosis:* is also called gametic meiosis and is found in animals and few lower plants. In this type of meiosis the meiotic division occurs just before the formation of gametes.

(b) *Intermediary meiosis:* In Intermediary meiosis the reductional division takes place at some intermediate time between fertilisation and the formation of gametes. This is also called sporic meiosis. This is characteristic for flowering plants.

(c) *Initial or zygotic meiosis:* occurs in some algae, fungi and some diatoms. Here the meiotic division occurs immediately after fertilisation.

Meiosis is a specialised kind of cell cycle that reduces the chromosome number by half. The exact replication and splitting of each chromosome in two identical parts and their subsequent separation into two cells would not ordinarily lead to any change in chromosome number between parent and daughter cells. In organisms where cells are always formed by sexual means, i.e. mitosis, the number of chromosomes would remain constant between generations. In sexually reproducing organisms where a zygote is formed by fertilisation of male and female gametes (haploid), the embryonic cells resulting from the fusion would have double the number of chromosomes than the parent cell, if no reduction in number of chromosomes occurred in the sex cell formation. But it is very clear that chromosome number always remains constant from generation to generation and so no such doubling occurs at any time. In the process of evolution, cells, i.e. gametic cells have somehow adapted to a mechanism of regularly reducing chromosome to half during sex cell formation. Meiosis is a simple process by which the chromosomes are separated during the formation of sex cells and their numbers reduced from diploid to haploid condition. Fertilisation then marks the event in which two haploid nuclei join to reform a diploid cell.

The number, as well as the size and shape of the chromosomes of a species, is called its karyotype and is usually constant. In diploid cells, each individual chromosome usually has a pairing mate called the homologue and both of them are called homologous chromosomes, one coming from each parent. In the absence of such homologous chromosomes the reduction division would result in random separation of chromosomes. This would result in haploid daughter cells but still lacking at least one essential chromosome. The reduction in chromosome number is accomplished by two sequential rounds of nuclear and cell division called Meiosis I and Meiosis II. which follow a single round of DNA replication. Each of these divisions is further divided into various stages as follows:

Meiosis I

Meiosis I, like mitosis, maintains after 'S' phase has been completed and the parental chromosomes have replicated to produce identical sister chromatids. The pattern of chromosome segregation, in meiosis I, is however, dramatically different from mitosis. During meiosis I, homologous chromosomes first pair with one another and then segregate to different daughter cells. Sister chromatids remain together, so completion of meiosis I results in the formation of daughter cells containing a single member of each chromosome pair consisting of two sister chromatids. The pairing of homologous chromosomes after DNA replication is only a key event underlying meiotic chromosome segregation, but also allows recombination between chromosomes of paternal and maternal origin. This critical pairing of homologous chromosomes takes place during an extended prophase of meiosis I which is divided into 5 stages. Leptotene, Zygotene, Pachytene, Diplotene and Diakinesis, on the basis of chromosome morphology.

Leptotene

The leptotene is first of the meiotic stages that differs from the previous interphase. The chromosomes first appear as long, slender threads, with many bead like structures called chromosomes, along the length. In some plants the chromosomes are clumped to one side of the nucleus. Such a condition is called synizesis. In some animals (many insects) they appear polarised with their ends drawn together toward that portion of the nuclear membrane close to the centriole. The initial association of homologous chromosomes is thought to be mediated by bass pairing between complementary DNA strands during this stage.

Zygotene

During this stage homologous chromosomes appear to attract each other and enter into a very close association in a zipper like fashion. This process is called synapsis. During this stage a zipper like protein structure called the synaptonemal complex is formed along the length of the paired chromosomes. This complex keeps the homologous chromosomes closely associated and aligned with one another. The pairing between homologous chromosomes is highly specific and occurs between all homologous chromosomes.

Pachytene

The Pachytene is a stage of progressive shortening and coiling of the chromosomes that occurs once zygotene has been completed. At this stage the two sister chromatids of a homologous chromosome are associated with the two sister chromatids of their homologous partner. The group of 4 chromatids is known as a bivalent or a tetrad. In this stage a series of exchanges of genetic material can occur or has already occurred between two non-sister homologous chromatids. The chromosomes may be linked at the sites of crossing over, giving the appearance of a typical chiasmata. The synaptonemal complex is also clearly seen.

Diplotene

At this stage, the synaptonemal complex disappears and each chromosome now acts as though it were repulsing its closely paired homologue, especially near the centromere. Distinctly visible separations occur between homologous chromosomes except for specific regions where an actual physical crossing over appears to have taken place between homologous chromatids. These are the chiasmas, which appear to be the only remaining force holding each bivalent together until metaphase.

Diakinesis

At diakinesis, coiling and contraction of the chromosomes continue until they are thick, heavy staining bodies. In this process the bivalents usually migrate close to the nuclear membrane and become evenly distributed. The nucleus disappears. The nuclear membrane slowly disintegrates and the bivalents rapidly attach themselves by their centrosomes to the rapidly forming spindles.

Metaphase I

In this phase the chromosomes reach their most condensed state and appear relatively smooth in outline. The chiasmata that had appeared during the diplotene now move towards the end of each chromosome. This process is called Terminalization. Terminalization leaves only the single terminal attachment between the formerly paired arms of homologous chromosomes. The kinetochores of sister chromatids are adjacent to each other and oriented in the same direction. The kinetochores of homologous chromosomes are pointed towards opposite spindle poles. Consequently, microtubules from the same pole of the spindle attach to sister chromatids, while microtubules from opposite poles attach to homologous chromosomes. This phase is characterized by the alignment of bivalent with chiasmata on the metaphase plate.

Anaphase I

Anaphase is characterised by the separation of chromosomes towards opposite poles. Anaphase I is initiated by disruption of the chiasmata at which homologous chromosomes are joined. The homologous chromosomes then separate, while sister chromatids remain associated at their centromeres. The pulling apart of the chromosomes by spindle fibres towards the poles gives them a typical 'V' shaped appearance. This is the phase where the recombined genetic segments in the chromosomes are pulled apart and the chromosome number is reduced to half in each cell thus being called the reductional division. The homologues are thus separated in this phase, resulting in the formation of a dyad in progeny. The greater the number of chromosomes a cell has, the more chances of recombination and the possibility of containing chromosomes from both parents rather than from one parent.

Telophase I

Events of this phase are quite similar to the telophase in mitosis. Once the dyads reach on the spindle poles, a nuclear membrane forms around them and the chromosomes pass into a short interphase before the meiosis II. Cytokinesis may or may not occur.

After telophase I generally cells enter a short interphase which is very short and chromosomes do not completely decondense, nor is a nucleolus formed completely. Soon the cell enters the second meiotic division, which contains all the phases as in mitosis.

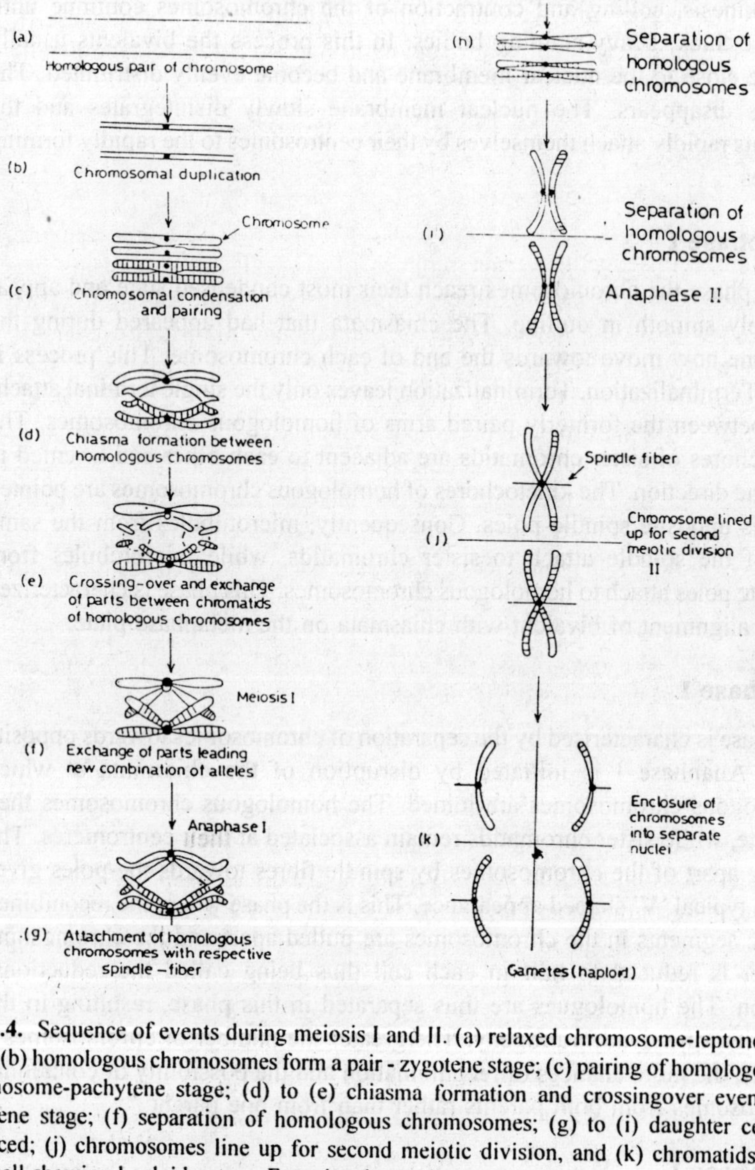

Fig. 6.4. Sequence of events during meiosis I and II. (a) relaxed chromosome-leptonene stage; (b) homologous chromosomes form a pair - zygotene stage; (c) pairing of homologous chromosome-pachytene stage; (d) to (e) chiasma formation and crossingover event - diplotene stage; (f) separation of homologous chromosomes; (g) to (i) daughter cells produced; (j) chromosomes line up for second meiotic division, and (k) chromatids in each cell showing haploid nature. Four daughter cells are produced.

Meiosis II

Meiosis II can be divided into prophase II, metaphase II, Anaphase II and Telophase II followed by cytokinesis. Prophase II shows dyads with two sister chromatids attached at their centromeres. Soon the spindles form as the nuclear membrane is disintegrated. The dyads rapidly bind to the spindle fibres by their kinetochores and are aligned on the equatorial plate in metaphase II. Anaphase II is characterised by the separation of the sister chromatids of the dyad, resulting in the formation of monads.

These monads soon move towards opposite spindle poles in Anaphase II, Telophase II and cytokinesis follow rapidly giving rise to four haploid cells from each initial diploid cell that entered meiosis.

SUMMARY

When a cell or an organism produces more cells or organisms by simple division of cytoplasm and the nucleus, the process is called asexual reproduction. Consequently exact replicas of the cells or organisms are produced. In many cases the process involves a mitotic division. This commonly occurs in all somatic cells of animals and plants. However, in prokaryotes the most widely used method of division is amitotic.

Many organisms begin their life as a single cell, the fertilised egg, which divides several times to form a multi cellular structure. These cells differentiate, and ultimately form an adult with full complement of organs and systems. Every cell grows and then divides into two daughter cells and before cell-division occurs, its components must have duplicated including the chromosomes and the cytoplasm so that each daughter cell receives full complement of the genetic material.

In most eukaryotic cells the actual process of cell division lasts for less than 1-hour, but the preparation for it may require a longer time during interphase. The overall events of the cell cycle are under the control of multiple gene products, which regulate the coordination of various events.

EXERCISE

1. What is the significance of interphase?
2. Explain in detail various stages of mitosis cell division.
3. What is the significance of meiosis?
4. Explain in detail various stages of meiosis.
5. Write in detail what do you know about cell cycle?
6. Write short notes on:
   (a) pachytene
   (b) leptotene
   (c) xygotene
   (d) diplotene

7

Cell Interaction and Integration

Cells are to survive in varying conditions; therefore, they should be able to get stimuli and respond to differences caused in the environment. Such responses are based on the integration of several components or systems mediated by its membranes, receptors, chemicals, effectors, pH differences, etc. They perform suitable actions that enhance the chances of survival and purposefulness. Cells create their own environment by secreting a thin layer around them. In favourable and unfavourable conditions, cells cannot be free of their environment.

In spite of adhering to their environment, cells retain their identity due to the plasma lemma limiting membrane.

In the course of evolution living cells searched new environments above the aquatic habitat and adapted themselves to the new terrestrial habitat locations. Their adaptive ability depended on the susceptibility of their genome to change. During the course of adaptation, acquisition of multicellular nature was proved advantageous over single cell structure; thus the cellular specialisation is the natural outcome.

Cells not only interacted with the environment but also with neighbouring cells in their immediate proximity. Like this cells created their own environment to become dependent on each other.

CELL SURFACE

Cell is bound by plasma membrane. Some regions of this are specialised to perform, absorption, secretion, transport, adhesion, communication and other physiological activities; according to these functions the membrane develops specialisation. In plant cell cellulose wall is present, whereas in animal cell external coating exists.

CELL JUNCTIONS

Side by side, plant cells are held together by middle lamella. These cells are further interconnected by five cytoplasmic strands passing through the cell walls and middle lamella; they are called plasmodesmata. (Fig. 7.1). In several animal tissues, cell membranes of adjacent cells are held together at several points, forming specific cell junctions, which are responsible for cell-to-cell attachment. These attachments may occur throughout the plasma membrane, but at certain places special junctions are formed to provide extra strength. These are of 3 types:

(i) Adhering junctions or septate junctions
(ii) Tight junctions, and
(iii) Gap junctions.

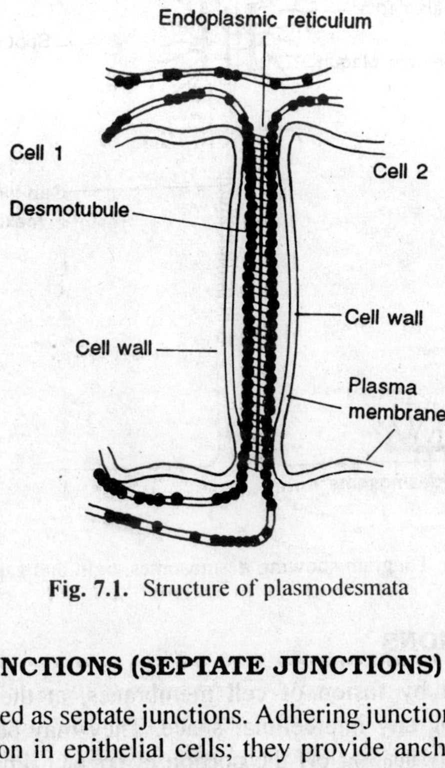

Fig. 7.1. Structure of plasmodesmata

ADHERING JUNCTIONS (SEPTATE JUNCTIONS)

They are also called as septate junctions. Adhering junctions are desmosomes which are common in epithelial cells; they provide anchorage to adjoining cells by strong points of attachments (Fig. 7.2). Adhering junctions are fine cytoplasmic tono filaments, which run across two parallel thickenings of cells, separated by intercellular spaces. At the basal surface of some cells, only half desmosomes are present, hence they are considered hemi-desmosomes, which prevent the cell from pooling off the underlying surface.

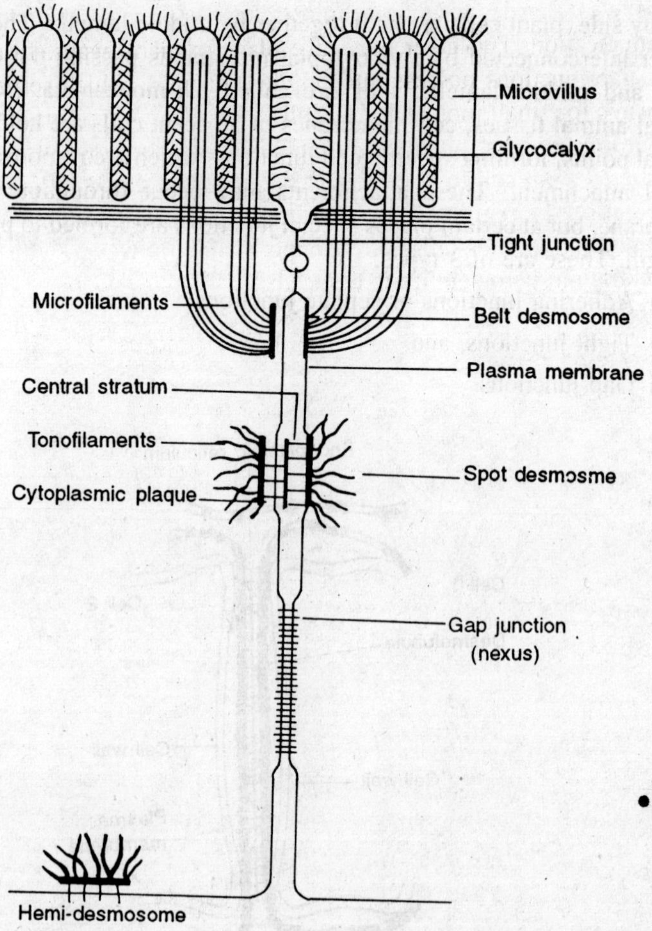

Fig. 7.2a Diagram showing desmosomes, tight and gap junctions

TIGHT JUNCTIONS

They are formed by fusion of cell membranes, at the point of contact, without possessing any intercellular space. They may be band like or belt like in form totally sealing off the portion of contact. Physiologically, tight junctions are important, as they prevent flow of materials across the epithelium in either direction, i.e. from lumen to basal region or vice versa. Second function of these junctions is maintaining cell polarity by functioning as a physical barrier to the mobility of integral proteins laterally (Fig. 7.3a).

GAP JUNCTIONS

They are also called as nexuses and are very simple junctions common in animal cells. (Fig. 7.3) Nexuses are tools for cellular interaction among themselves. Besides forming cellular adhesion, they also help in intercellular communication. The latter may take place in many ways. In a majority of cells, gap junctions possess intercellular channels, which serve as transport pathways for small ions and molecules (Fig. 7.4). Therefore, low molecular weight ions, sugars, vitamins, hormones and other molecules can be easily distributed from cell to cell. This is called as metabolic coupling. Metabolic cooperation is regulated by calcium-ion concentration. With low Ca concentration, permeability of channels is high; however, with increased Ca concentration, permeability is reduced.

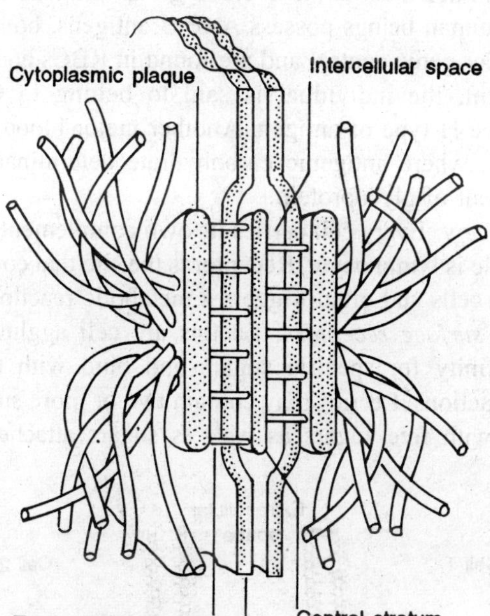

Fig. 7.2b Diagrammatic representation of a spot desmosome

Next type of communication through gap junctions is through electrical coupling. Gap junctions help as electrical pathways for commu ̀ations. Gap junctions normally have low resistance but the impulse conduction is very quick. Electrical coupling is found in embryonic cells, cardiac cells and liver cells.

Proper functioning of a cell is possible only by proper interaction and integration of the cell.

Plasmalemma in many cells is surrounded by cell coats, which serve several functions. Basically, cell coats are protective in function; added to this cell coats contain enzymes, that possess sites for interaction of signals from other cells. Some of its important functions are mentioned below:

Cell-surface carbohydrates serve for interaction: Plasmalemma contains many carbohydrates existing in conjugated form. They attach either to lipids or to proteins of the plasmalemma facing the exterior; carbohydrates attached to proteins form glycoproteins and those attached to the lipids give rise to glycolipids. Carbohydrates comprise D-galactose, D-mannose, D-fucose, N-acetylmuraminic acid, N-acetylgalactosamine and N-acetyl glucosamine. Surface proteins and carbohydrates together give rise to surface antigens. Antigenic specificity is mainly due to the arrangement and sequence of oligosaccharide chains, which is cell specific. Orientation of glycoproteins also help in keeping contact with external world of the cell. Cell adhesion to its neighbouring cells is also due to surface carbohydrates.

On studying in detail about ABO blood group antigens of erythrocytes, it is evident that human beings possess A or B antigens, both or none. These antigens are under genic control and are found in RBCs and other tissues. If antigen is absent, the individual is said to belong to O group. These erythrocytes have H type of antigen. Another major blood group system is the MN system, where antigenic carbohydrate determinant is attached to protein component of glycoprotein.

Receptor concept: Every cell carries its own complement of receptors, but the good example is lymphocyte. Receptor is the site that communicates with the surrounding cells and often triggers a metabolic reaction.

Lectins with surface receptors: Lectins are cell agglutinating proteins; they posses affinity for specific sugars and bind with them, producing agglutinating reactions. Lectins may contain two or more sugar-binding sites and may bind with free sugars as well as sugars attached to the plasma membrane.

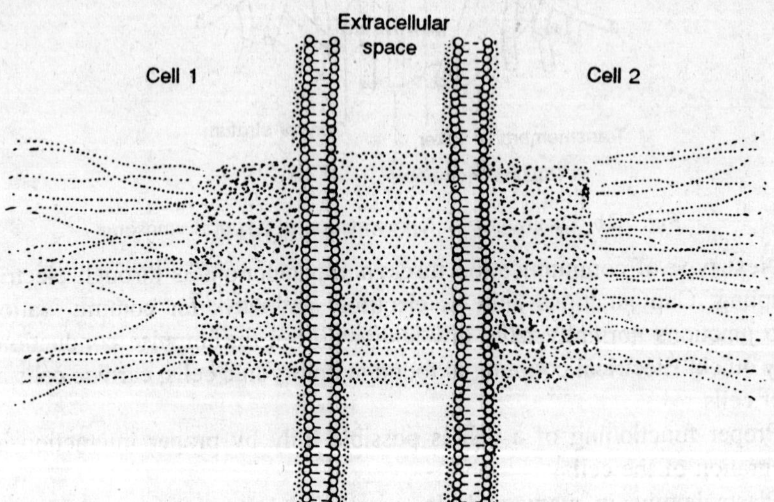

Fig. 7.2c. A magnified view of desmosome Adhering type of junctions

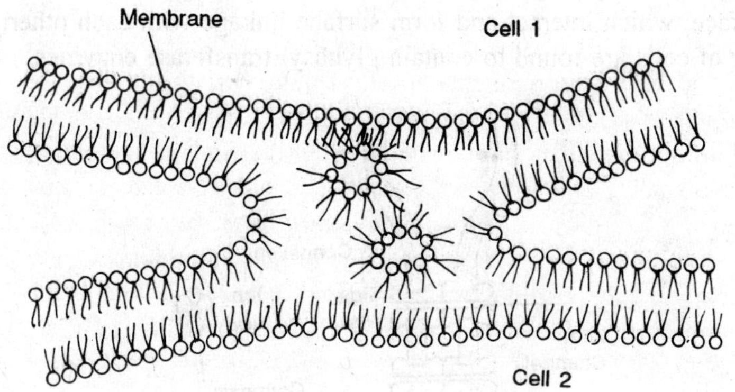

Fig. 7.3a. A tight junction formed by the fusion of adjacent cell membranes

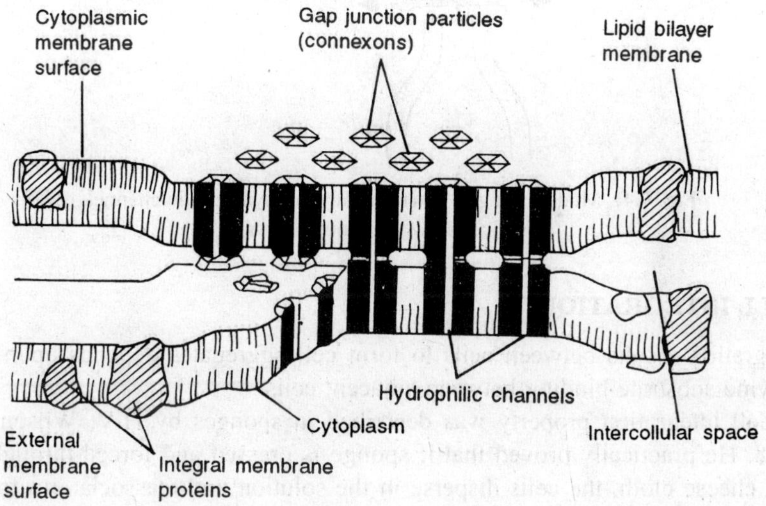

Fig. 7.3b A diagrammatic representation of gap junction common in animal cells

When many lectin molecules are involved, they give rise to cross linking of cells, resulting in agglutination (Fig. 7.5). Behaviour of lectins towards surface sugars is similar to that of antigen-antibody binding.

Cell to cell interaction: This is clearly a property of cell surface, where membrane carbohydrates are involved in the mechanism. For example, Lymphocytes are produced in bone marrow from the stem cells, and migrate to the primary lymphoid organs where they differentiate. Lymphocytes that differentiate in thymus are called T-lymphocytes and those differentiating in bursa are called B-lymphocytes.

Cell to cell interaction results in attachment between cells, leaving nil or very little intercellular spaces. Roseman propounded hypothesis saying that adhering cells carry polysaccharides and the enzyme glycosyl transferase on the surface, which interact and form surface linkage with each other; many number of cells are found to contain glycosyl transferase enzymes.

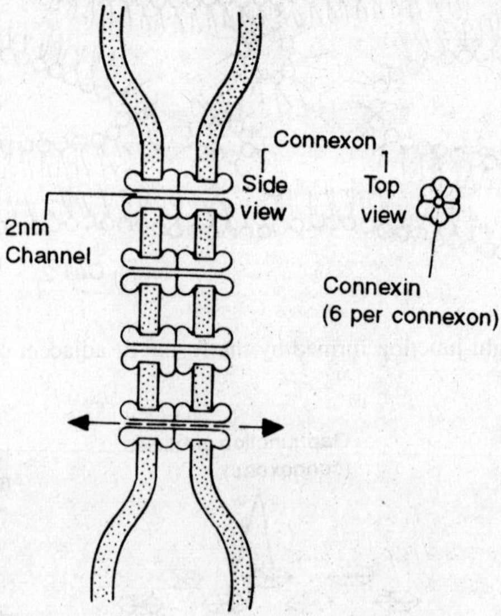

Fig. 7.4. Model of a gap junction showing hydrophilic channels which allow nutrients to pass

CELL INTEGRATION

Integration results between cells to form cell aggregates depends upon an enzyme substrate binding between adjacent cells.

Cell integration property was described in sponges by H.V. Wilson in 1908. He practically proved that if sponge is pressed and forced through a fine cheese cloth, the cells disperse in the solution and reassociate to form aggregates.

Cell integration appears to be due to three factors:

(i) Aggregating cell should be motile,

(ii) Some integration factor is involved,

(iii) Divalent cations (Ca) are essential to stabilise the aggregation factors.

In multicellular organisms cells of tissues have an inherent tendency to remain in permanent association with each other. If these cells lose their adhesive property, they gain the capacity of uncontrolled growth, develop malignancy and may also distribute themselves in other tissues.

Form of communication used during the phase of integration in Dictyostelum discoideum, Slime mould is very interesting. Its life cycle consists of:

a) growth phase,

b) integration phase &

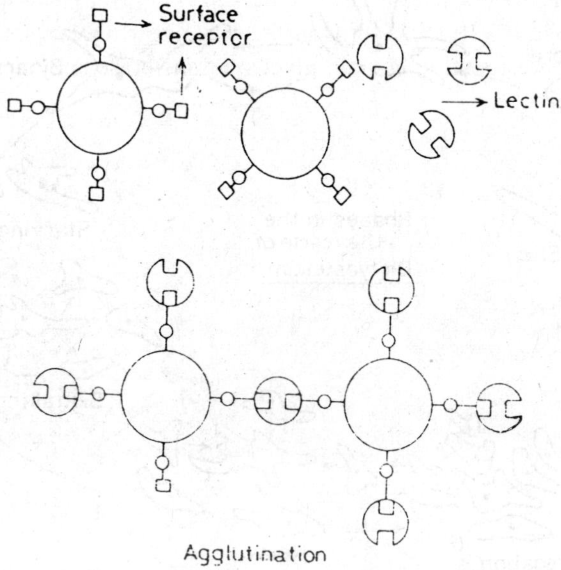

Fig. 7.5. Lectin molecules are involved in cross linking of cells, resulting in agglutination of cells

c) development phase.

Growth phase is a free living dividing stage, in which organism is amoeboid which feeds on bacteria by phagocytosis and divides by binary fission. When the organism is starved, several amoebae integrate to form a multicellular mass in response to pulses of an attractant, acrasin. Acrasin in D discoideum is cyclic adenosine monophosphate (cAMP) Integration phase is followed by development phase, in which mass of cells undergo morphological changes from sling like bodies to saucer shapes (Fig. 7.6).

SIGNALS DURING CELL INTEGRATION

On starvation some cells produce slow rhythmic pulses of cAMP at a frequency of about one pulse every 10 minutes. This frequency slowly increases to one pulse every 3 minutes. Amoebae start integrating and induce other cells in the population to produce cAMP receptor proteins on the cell surface of rest of the population to induce the amoebae to sense the pulse signals.

Oscillating pulsed signals of cAMP are recorded which correspond to pulses of cAMP synthesis and emission from all the cells. Such a pulsed emission is attributed to an internal metabolic oscillator, generating a rhythmic change in the activity of adenyl cyclase, the cAMP synthesising enzyme. This oscillator is believed to be a cytochrome system of mitochondria. The amoebae responding to cAMP pulse start moving towards chaemotaxis, which is the source aided by calcium ions.

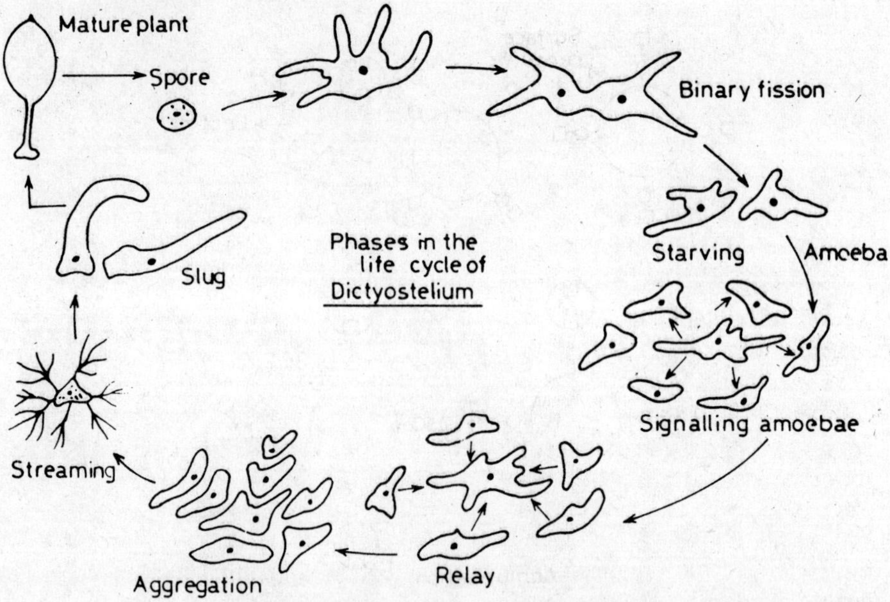

Fig. 7.6. Life cycle of the slime mould. *Dictyostelim discoideum* in which aggregation of amoeboid cells takes place

SUMMARY

It is a universal phenomena of different types of cells, tissues, organs and systems with specialised rọle to perform in biological organisms. Proper functioning is possible only by proper interaction and integration of the cell which is possible only through proper recognition of the cell by the cell. The correlation of glycoproteins helps in making contact with external world of the cell is nothing but interactions and integrations which lead to adhesion. Adhering cells carry polysaccharides type starvation some cells start producing slow rhythmic pulses of CAMP. The pulse signals travel short distance from the source and get destroyed by phosphoesterase enzyme released by starved cells and get bounded to the CAMP receptors. The CAMP pulse starts moving towards source (chemotaxis) and even cooled by ca ions. The CAMP bonding to the receptors sites triggers the release of calcium ions which activates actomycin protein inducing the cell movement.

EXERCISE

1. Explain in detail about cell interaction and cell integration.
2. Explain how cell surface carbohydrates serve for cell interaction.
3. Write short notes on:
   (a) Adhering junctions or Septate junctions.
   (b) Tight junctions.
   (c) Gap junctions.

8

Cell Locomotion

Cell locomotion may involve movement of the entire cell or a portion of it, to the benefit of the organism. This is a biological phenomenon displayed to enable many roles such as feeding, digestion, reproduction, circulation and protection. Flagella, cilia and particularly a muscular system are the means of locomotion for a cell or a unicellular organism.

Biological movements are also performed by cytoplasm of the cells such as cytoplasmic streaming in plant cells or cyclosis in Amoeba.

AMOEBOID LOCOMOTION

As there is no exterior cell wall in animal cells, finger like blunt protrusions from the cell may project, which help in locomotion and also feeding.

In Amoeba, which is a unicellular Protozoan, such protrusions are called pseudopodia which are formed at any point on the surface. Pseudopodia are temporary processes formed when an Amoeba is moving on a solid substratum.

It is suggested that actin filaments and myosin molecules are found in Amoeba. When they interact, actin-myosin complex is formed, providing the basis for locomotion. Due to this interaction, contraction is possible with continuous hydrolysis of ATP.

Various hypotheses have been postulated for explaining the formulation of pseudopodia and locomotion in Amoeba. Amoeba has a layer of ectoplasm that is more rigid and gel-like, whereas endoplasm is diluted and sol-like. When ectoplasm contracts in a certain region, the liquid endoplasm streams towards the region, which can be observed under a polarised microscope. When the endoplasm is propelled in the ectoplasmic tube, it also changes its viscosity and becomes rigid. Automatically contraction is created in the direction opposite to that of streaming of endoplasm.

According to Mast, during the formation of pseudopodium, the plasma membrane gets attached to the substratum followed by a local and partial liquefaction in the plasmagel at that point. Rest of the plasmagel flows out in that area to produce a bulge (Fig. 8.1). At the posterior end, contracting plasmagel is converted into plasmasol and anteriorly, an ectoplasmic tube is continuously regenerated by gelation of plasmasol.

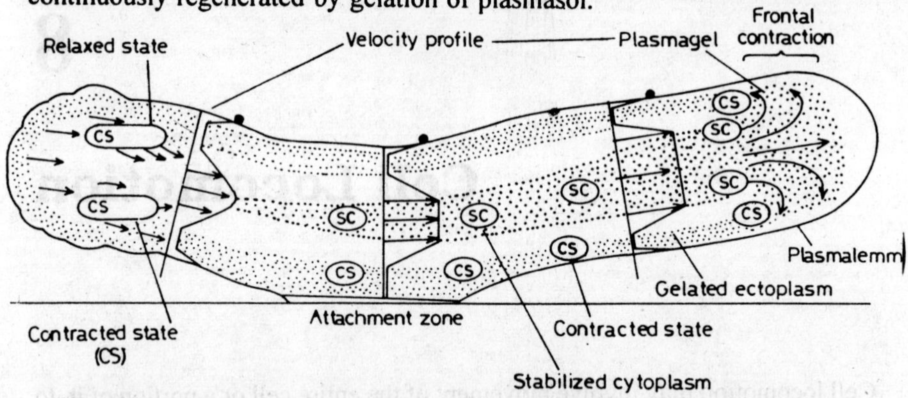

Fig. 8.1. A proposed scheme to show amoeboid locomotion. The cytoplasm on reaching the frontal end undergoes contraction and merges with the plasmagel which moves posteriad as ectoplasmic gel. The gel is then converted to sol which is then converted into relaxed state

The sol-gel interconversions are actually the contraction-relaxation events, which are enforced by osmotic pressure and other ionic changes. It is considered that ca^{++} ions play an important role in regulating the amoeboid movements. Free ca^{++} ions in the plasmasol induce contraction and bring about conversion of sol into gel.

Flagellar and Ciliary locomotion

With the help of specialised locomotor organelles on the surface, viz., flagella and cilia, many unicellular organisms exhibit locomotion.

Flagella are long, slender, thread-like projections from the surface of the cell. The base of the flagellum is anchored in the cytoplasm on a motion-controlling kinetosome region or granule. A cell usually has one flagellum but in many cases there may be more anchored in the kinetosome granules.

Cilia are shorter and each cilium has its own kinetosome at its base (Fig. 8.2). A cilium is about 1 to 15μ long, whereas a flagellum can be about 150μ long.

Cilia and Flagella are made up of a central core of cytoplasm surrounded by a double membrane which is actually the extension of the cytoplasm. Inside the cytoplasm pairs of elongated fibres are arranged along the periphery and two unpaired elongated fibres are present in the centre. In all there are 9 paired and 2 unpaired fibrils attached to a basal plate.

The cilium originates from a basal body or kinetosome which lies embedded in the ectoplasm. The tip of the flagellum or cilium tapers to a point where the number of fibrils are reduced. In certain cases the basal region contains a

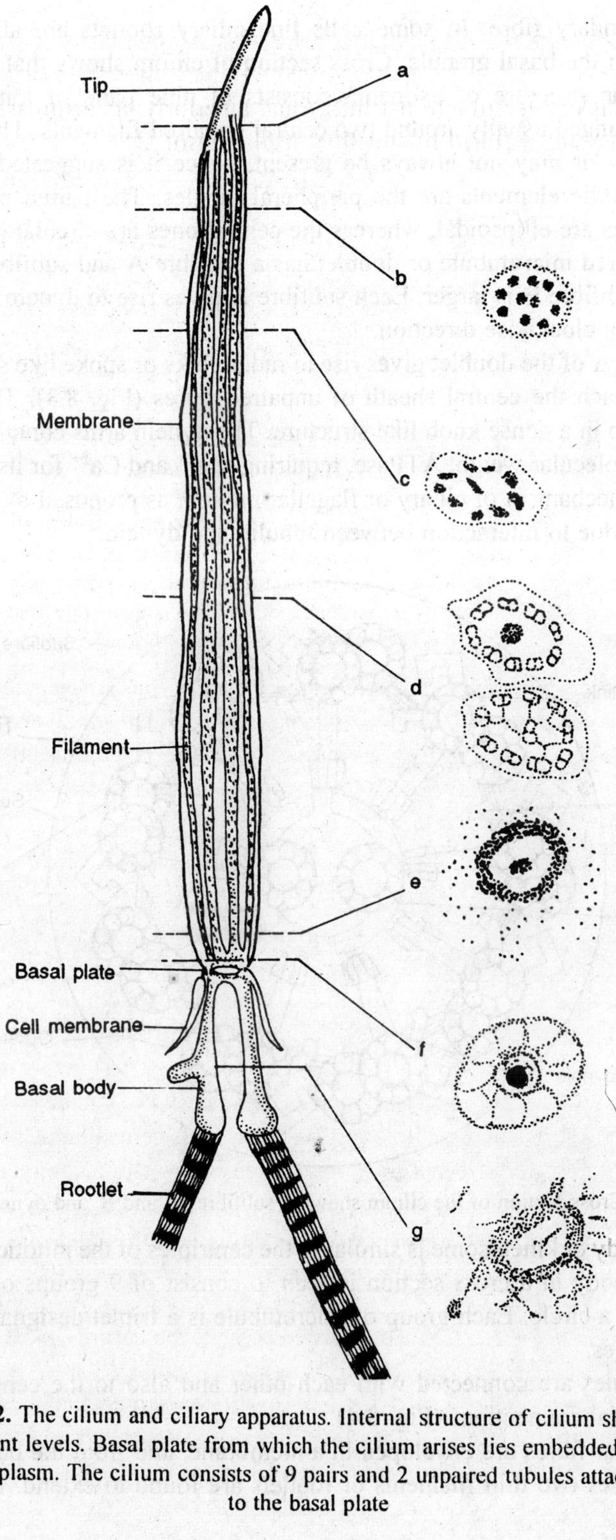

Fig. 8.2. The cilium and ciliary apparatus. Internal structure of cilium shown at different levels. Basal plate from which the cilium arises lies embedded in the ectoplasm. The cilium consists of 9 pairs and 2 unpaired tubules attached to the basal plate

single secondary fibre. In some cells fine ciliary rootlets are also found arising from the basal granule. Cross section of cilium shows that the axial microtubular structure of axonema consists of nine pairs of longitudinal tubules, arranged usually around two central unpaired filaments. The central tubules may or may not always be present, hence it is suggested that the essential motile elements are the peripheral tubules. The paired peripheral microtubules are ellipsoidal, whereas the central ones are circular.

Each paired microtubule or doublet has a subfibre A and subfibre B, out of which subfibre B is larger. Each subfibre A gives rise to dynein arms, all oriented in a clockwise direction.

Subfibre A of the doublet gives rise to radial links or spoke like structures and they reach the central sheath of unpaired fibres (Fig. 8.3). The radial links end up in a dense knob like structure. The dynein arms contain dynein or a high molecular weight ATPase, requiring Mg^{2+} and Ca^{2+} for its activity. The basic mechanism of ciliary or flagellar motion, as proposed by Gibbons in 1977, is due to interaction between tubulin and dynein.

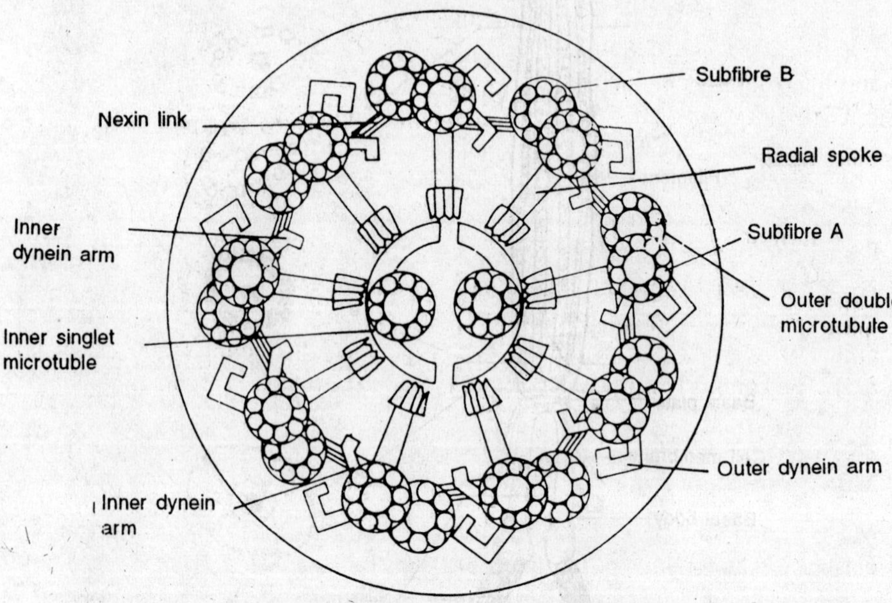

Fig. 8.3. Cross section of the cilium showing subfibres A and B, and dynein arms

Basal body or kinetosome is similar to the centrioles of the mitotic spindle. The basal body in a cross section is seen to consist of 9 groups of tubules arranged in a circle. Each group of microtubule is a triplet designated A, B and C tubules.

The tubules are connected with each other and also to the central fibre through radial connections (Fig. 8.4).

All tubular fibres are enveloped in a membrane, and from the basal body in many cases two thin filaments or rootlets are found to extend. From the basal plate of the cilium two dense perpendicular processes arise which seem to originate from the triplets. These pro-cesses have been termed the basal feet, which are composed of microfilaments.

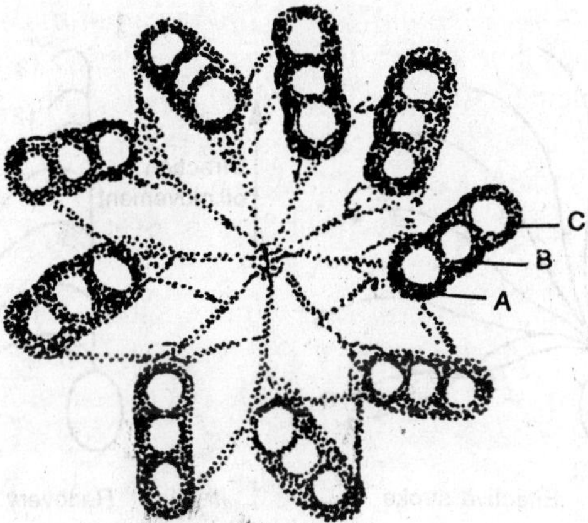

Fig. 8.4. Cross section of the basal body of cilium (highly magnified)

Ciliary Motion: Movement of cilia is coordinated when they beat together, and the rhythm is called isochronal. But certain times each cilium moves a fraction of a second after the preceding one, producing a wave-like movement and this rhythm is called metachronal.

Ciliary movement is studied in two parts: the effective stroke caused by the simultaneous contraction of five (5) of the paired microtubules, causing the cilium to bend like a hook; the second part is the recovery stroke, which is slower than the effective stroke and is supposed to be caused by the contraction of the other four tubular filaments. In the recovery stroke the cilium is not stiff, but exhibits a contractile motion from base to tip (Fig. 8.5). The central two filaments act to extend support. However, their role is not clear.

Muscle movement

Muscles and muscle cells have properties of contraction, expansion and elasticity. The function of muscle cell is closely related to its structure, hence a brief account of different types of muscles is necessary. There are (3) three types of muscle cells; they are:

(i) Striated muscles,
(ii) Smooth muscles and,
(iii) Cardiac muscles.

Structure and function of striated muscle cells: These cells are composed of multinucleate muscle fibres. Each fibre consists of a semi-fluid sarcoplasm, containing many longitudinal myofibrils, called fasciculus. The membrane enclosing each fibre is called sarcolemma, beneath which are located numerous nuclei is scattered. A live myofibril transparent, shows cross-striations with alternating light and dark bands, under polarised microscope.

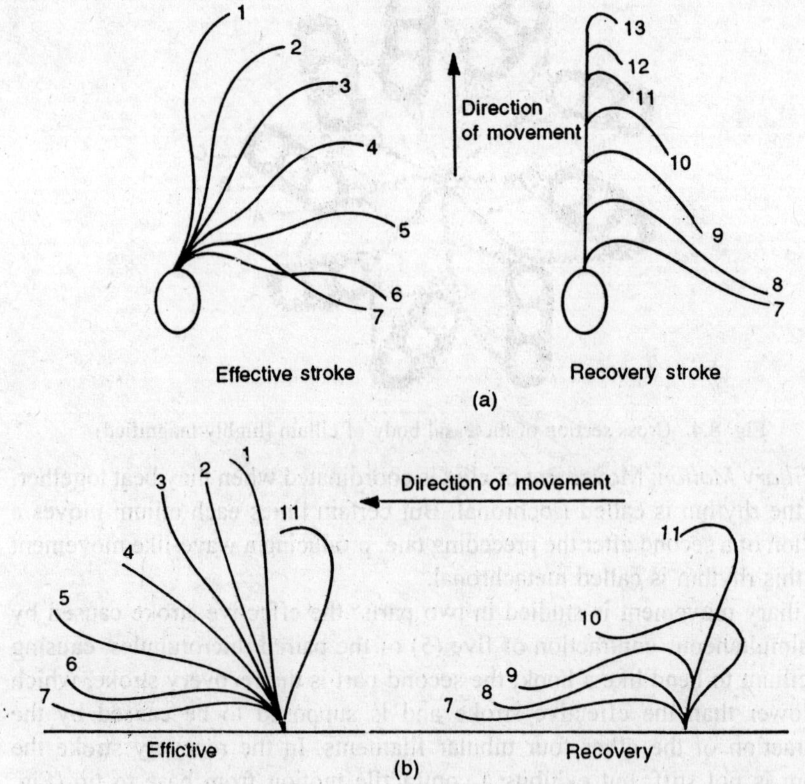

Fig. 8.5. Flagellar and ciliary movement: (a) Flagellar movements during effective and recovery stroke; (b) successive ciliary motions produced by bending movements. Bending motion is produced by making and breaking of cross bridges between adjacent microtobular doublets

Myofibrils are basic contractile units which in a relaxed state show dark and light bands due to the concentration of myofilaments. (Fig. 8.6).

Structure and function of smooth muscles cell

They are also called unstriated or involuntary and are devoid of any cross-striations. They are spindle shaped cells with long-tapering ends and a centrally placed large nucleus. They are generally found in the wall of internal organs such as digestive tract, respiratory passage, urinary bladder, arteries and veins. They are slow in their contractile behaviour and do not come under voluntary control. Although these cells contain actin and myosin proteins, their mechanism of contraction is still not completely understood (Fig. 8.6).

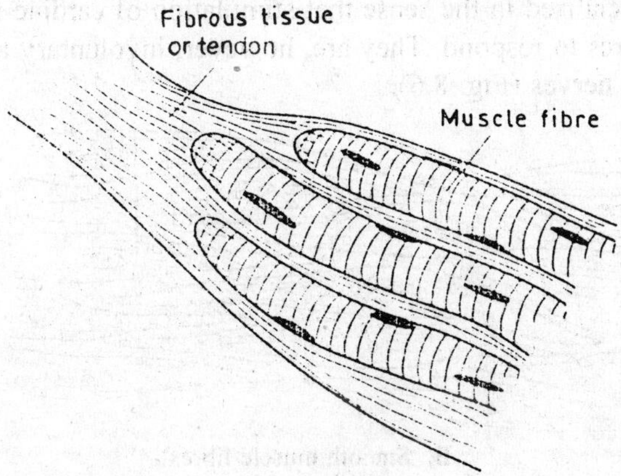

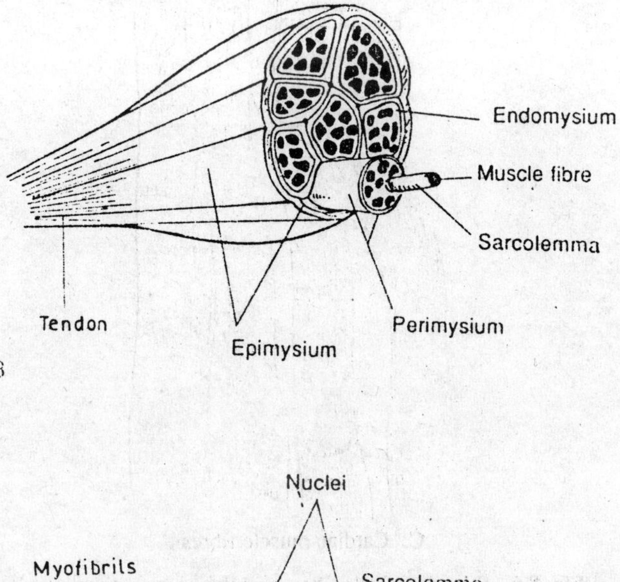

B

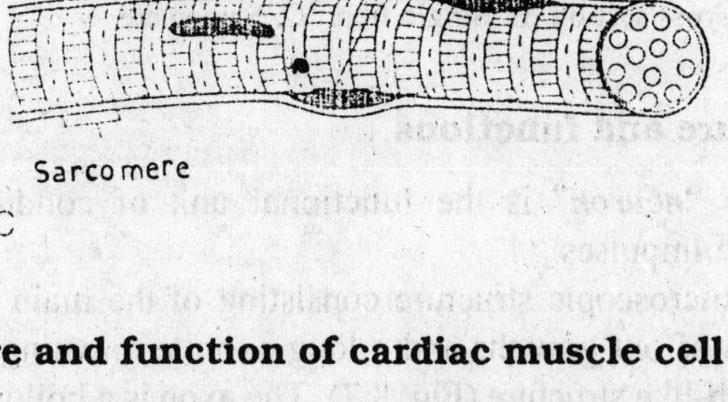

C

Structure and function of cardiac muscle cell

These cells are found in the heart only. They are made of striated multinucleate fibres, resembling skeletal muscles in many respects. These muscle fibres are arranged in a syncitial fashion and do not appear to be fused with each other. These are specialised in the sense that stimulation of cardiac muscle causes all muscle fibres to respond. They are, however, involuntary and innervated by autonomic nerves (Fig. 8.6).

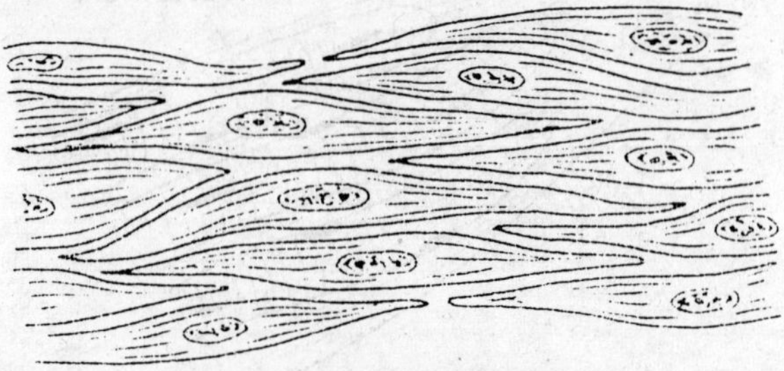

B. Smooth muscle fibres

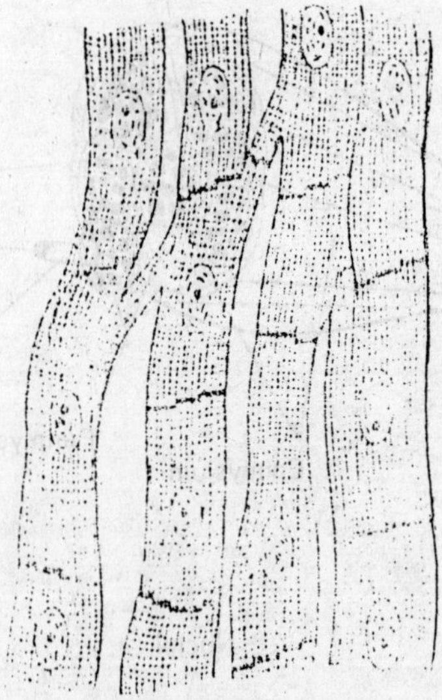

C. Cardiac muscle fibres

Fig. 8.6. Striated muscle fibres and their connective tissue Longitudinal section with some connective tissue approved. A. Cross section B. Single fibre C. Sarcomere

Nerve cell structure and functions

A nerve cell or the "*neuron*" is the functional unit of conduction and transmission of nerve impulses.

The neuron is a microscopic structure consisting of the main cell body, dendrites in the form of out growths and a long axon terminating into axon fibres ending in a bulb-like structure (Fig. 8.7). The axon is a hollow cylinder filled with axoplasm, which differs in chemical composition from the surrounding fluid. It is along this axon an impulse travels. The main cell body with its nucleus is needed to generate an impulse. The axon can be distinguished from the dendrites as it is longer and carries neurosecretory vesicles. Neuron is covered by a myelin sheath which acts as an insulator (Fig. 8.7).

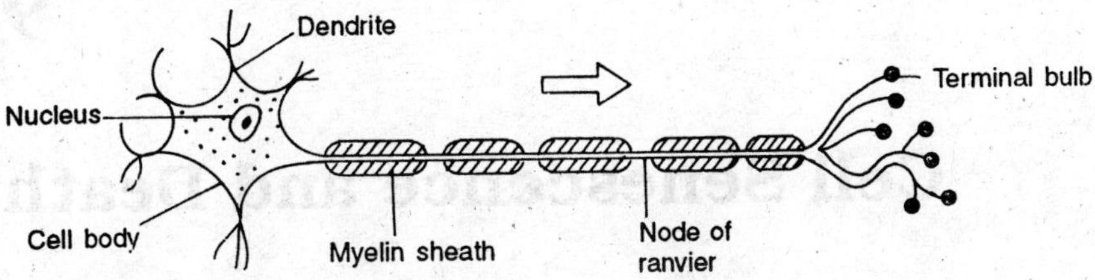

Fig. 8.7. Structure of a neuron show myelinated axon terminating in axon fibres

The neuron when excited develops an action potential, which is a result of profound changes in the electrical property of the nerve membrane. The events that follow excitation are conduction of impulse and the response evoked by the stimulus.

SUMMARY

Cell locomotion may involve movement of the entire cell or a portion of it, to the advantage of the organisms. It is a phenomenon that is displayed to accomplish many roles such as feeding, protection, digestion, reproduction, and circulation. Flagella, cilia and particularly a muscular system are the means of motion for a cell or an organism, and this motor ability is used mainly to capture food in two major ways. Biological motions are also performed by cytoplasm of the cells such as cytoplasmic streaming in plant cells or cyclosis in Amoeba. The cytoplasmic motion is a great aid in the transport of materials. Once the capacity of motion is acquired, it may serve not only in feeding but also in protecting them from potential dangers.

EXERCISE

1. Discuss various types of locomotary organelles, and their utility to organism.
2. Describe the ciliary apparatus and the mechanism of ciliary motion.
3. Explain in detail amoeboid locomotion with supporting hypotheses.
4. Explain in detail about different muscle cells their structure and function.
5. Write Shortnotes on:
   (a) striated muscle.
   (b) smooth muscle.
   (c) cardiac muscle.
   (d) nerve cell.

9

Cell Senescence and Death

Cell Senescence can be defined as a process occurring in all members of population, after maturity, involving progressive decline in vital capacities of the organism terminating in death. Senescence is also considered as death.

The science that studies biological causes of senescence is called gerontology.

Senescence (Ageing) follows cessation of growth. At the cellular level senescence can be studied on the basis of three processes:

(a) Decline in the final efficiency of non-dividing highly specialised cells such as neurons and muscle cells to the possible extent.

(b) Progressive stiffening with age of the structural proteins – such as collagens.

(c) Limitation imposed on cell division as revealed by the studies on fibroblast producing collagen and fibrion.

Mechanisms of Cell Ageing: It can be suggested that normal cells have a finite capacity for replication, and this finite limit is rarely reached in vitro but is, of course, demonstrable in vitro. The functional losses that occur in cells prior to their loss of division capacity produce age changes.

According to L. Orgel (1963), cellular ageing results from impaired specificity of the translation step in protein synthesis. This hypothesis has been confirmed to a great extent.

THEORIES OF SENESCENCE

Many theories related to cell senescence depending upon the different types of nucleic acid and nucleoprotein structures are described.

A. *Change in Nucleic Acids Quantity*: Loss of DNA or RNA per cell per organ could explain decreasing functional efficiency with increasing age. Histological findings point to progressive loss with age of certain irreparable types of cells.

B. *Mutation theory*: The effect of mutations is to bring about the synthesis of faulty messenger RNA and thus, in turn, faulty proteins which are unable to fulfil their biological functions or can do so only imperfectly. Due to faulty proteins, impaired specificity of the enzymes are formed in the translation mechanism, according to Orgel.

There are many ways in which protein synthesis - information content can be changed, falsified or diminished. Results of many biologists' observations are mentioned below:

(1) In ageing cell replacement of defective molecules of metabolic DNA becomes impossible, therefore defective molecules accumulate; as a result of many faulty DNA molecules in the cell, functional impairment takes place.

(2) Ageing is attributed to loss of repetitive information.

(3) The number of methyl group (5' methyl cytosine) in DNA decreases with age, thus DNA information content gets modified and affects protein synthesis.

C. Modifications in protection regulatory mechanisms: According to Orgel, ageing of cells may ensue as a result of cumulation of transcription and translation errors in protein synthesis. Errors leading to reduced specificity of an information handling enzyme lead to an increasing error frequency. Such processes are cumulative. It is a question whether cellular ageing may be due to failure or switchover regulatory processes. Could there be an ageing programme? As an answer to this DNA pool in old age has been observed, and a hypothesis was formulated that each species has species specific programme that serves to maintain and prolong useful life. It is also proven that there is a shift from, arginine-rich to a lysive rich histone composition in the liver of old animals.

FREE RADICAL THEORY OF AGEING

Ageing process may be divided into two categories of cumulative degradative changes.

(i) wide damage produced by a variety of means, such as autoimmune reactions, ionizing radiations and smog.

(ii) alterations in the so called biological clocks-determine the maximum life span of an individual cell.

Free radical reactions lead to a variety of products

Initiation	RH+O	-cu-	R+H O
Propagation	R+O	-	RO
	RO+RH	-	R+ROOH
Termination	R+R	-	R:R

The high reactivity of free radicals is due to the presence of a free electron. Due to the magnetic moment, associated with a free electron, free radicals can be detected at low concentration using the technique of electron spin resonance (ESR) spectroscopy.

Free radical reactions rate, involving molecular oxygen is accelerated by catalysts such as copper, iron and manganese; and inhibited by antioxidants such as vitamin E, butylated hydroxytoluene, and 2-mercapto ethylamine (2-MEA) which are capable of removing intermediate free radicals. These compounds are expected to minimize deleterious effect of free radicals. E-vitamin is a natural antioxidant which has a modest beneficial effect on the life span as it decreases the rate of free radical formation.

SENESCENCE AND IMMUNOLOGICAL SURVEILLANCE

Senescence process is basically related to the fate of the less differentiated mesenchymal cells responsible for the functions of defence and repair with lymphocytes and fibroblasts. Like other somatic cells, these cells are prone to mutations. Though genetic mutations in the somatic cell may modify any cellular activity, they cause two major changes.

(a) Against the changed cells, altered antigens generate immune responses and

(b) Mutation within lymphycytes induces a change in torlerance to self components.

SOMATIC MUTATION HYPOTHESIS

Every living species has certain life span, beyond which genetically controlled programmes become less effective until death. Most vulnerable material to age related damage is the genetic material, i.e. DNA. Changes in the DNA base sequence alter the template and, as a result, affect the regulatory and metabolic capacities of the cell. Cell genome damage may occur in several ways, such as radiations or multiplying errors caused by unmatched nucleotides. Somatic mutation continuation involves error in the structure of enzymes connected with protein synthesis, causing an irreparable damage that will lead to failure of cell mitosis.

THYMUS FUNCTION

Thymus is a very important "Biological Clock" which allows phenotypic expression of genetically decided age. It is a known fact that serum antibodies are produced only by B-lymphocytes and most natural antigens can stimulate antibody production exclusively by the cooperation of T and B immunocytes. T-immunocytes (thymus determined lymphocytes) without B-cell co-operation will lead to delayed hyper sensitivity, for homograft rejection and for medical examination important segment of immunity against viral and mycobacterial infections. Immunological inadequacies of old age involves weakness of T-cell rather than B-cell function. So, due to losing flesh of thymus in middle age, T-cell loses to incorporate effective immunocyte clones against antigens which were previously not encountered.

IMMUNE SURVEILLANCE

Somatic mutations are neither continuous nor heritable changes in a cell line. The development of new plastic change, benign or malignant, must be considered as somatic mutation. Such modifications are sometimes involved with some degree of antigenicity.

Common types of cancer can be considered as old age disease in which any directly influencing carcinogenic agents are very common to have influence on incidence. Senescence may be presumed to act in two ways:

(1) By allowing time for any mutation to occur.

(2) By a progressive weakening of immune surveillance with age.

SENESCENCE OF CONNECTIVE TISSUE

Gross age related changes are observed in connective tissues of the body. All the forms of connective tissues, viz., dermis, tendons, cornea, vascular walls, cartilages, bones, etc. are derived from embryonal mesenchyma. All these connective tissues possess large intercellular spaces filled with collagen, elastin, proteoglycans and structural glycoproteins, collectively considered as intercellular matrix macromolecules (IMM).

In the beginning of embyyonic life proteoglycons and structural glycoproteins participate actively in differentiation process and later this role is taken over by collagen and elastin. In adult life, collagen functions whenever necessary, whereas elastin is suppressed. Collagen synthesis reduces rapidly at an advanced stage.

SENSCENCE OF ELASTIN TISSUE

Elastin fibres play an important role in elasticity and normal tone of skin and blood vessels. Quantity of these elastic fibres deteriorates with age in skin, blood vessels, particularly during arteriosclerosis and ageing disease. Degradation of elastic fibres start usually at a relatively young age, but at the age of 45 years, the degradation accelerates.

AGEING AND TERMINATION OF SYNTHESIS PROGRAMME

Translation and transcription regulations at the genetic level determine the speed of ageing process, and if this control is upset, the cells may synthesise wrong molecules or stop synthesising at all. This process may be either due to exhaustion of genetic programme or to the collection or errors.

Further, another mechanism may also contribute to senescence of ageing; which is dependent upon the cellular micro and macro environment, on nutrition, on physical exercise and balanced diet etc. Avoidance of regular exercise damages healthy functioning of muscle tissues.

"APOPTOSIS"

Multicellular organisms, homeostasis is maintained through a balance between cell multiplication and cell senescence(Cell Death). Physiologically cell death occurs mainly through a process apoptosis.

Recent evidences suggests that alterations in cell survival contribute to pathogenesis of a variety of human diseases including cancer, viral infections, autoimmune diseases, neurodegenerative disorders and AIDS. Treatments established to alter the apoptotic threshold may cure some of these diseases (Thompson 1995).

Apoptosis is a kind of programmed cell death which plays an important role during development, homeostasis, and in several diseases like cancer, acquired immuno deficiency syndrome (AIDS) and neurodegenerative disorders. It operates through the activation of a cell intrinsic suicide program. Its important machinery appears to be present in most mammalian cells at all times, but the activation of the suicide program is regulated by many different signals that originate from both intracellular and extracellular milieu (stellar, 1995). Some components of apoptotic program are established among worms, insects, and vertebrates (Stellar 1995).

Homeostasis in higher organisms is regulated not only by the multiplication and differentiation of cells but also by cell senescence. Cell senescence during embryo genesis, metamorphosis, endocrine dependent tissue atrophy, and normal tissue turnover is termed programmed cell senescence. Most programmed cell senescence occurs by apoptosis which includes condensation and segmentation of nuclei, cytoplasm and extensive fragmentation of chromosomal DNA into nucleosome units.

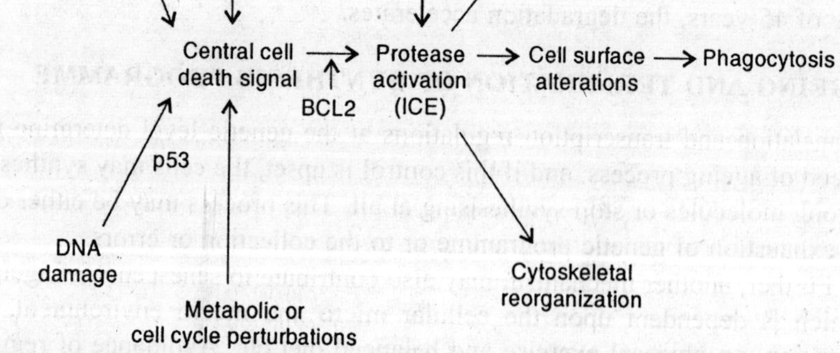

Fig. 9.1. Model for the regulation of apoptosis. The major end point of apoptotic cell death is the removal of the dying cell by phagocytosis (After Thompson, 1995).

Most animal cells are capable of self destruction by this intrinsic cell suicide program. Execution of this apoptosis is often associated with morphological and biochemical changes. During apoptosis, nucleus and cytoplasm condense and the dying cell fragments into membrane bound apoptotic bodies, which are quickly phagocytosed and digested either by macrophagus or by neighbouting cells. Dead cells are thus quickly removed, leakage of their poisonous or dangerous contents is avoided. Apoptosis is different from necrosis which is a pathological form of cell death that results from cellular injury. Apoptosis is usually associated with the activation of nucleases that degrade the chromosomal DNA first into large (50-300 kilobases) and subsequently into very small oligonucleosomal fragments.

Apoptosis is very important for the development and homeostasis of metazoan animals. Besides the beneficial effects of cell death, the inappropriate activation of apoptosis can contribute to several diseases, including AIDS and ischemic stroke (Fig. 9.1).

Certain cysteine proteases are involved in causing apoptosis. The actual cause and mechanism of apoptotic death is still unknown. But there are several similarities between apoptosis and the cell cycle implying that apoptosis and mitosis may be mechanically related or even coupled. There is also an extreme view of this idea that apoptosis may be an aberrant type of mitosis.

SUMMARY

The most important inviolable law that governs biological events of living cell is senescence or ageing. The process of ageing is an inevitable biological event that can neither be slowed nor stopped. The science that studies biological causes of senescence is called gerontology. Ageing processes are progressive and not reversible under physiological conditions. In living systems ageing occurs at all levels—from macromolecules to the intact cell of animals and plants.

EXERCISE

1. Define Cell senescence; Explain various processes by which it can be studied.
2. Explain the mechanism of cell senescence and various theories of senescence.
3. Write in detail about Apoptosis.
4. Write Shortnotes on:
   (a) Free radical theory of aging.
   (b) Senescence and immunological surveillance.
   (c) Somatic mutation theory.
   (d) Thymus function.

10

Cell Differentiation in Plants and Animals

Cells and cytoplasmic contents may become so organised as to perform specialised functions that other cells do not perform. The change by means of which such specialisation is achieved is known as differentiation. The event of differentiation is accompanied by changes which occur in embryonic tissues of plants and animals. Zygote is a unicellular stage which develops as a highly organised individual after several divisions. Cleavages or divisions of zygote forms millions of cells, each of which has some kind of functional specialisation resulting in the formation of specific organs and systems. Through metamorphosis or development of multicellular structures, varied tissues are established.

FACTORS CONTROLLING DIFFERENTIATION

There are two factors which determine the future of a given cell, viz., internal-localised within the cells, and external which depends upon the influences from outside the cell.

Role of Internal Factors: Development and specialisation of embryonic cells are dependent on internal factors. For example, the egg of sea urchin exhibits a definite polarity, the animal and vegetal poles comprising animal – Vegetal, axis. A definite metabolic gradient exists along this axis. The first division plane of the sea urchin egg is along the animal-vegetal axis. When the two resulting cells are separated and maintained under optimal conditions, each cell develops into a sea urchin larva, because the cytoplasmic contents are identical in each cell. The second division is at right angle to the first one, forming four cells; on separating and allowing to develop in optimal conditions each cell develops into a larva. The third division is perpendicular to the animal-vegetal axis, resulting in an eight celled stage, with four cells in each pole. The four cells of the animal pole are different from the cells at the vegetal pole in their cytoplasmic contents on isolation. They do not develop into normal larvae. Thus it can be concluded that the embryonic development depends upon the cytoplasmic internal factors distributed along the animal-vegetal axis.

Role of external factors: External factors also influence embryonic development. For example the differential oxygen availability and the opportunity for waste removal which exists among cells at the surface of an embryo and those that lie more deeply result in different metabolic and divisional rates. Added to this there are cluster of cells which exert influence on other cells. Such cells act as organisers which participate in differentiation of cells into specific organs.

Totipotency: A mass of cells produced due to zygotic repeated divisions are totipotent, i.e., each cell in its early embryonic stage has the entire genetic potential to develop into an individual. As the embryo grows further, cells are differentiated along some predetermined path subject to the influence of genes. When the cells of the differentiated embryo are separated and in a suitable medium under optimal conditions, each cell does not develop into a whole organism.

However, the cultured cell develops into a predetermined path, i.e., the cell predetermined to develop into a ear will develop into a ear only. It can be concluded that the cell destined to develop into a ear has the potential to develop a ear. Such a cell is pluripotent and it cannot evidence totipotency.

Weismann suggested that differentiation is due to the expression of certain chromosome segments into different tissues. He assumed that complete chromosome complement was retained by those cells which were supposed to give rise to gametes in order to provide genetic continuity. This theory cannot be accepted, since the somatic cells of all organisms have the same chromosomal complement.

Accepting the fact that genetic equipment of every somatic cell is identical, Biologists have formulated Plasmagene concept. Plasma-genes are gene-like elements, which are synthesised in the nucleus under special conditions and later released in the cytoplasm. Due to lack of evidence, plasmagene theory was not acceptable to cell Biologists.

Genes control differentiation: Modern Biology advance techniques could prove that cell differentiation was under control of gene function. Every tissue or cell has a specific protein and a specific complement of enzymes which are produced by genes. Though genes do not produce these proteins and enzymes directly, they are produced through an intermediate – the messenger RNA. Hence, the process of cell differentiation is due to the cell's ability to synthesise different kinds of proteins.

Effect of Environment on differentiation

Basal layer cells of plant epidermis are cytologically and biochemically not specialised. These stem cells are actively proliferating and respond to their environment in a variety of ways. Some basal cells differentiate into keratinizing cells and get separated from the remaining layer. But on exposure to Vitamin A, cells differentiate into mucus-secreting cells. Wilde (1961) has reported that a single isolated cell, when kept in a nutrient medium, did not differentiate at all, but when two cells were kept in the medium, one of them underwent differentiation. This clearly shows that a neighbouring cell provides a microenvironment to initiate cell differentiation. The metabolic exchange between the cells and the medium result in altering the medium for differentiation of isolated group of cells.

Levels of differentiation

Cell differentiation can be studied both at cell level and population level. The cellular level involves progressive morphological and biochemical changes; differentiation at population level involve specialisation of two or more cells in a multicellular organism, which may follow different paths. In this type, cells undergo cytoplasmic, nuclear and biochemical changes to establish different development patterns.

Differentiation has been recognised as a four-stage process:

1. Non differentiated stage of the stem cell represents a condition in which the cells produce specific proteins prior to morphological changes set in.
2. Pre-differentiated cell represent extensive proliferative stage to produce vast population. Some are differentiated as specialised cells and others remain as memory cells with arrested growth.
3. The differentiated stage is distinct from other cells. In this the cells have cell-specific proteins.
4. In the differentiated cells modulation may take place in response to extra cellular factors, viz., hormones, pH and temperature.

Differentiation is an *Irreversible process:* In the majority of cases differentiation is an irreversible process.

A mature sperm cannot revert back to spermatocyte and a red blood cell cannot go back to its original stem cell.

In plant cells the situation is different. A differentiated plant cell can retrace its path and dedifferentiate. In many woody plants, the meristematic cells of the shoot and root give rise to greater number of meristematic cells, xylem and phloem tissue and cortical cells. The cortical cells generally form the filling tissues of root and stem, but sometimes take over the function of storage cells. In a two-year-old woody plant the cortical cells redifferentiate to become embryonic cells and develop into new meristematic tissue, called cork cambium. The cork cambium may further give rise to many cambium cells that differentiate into cork parenchyma and cork cells.

This situation is not clear in animals; for example if a salamander's organ like tail or limb is lost, a mass of undifferentiated cells divide mitotically to differentiate into bone, muscle, nerve cells, epidermis, capillaries, etc. for replacing the lost part. This process is called regeneration. Higher organisms, however, do not have regenerative potential.

MECHANISM OF DIFFERENTIATION

The embryonic cell, zygote, has the potential to develop into many cells; some of them remain undifferentiated and persist throughout life but may be utilised when called upon to produce specialised types. Various molecular mechanisms responsible for differentiation are as follows:

- constant genetic information,
- partial shielding of Genome to Transcription,
- mRNA selection for Translation, and
- Amplification of gene in differentiation

CONSTANT GENETIC INFORMATION

Cells do not lose their potential after differentiation. For example, a single cell derived from the pith callus of carrot grows into a whole plant. The cells undergo vigorous superficial cell division resulting in a uniform parenchyma; after some time the callus begins to differentiate. The Parenchyma has various tissue forming properties. Successive differentiation leads to the formation of vascular bundles and other tissues, leading to the growth of an entire plant. This experiment shows that from a single cell, by repeated cell divisions, a set of tissue are formed to produce a total plant.

PARTIAL SHIELDING OF GENOME TO TRANSCRIPTION

In certain in vitro experiments conducted with athymus DNA, in which RNA Polymerase was added to synthesise RNA on DNA templates, analysis proved that RNA produced in vitro represented only 10% of the transcribed DNA. This reveals that the transcription is restricted to some parts of the genome, and the kinds of mRNA produced by a cell are regulated by the amount of DNA available for transcription.

This reveals that the DNA available to transcription in each tissue is different in its polynucleotide sequence and the balance DNA is not available for the purpose.

mRNA SELECTION FOR TRANSLATION

It is understood that synthesis of proteins depends upon the DNA information that has been successfully transcribed to mRNA. Some selection process operates to inhibit or permit mRNA message for protein synthesis.

An example of translational control: During fertilisation, sea urchin eggs show increase in protein synthesis rate. Little quantity of messenger RNA is synthesised just after fertilisation. Gross and Co-Biologists suppressed the RNA synthesis by using actinomycin D, an antibiotic, to prove that increase in the rate of protein synthesis takes place in the absence of RNA synthesis. In the next experiment sea urchin eggs were divided into two halves, the non nucleated half synthesised as much protein after activation as did the nucleated half after fertilisation. If the message is not translated, it could be due to non-availability of some components other than mRNA, such as ATP, transfer RNA etc. If in presence of all components translation does not take place, it can be attributed to the masking of mRNA. It is thus concluded that mRNA species are selected for protein synthesis and those not needed are destroyed.

AMPLIFICATION OF GENE IN DIFFERENTIATION

Gene amplification is an important phenomenon which has recently been established in animal development. Gene amplification should be achieved by selective replication of required genes, so that some genes proportionally increase, while the amount of DNA remains stable. This phenomenon is very important as it contributes to adjustment of gene activity in a cell by changing the proportion of some genes.

This phenomenon has been demonstrated in the giant polytene chromosomes of salivary glands of certain insect larvae at a particular developmental stage. DNA synthesis establishes at specific bands in giant chromosome, which has an ability to hybridise with labelled thymidine, expressing that H-thymidine is incorporated into DNA regions of chromosome where there is no addition in DNA content.

CYTOPLASMIC FACTORS IN DIFFERENTIATION

Cytoplasmic factors affect gene expression during differentiation. Briggs and King have worked out techniques of nuclear transplantation in animals which are successfully being applied to several species of amphibia. For example, thenucleus of a specialised intestine epithelial cell was transplanted into an enucleated egg of Xenopus leavis, frog; subsequently, a normal frog was produced (Fig. 10.1). This reveals that in acquiring specialisation, the intestinal epithelial cell did not lose genes. Nor wasits activity lost forever. This experiment also concludes that the genes present in a living cell exhibit differential activity.

If a nucleus from one kind of cell is transplanted to the cytoplasm of another type of cell engaged in a different kind of synthetic activity, always the cytoplasm only determines the nature of nuclear activity. Therefore, it can be presumed that factors controlling gene activity are located in cytoplasm. Nuclear transfer experiments also prove that the cytoplasmic proteins enter the transplanted nuclei at the time of gene activity. It also indicates that an

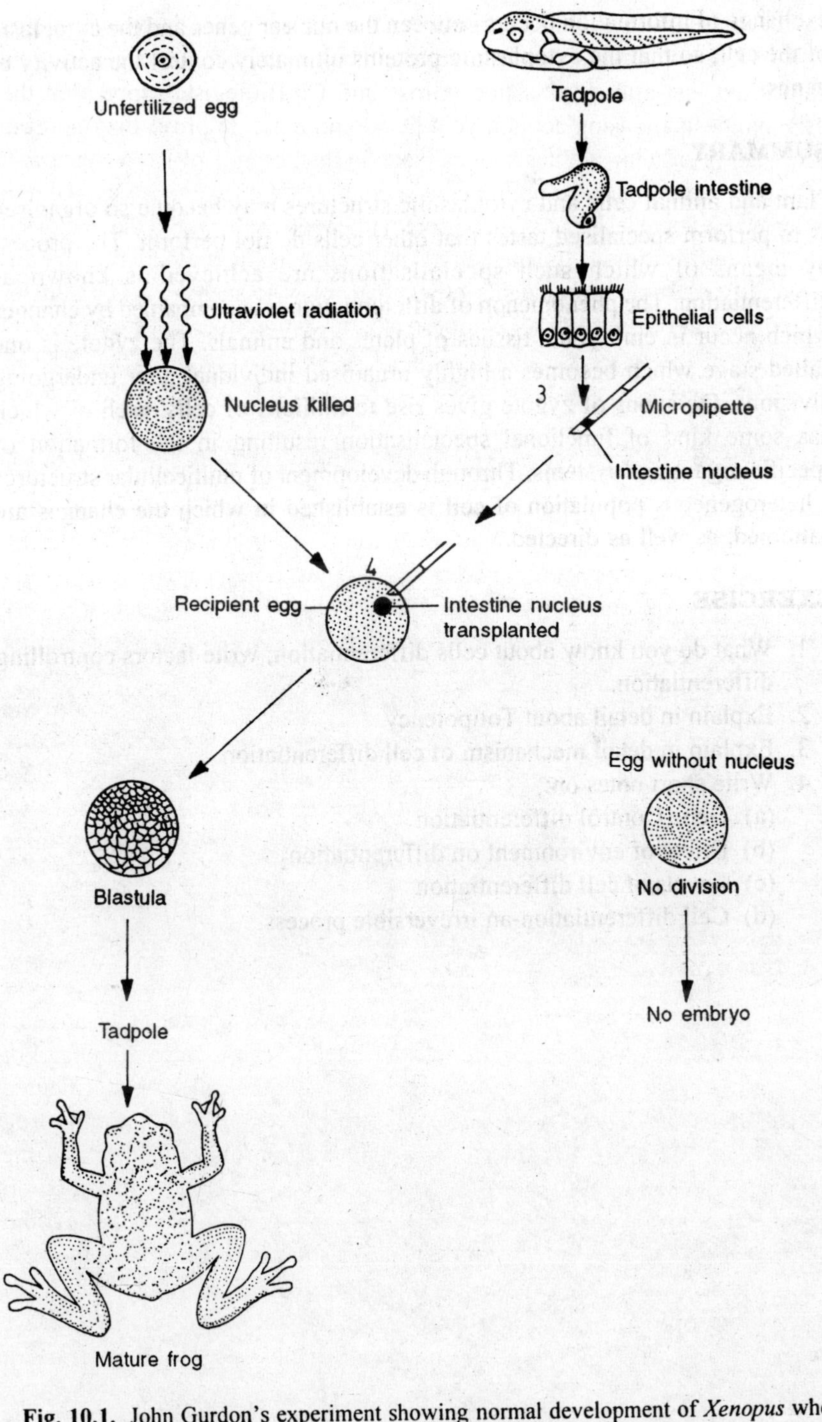

Fig. 10.1. John Gurdon's experiment showing normal development of *Xenopus* when the differentiated intestinal nucleus was transplanted in an enucleated egg by ultraviolet irradiation. enucleated eggs do not divide.

exchange of information occurs between the nuclear genes and the cytoplasm of the cell, so that the cytoplasmic proteins ultimately control the activity of genes.

SUMMARY

Plant and animal cells and cytoplasmic structures may become so organised as to perform specialised tastes that other cells do not perform. The process by means of which such specialisations are achieved is known as differentiation. The phenomenon of differentiation is accompanied by changes which occur in embryonic tissues of plants and animals. The zygote is one called stage which becomes a highly organised individual after undergoing divisions. Divisions of zygote gives rise to millions of cells, each of which has some kind of functional specialisation resulting in the formation of specific organs and systems. Through development of multicellular structures a heterogeneous population of cell is established in which the changes are patterned, as well as directed.

EXERCISE

1. What do you know about cells differentiation; write factors controlling differentiation.
2. Explain in detail about Totipotency.
3. Explain in detail mechanism of cell differentiation.
4. Write short notes on:
   (a) Gene Control differentiation
   (b) Effect of environment on differentiation
   (c) Levels of cell differentiation
   (d) Cell differentiation-an irreversible process.

Part B
Genetics

11

Nature of Genetic Material

NUCLEIC ACIDS

The nucleic acids are divided into two categories, the RNA and DNA. In different organisms the genetic information is stored either within the DNA or the RNA. Example, *Bacterial Viruses* (Bacteriophages) ϕX174, λ, T2, T4, have DNA while F2, Ms2, R17, and QB have RNA as genetic material.

Animal Viruses

Simian virus 40, mouse polyoma and rabbit papilloma, herpes simplex, adenovirus have DNA. Rous sarcoma in fowl, poliomyelitis, influenza, reo virus have RNA.

Plant Viruses

TMV and Tomato bushy stunt virus have RNA.

The Nucleic acids consists of nucleotides as their building blocks, where, they are made up of nucleoside and the phosphate group. The nucleosides are further made up of pentose sugar + nitrogenous base.

Pentose in RNA is ribose while in DNA it is a 2′ Deoxyribose. But the pentose occurs in the nucleotide in β-furanose form. The phosphate group is esterified usually at 5′ position and in the cell pH has a –ve charge. The nitrogenous bases are linked to the 1′ carbon by N-glycosyl linkage.

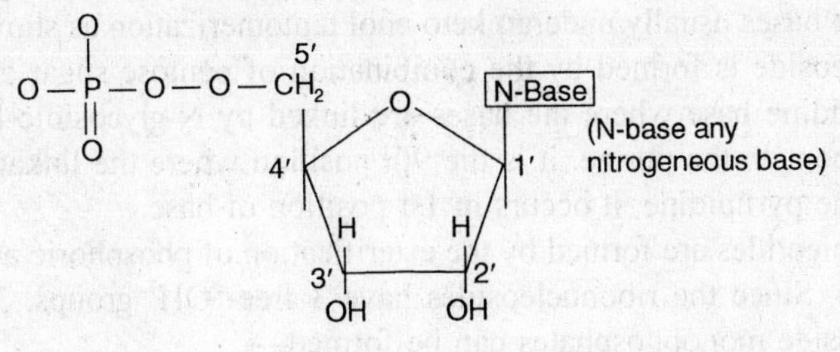

Nitrogenous bases are the derivatives of 2-parent heterocyclic compounds namely pyrimidine and purine. Derivatives of pyrimidine are cytosine, uracil and thymine, where the RNA consists of cytosine and uracil while DNA has thymine and cytosine. The purine derivatives are adenine and guanine both of which occur in DNA and RNA.

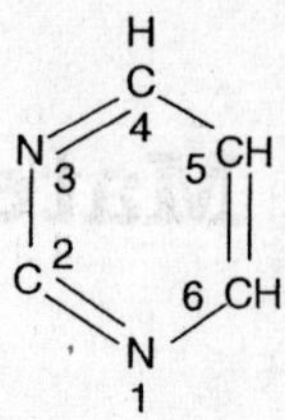

Pyrimidine

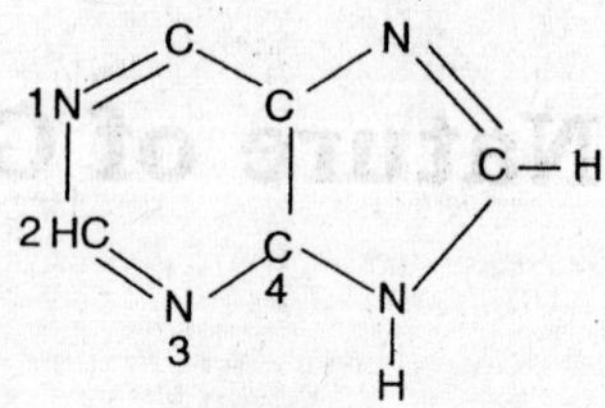

Purine

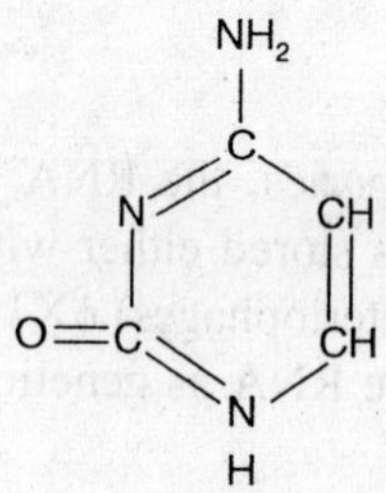

Cytosine

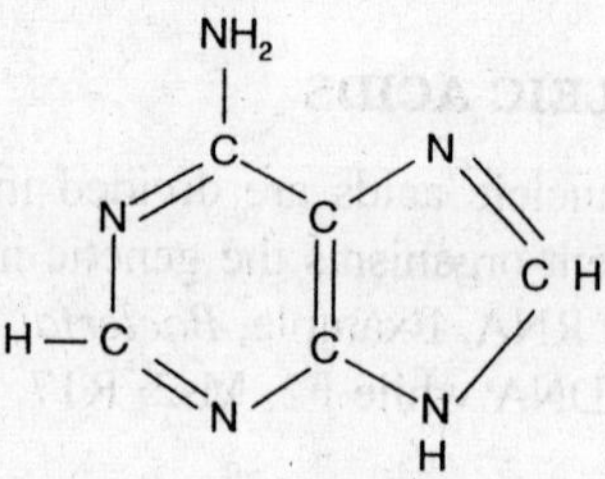

Adenine

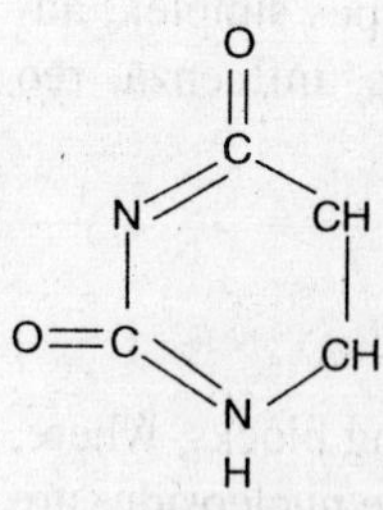

Uracil

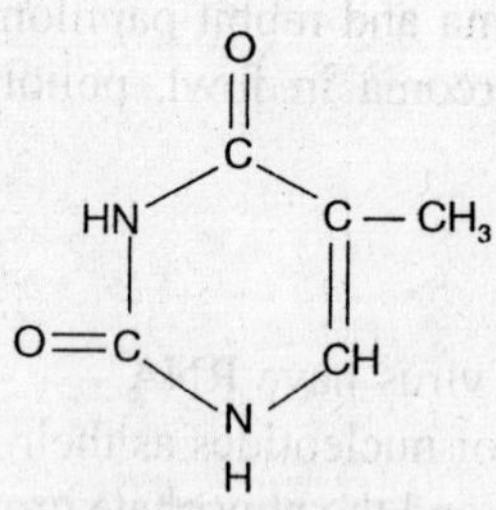

Thiamine

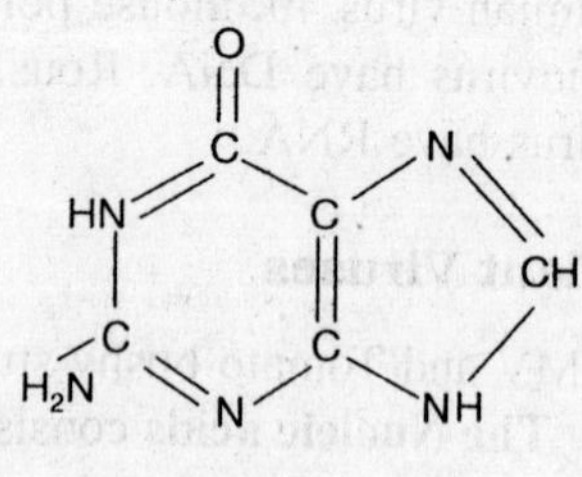

Guanine

only $-CH_3$ is the difference

Apart from these major bases, there are some minor bases, which are methylated forms of the principle base and in some viral DNAs they may be methylated or glucosylated. Such bases have specific roles as signals. The pyrimidine bases usually undergo keto enol tautomerization as shown below.

A nucleoside is formed by the combination of pentose sugar and purine and pyrimidine base where the bases are linked by N-glycosidic linkage to the 1′ carbon. In the purine, it is the 9th position where the linkage occurs, while in the pyrimidine, it occurs in 1st position of base.

The nucleotides are formed by the esterification of phosphoric acid on the nucleoside. Since the ribonucleosides have 3 free 'OH' groups, 3 possible ribonucleoside monophosphates can be formed—

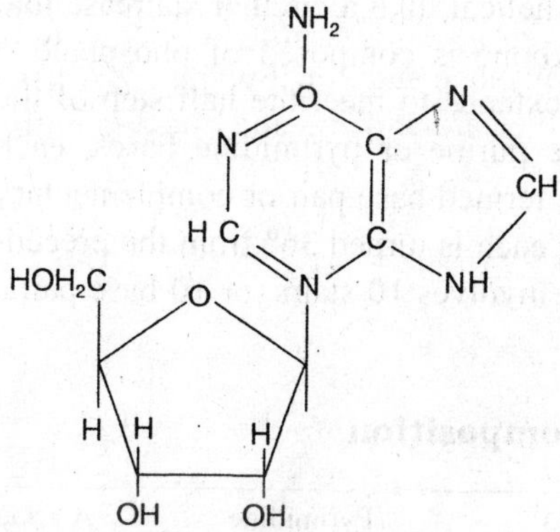

Adenosine

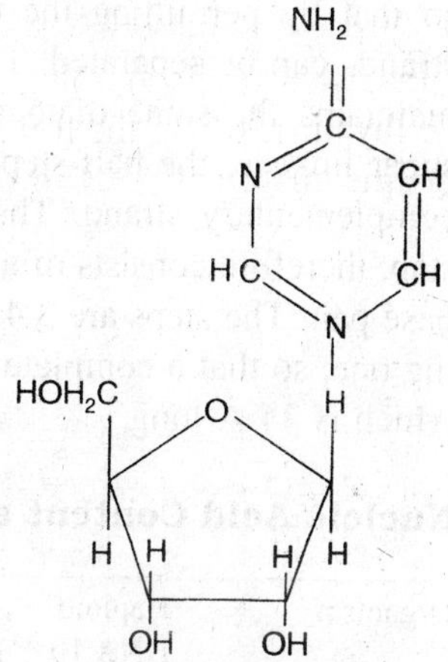

Cytidine

Adenosine 3′-Monophosphate, Adenosine2′-monophosphate and Adenosine 5′-monophosphate. The ribonucleosides at the 5′ phosphate may be further phosphorylated to yield the 5′ di and triphosphates. Eg.: AMP, ADP and ATP.

However, no such formation is seen with the deoxyribonucleoside which has only the 3′ and 5′ positions available for esterification. They also form mono, di, tri, phosphates.

Structure of Nucleic Acids

Nucleic acids are polymers made up of 100, ... 1000, ... many nucleotides coupled together by phosphodiester bond. In the case of DNA, the 4 carbon has no 'OH' group and the only availability for internucleotide linkages are with the 3′ and 5′. However in the case of RNA, which has the 'OH' group at the 2′ position, it is possible to postulate 2′, 5′ linkage in addition to 3′, 5′ linkage but result indicates, even in RNA it is the 3′, 5′ linkage.

Chargaff *et al.* showed that all the nucleotide bases were not present in equal amounts and that the ratio varied between species. The experiments conducted by him in conjunction with improved techniques of isolation suggested that the DNA was not a single molecule but made up of repeated tetranucleotides sequences. As a rule, Chargaff had shown that the purine bases (A + G) equalled to the sum total of pyrimidine bases. There was an equivalence between bases carrying amino group at the 6th or 4th position namely A and C, and the bases carrying keto groups namely T and G; the ratio of A to T, G to C was close to unity in various eukaryotic species. X-ray studies on DNA by Wilkins, Franklin and others indicated a well-organized multiple stranded fibre about 22 Å in diameter, characterised by groups spaced 3.4 Å and a repeating unit every 34 Å. Taking into account the points, Watson and Crick proposed a double-helical structure for DNA. According to them, the DNA molecule is 2-stranded and coiled like a rope, so that by permitting the ends to revolve freely, the two complementary strands can be separated. The coiling is helical, like a circular staircase that maintains the same diameter. The backbone is composed of phosphate + sugar linkage, the half-step of a strand extends to meet the half step of the complementary strand. There are single purine or pyrimidine bases, each step, therefore consists of a pair of base, termed base pair or complementary base pair. The steps are 3.4 Å apart, and each is turned 36° from the preceding one, so that a complete turn of 360° involves 10 stairs, or 10 base pairs, which is 34 Å long.

Nucleic Acid Content and Base Composition

Organism	Haploid DNA 10^{-12} g (pico gram)	Purine (A.G.)		Pyrimidine		A + G/
		A	G	T	C	T + C
Man	3.2	31	19.1	36.5	18.4	1
Cattle	3	28.1	22.2	21.2	22	1.03
Fruit fly	0.18	27.3	22.5	27.6	22.5	1
Corn	7.5	25.6	24.5	25.3	24.6	1
E. Coli	0.0047	24.6	25.5	24.3	25.6	1
Herpes (simplex)	0.00011	13.8	37.7	12.8	35.6	1.06
Lambda	0.000055	26	23.8	25.8	24.3	0.99
174 (single stranded)	0.00000284	24.7	24.1	32.7	18.5	0.95
TMV	0.0000033	29.3	25.8	26.8	18.1	1.23

The rotation of this double helix around its axis generates one major (wide) and one minor (narrow) groove. In each complete turn, the minor groove may bind to some Histone proteins while the major groove binds to nonhistone proteins. The base pairs are held by 'H' bonding as shown in the figure.

Different Conformational forms of DNA

Form	Forms of DNA	Pitch (nm)	Residues per turn	Inclination of base pair from horizontal
A	Na-salt 75% RH.	2.8	11	20°
B	Na-salt 92% RH.	3.4	10	0°
C	Li-salt 66% RH.	3.1	9.3	6°

Through the basic model put forward by Watson and Crick is close to the accepted structure of the DNA molecule in solution. The refined X-ray diffraction studies of Wilking *et al.* indicate three possible structures of DNA fibres. The B structure comes close to the original Watson–Crick model.

The A and C conformations are also right handed but differ in the pitch and in the number of bases per turn. In both A and C forms, the bases are tilted. Their biological significance is not clear but the A conformation is believed to be close to the structure adopted by double-stranded RNA and by DNA-RNA hybrid. Because of the presence of the extra 2-OH group, RNA appears to be unable to adopt the B confirmation. Thus, during the formation of the template for making RNA, the DNA molecule must adopt the conformation.

Various organic solvents also force the DNA to change its conformation. Adjacent bases are related to each other by roll, slide and twist and the A, B and C forms are simply structures, which exist across a continuation of values. The bases may slide relative to each other with the values of slide (σ) ranging from –0.1 nm to +0.2 nm. This value depends on the actual bases involved and so a stretch of DNA may have regions of A, B and C from DNA.

Similarly, the helical twist angle is highly dependent on the particular sequence and varies from 27.7° (A_PG) to 40.0° (G_PC). B. DNA is not therefore a smooth, regular double helix but rather shows an irregularity dependent on the base sequence. The alteration of the helical twist angle results, partly from a rolling of adjacent base pairs over one another along their long axes *i.e.,* along a line from C_8 of purine to the C_6 of the pyrimidine.

When the sequence of bases in DNA shows a regular pattern, the structure of DNA undergoes dramatic changes. When there is a region of DNA containing alternating Pu and Py, it can lead to wrinkled form of DNA. In more extreme cases to a complete breakdown of the B form of DNA and conversion to Z-DNA. The wrinkled DNA has deeper minor grooves which could lead to changes in the interactions of the duplex with cations and water molecules. It may also affect DNA protein interaction which could be the reason why the lac repressor binds more strongly to poly [d(A-T)] poly [d (A-T)] than to DNA of general sequence since in the normal recognition site (Lac operator) three fourths of the bases are arranged as alternating Pu and Py, thus taking up the wrinkled configuration.

If instead of alternating Pu and Py, DNA has a Pu-Py dinucleotide followed five bases later by a py-pu dinucleotide (*i.e.,* PuPy abc PyPu def PuPy ghi PyPu), then the DNA would be bent. In a similar manner, runs of As appropriately placed, lead to the formation of a bent molecule.

Under conditions of high salt concentrations, DNA with an alternating Pu-Py sequence, ex: poly (d A-T) tends to form a left handed double helix known as Z-DNA which has 12 base pairs per turn in solution. It may also be formed in supercoiled DNA molecules where the region of left-handed DNA serves to relax the tension of the supercoiled molecule.

The secondary structure of DNA is a dynamic one, wherein the double helix is not a totally fixed or rigid molecule but undergoes considerable internal deformation in a continuous manner. The swinging of small segments of the double helix is referred to as *breathing of the DNA* (in tritium exchange experiments). Sequences, which breathe most readily have been calculated, where the opening of a base pair is determined by the stability of the base pair and its nearest neighbour 5′ Pu-Py dinucleotide pairs are stabler than 5′ Py-Pu dinucleotide pair.

Denaturation

When double-stranded DNA molecules are subjected to extremes of temperature or pH, the hydrogen bonds of the double helix are ruptured and the two strands separate. The DNA is said to be denatured and forms a random coil. When heat is used as the denaturant, the DNA is said to melt and the temperature at which the strands separate is known as the transition temperature or melting temperature (Tm).

The nitrogenous bases of a polynucleotide absorb light at 260 nm, but in a double-stranded DNA, this absorption is partially suppressed due to coupling between the transition of neighboring chromophores (*i.e.*, interaction of π electrons of the bases), which is due to stacking of the bases. When the duplex melts, the hydrogen bonds break and the bases unstack; as a result, the absorption at 260 nm rises by 20–30%. This is known as *hyperchromic effect* which is used to monitor melting of DNA.

Along with this, the transition is accompanied by a change intensity of the DNA, the single-stranded molecule being more dense than the corresponding duplex.

The nature of melting transition is affected by:

(a) The (G + C) content of DNA

Higher the G + C content of DNA, more stable will be the molecule and therefore higher will be its melting temperature.

(b) The nature of solvent

In low concentration of counter ion, denaturation occurs at relatively low temperatures and over a broad range, while at high concentration of counterion the Tm is raised and the transition is sharp.

(c) The nature of DNA

Most DNA have varying G + C and A + T regions occurring as mosaics, which results in the two strands being held together at G + C regions. This region thus allows the reannealing of the broken H-bond on lowering the temperature.

Renaturation of DNA

When 2 DNA strands are returned from the extreme conditions they may reassociate to reform a double helix. Simple DNA molecules reanneal correctly and instantaneously than complex DNAs, since to reform the double helix they must align themselves perfectly. It depends upon the concentration of DNA and the time allowed. If the solution is quenched at 4°, it limits diffusion and prevents the separation of any DNA strand which have mismatched producing solutions of denatured DNA in case of complex molecules.

The complexity of a DNA sample is reflected in the time it takes a solution of DNA of given concentration to reanneal. The reassociation can be followed spectroscopically or by taking advantage of the fact that a duplex DNA binds more strongly to *hydroxy apatite* than a single stranded DNA. Another method makes use of S_1 *nuclease*, an enzyme that preferentially digests single-stranded DNA.

Buoyant Density of DNA

Physical properties of DNA are strongly influenced by the percentage of G + C, which also influences the buoyant density of DNA. In concentrated CsCl solutions G + C rich DNA has a higher buoyant density than A + T rich DNA.

Isolation and Separation of DNA

Phenol extraction is used widely in the isolation of both DNA and RNA which are retained in the aqueous phase while the denatured protein collects at the interface between this and the phenol phase which contains the lipids. RNA can be removed from the DNA by using pure pancreatic ribonuclease or by *isopycnic ultracentrifugation* in a gradient of CsCl. A density gradient is established utilising the buoyant densities of the molecules to be separated. Since an equilibrium is established, it is also sometimes known as *equilibrium ultracentrifugation.*

The buoyant density of double-stranded DNA varies with the G + C content, however RNA has a much higher buoyant density (about 1.9 g/ml) than double-stranded DNA (about 1.7 g/ml) that which in turn is higher than that of protein (about 1.3 g/ml). Single-stranded DNA has a slightly higher buoyant density.

Bacterial plasmids can be separated from chromosome DNA using the Isopycnic ultracentrifugation. Chromosomal DNA is linear while plasmids are circular. The difference is exploited by addition of saturating concentrations of intercalating fluorescent dye, *ethidium bromide,* where intercalation requires that the DNA strands be forced apart with a concomitant decrease in buoyant density and partial unwinding of the double helix. The unwinding of the double helix is hindered in the closed-circular plasmid molecule with the result that they bind less ethidium bromide and have a higher buoyant density than linear chromosome.

Chemical Reactions

Some of the important reactions of the DNA are:

(a) Strong mineral acids lead to depurination of Nucleic acid which leads to the conversion of sugars into furfural derivatives which give specific colour reactions with orcinol (RNA) or diphenylamine (DNA).

(b) Nitrous acid reacts with amino groups to convert them into hydroxyls. Thus converting

Cytosine ⟶ Uracil

Adenine ⟶ Hypoxanthine

Guanine ⟶ Xanthine

Eukaryote DNA

Eukaryotic DNA is contained in a relatively small number of chromosomes. No direct correlation can be made between the amount of DNA in the nucleus and the number of chromosomes in which it is contained. Somatic cell contains diploid number of chromosomes. The characteristic number and morphology of the chromosomes in a particular cell is known as the karyotype of that cell.

DNA Content and Chromosome Number

Organism	Haploid DNA picograms	Content base pairs	Haploid chromosome number
Simian Virus 40	0.000006	5.3×10^3	1
Herpes simplex	0.00017	151×10^3	1
E. Coli	0.005	4.5×10^6	1
Drosophila melanogaster	0.17	0.15×10^9	4
Homo sapiens	2.7	2.4×10^9	23
Xenopus laevis (toad)	4.2	3.8×10^9	38
Triturus Cristatus (newt)	35	31.5×10^9	12

The amount of DNA is at least an order of magnitude in excess of that required for the known gene coding capacities of the cells where the bulk of the DNA is not expressed.

The minimum size of genome *i.e.,* amount of DNA per cell (C-value) increases with the stage of evolutionary development. However, certain amphibia have a 'C' value more than man, even more than other amphibia, the reason being not clear. This phenomena is known as the *C value paradox*. The species with greater amount of DNA should have the advantage of greater cording potential and the disadvantage of the requirement to replicate very large amounts of DNA prior to cell division.

Eukaryote DNA is divided into three frequency classes of Unique DNA, moderately repetitive DNA and highly repeated DNA with a considerable overlap between the three categories.

Unique DNA

It comprises about half of the total haploid DNA content and consists of the sequences coding for most enzyme functions for which there is only one or a small number of genes per haploid genome. In contrast to the prokaryotic genes which occupy a single uninterrupted sequence of DNA, the majority of eukaryotic structural genes have extra sequences inserted into the middle of the gene. These intervening sequences are called *introns,* which may be small and single as in the case of the gene for tyrosine suppressor tRNA or large and multiple. These intervening regions are transcribed into RNA and then removed in the nucleus to produce the final mRNA. Such intervening sequences appear to be less frequent in the lower eukaryotes. Majority of structural genes display this phenomenon, and is not universal. The significance of introns is not properly understood and has been suggested to reflect the continual process of evolution within the eukaryotic chromosome.

Repetitive DNA includes all the rest of the DNA. The sequences represented in this group are generally thought to be those coding for proteins which form major structural components of the cell, example, genes for rRNA, tRNA, histones etc.

Satellite DNA

It represents highly repeated sequences of which there may be a million or more copies per haploid genome which are usually quite short and are arranged in tandem arrays. The name relates to the method of its isolation on CsCl buoyant density gradient of sheared DNA, where it forms a satellite band separate from the main DNA band due to its differing content of A and T residues.

The distribution of satellite DNA among chromosomes varies, some chromosomes have virtually no satellite while others (especially y chromosome) are largely composed of satellite sequences. It appears to be concentrated near the centromere of the chromosome in the *heterochromatin fraction*. The function of satellite DNA is not known properly. It was thought that satellite DNA was not transcribed since its corresponding RNA was seldom isolated but occasional cases of satellite transcription have been reported. Its transcriptional inactivity lies in with its localization in heterochromatin.

Interspersed Repetitive DNA

It differs from the satellite DNA in that, in it the sequences are not clustered but dispersed singly throughout the genome. They are of two classes.

(a) Short interspersed nuclear elements (SINES) consisting of repeats shorter than 500 bp and (b) the long interspersed nuclear elements (LINES).

Example of SINES is the Alu family in human cells (it derives its name from the presence of a single site for the restriction enzyme Alu I in the repeat). The Alu sequence is transcribed and it is believed to have become dispersed throughout the genome by reverse transcription and reintegration, therefore it is termed as Retroposon.

Foldback DNA or Palindromic DNA—these sequences comprise 3–6% of eukaryotic DNA. The size range is from 300 to 1200 bp.

5′ TGCGATGC : CGATCGCA 3′

3′ ACGCTACG : CGTACGCT 5′

Chromatin Structure

In the interphase, the chromosome is referred to as chromatin. Earlier chromatin was loosely defined and subdivided into two classes – Euchromatin and Heterochromatin. Heterochromatin comprises the dense readily stained areas of the nucleus or chromosome and was thought to represent inactive chromatin. The Euchromatin is more loosely packed, which was thought to represent the transcriptionally active material.

Chromatin consists of DNA, RNA and proteins. In general, the amount of protein is equal to or greater than the amount of DNA while the amount of RNA is comparatively small.

The protein content of chromatin can be further subdivided into *histone and nonhistone proteins.*

Histone

The histones are basic proteins of low molecular weight. Each molecule consists of a hydrophobic core region with one or two basic arms. They are classified into five types namely H_1, H_2A, H_2B, H_3 and H_4.

H_1 is very lysine-rich protein of about 216 amino acids which shows a high degree of sequence conservation among Eukaryotes particularly at the central a polar regions. H_2A and H_2B are even more highly conserved and are known as lysine rich histones. The most conserved of all are the Arg-rich histones H_3 and H_4.

The effect of variations in sequence of histones on chromatin structure is not clear. However, the histones may be methylated, phosphorylated, etc., and thus affect the interaction of histones with each other or with the DNA.

In sperm cells, histones are replaced by other small basic proteins known as *protamines.*

In the absence of DNA, the 'core' histones associate with each other, the predominant species being a homotypic tetramer in the case of H_3 and H_4, a dimer in case of H_2A, H_2B.

Nonhistone Proteins

They occur in approximately equal amounts to histones. Some are the enzymes involved in replication and transcription, other resemble the histone in being of low mol. wt. and are known as high mobility group of HMG proteins. They are also basic like the histones and play a structural role, but differ from histones in being only loosely associated with chromatin.

Nucleosome

The DNA of human cell is of the order of 1 m in length and must be condensed in a cell nucleus whose diameter is of the order of 10 µm. Thus, packing is essential which should also maintain accessibility during replication. This is possible by a packaging mechanism involving histones and some other chromosomal proteins. The complex of DNA–histone forms the beads called nucleosomes in which the double-stranded DNA forms the string.

Each nucleosome is composed of two molecules each of histones H_2A, H_2B, H_3 and H_4 and about 200 bp of DNA. The H_2A, H_2B, H_3 and H_4 molecules form a protein complex called the histone octamer around which the DNA is wrapped. About, 146 bp of DNA are in close contact with the histone octamer to form a nucleosome core particle. The DNA between each core particles is called linker DNA, it is about 54 bp long. Histone H_1 is bound to the linker DNA and to the nucleosome core particle. H_1 is responsible for higher order chromatin structures.

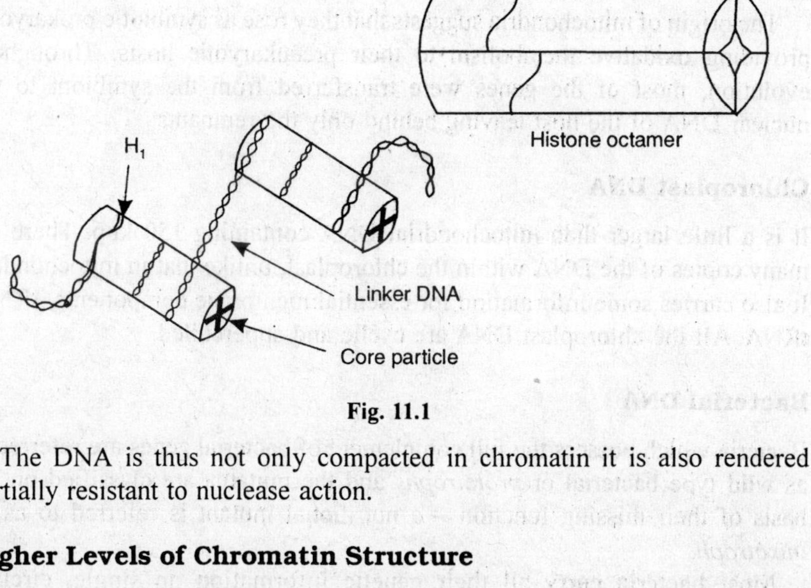

Fig. 11.1

The DNA is thus not only compacted in chromatin it is also rendered partially resistant to nuclease action.

Higher Levels of Chromatin Structure

The packaging of DNA into nucleosome accounts for a 10-fold reduction in chromosome chain length. Further reduction comes from higher levels of DNA packaging. The beads on a string structure is itself coiled into a solenoid to yield the 30-nm fiber. It forms when every nucleosome contains a molecule of the H_1. The adjacent molecules of H_1 bind cooperatively bringing the nucleosomes together into a more compact and stable form of chromatin. This achieves a further four-fold reduction in the length.

The 30-nm fibers are finally themselves attached to a nonhistone protein scaffold that holds the fibers in large loops.

There may be as many as 2000 such loops on a large chromosome. Such packaging yields chromosomes that are 5–10 μm in length and 1 μm in diameter. The organization of the DNA into loops accounts for the remaining 200-fold condensation in the length of the DNA.

Extra Nuclear DNA

(a) Mitochondrial DNA

It is found in cyclic, double stranded super coiled molecules with an exception of linear mitochondrial DNA in paramecium.

Mammalian mitochondrial DNA are not packaged into nucleosomes. These are about 15 kbp long and can therefore code for 15 to 20 proteins. Yeasts have 75 kbp, plant mitochondrial DNA are still longer.

Mitochondrial DNA can code only for a small proportion of mitochondrial proteins, yet it is described as a "lesson in Economy". Introns are absent in mammalian mitochondrial genes but present in some yeast genes. There are genes for rRNAs, tRNAs and some other proteins of the ETC present on the mitochondrial DNA.

The origin of mitochondria suggests that they rose as symbiotic prokaryotes providing oxidative metabolism to their preeukaryotic hosts. Throughout evolution, most of the genes were transferred from the symbiont to the nuclear DNA of the host leaving behind only the remnants.

Chloroplast DNA

It is a little larger than mitochondrial DNA containing 150 kbp. There are many copies of the DNA within the chloroplast, unlike that in mitochondria. It also carries some information for essential membrane components, rRNA, tRNA. All the chloroplast DNA are cyclic and supercoiled.

Bacterial DNA

Bacteria which possess the full complement of bacterial genes are referred to as wild type bacterial or *prototrophs* and the mutants are classified on the basis of their missing function – a nutritional mutant is referred to as an *auxotroph.*

Most bacteria carry all their genetic information on single, circular chromosome, others also possess small extra chromosomal elements known as plasmids, which carry genes for drug resistances.

Chromosome of *E. coli* for example, is a single cyclic duplex molecule of 4.5 million basepairs having an effective circumference of 1 mm, which

Name of animal	Body character	Dominant character	Recessive character
4. Fruit fly (Drosophila)	Eye colour	Red	White
	Wings	Flat and yellow	Curled and white
	Body colour	Grey	Black
5. Sheep	Wool	White	Black
6. Sunflower	Stem	Branched	Unbranched
7. Nettle	Leaf margin	Serrated	Smooth
8. Cotton	Lint colour	Coloured	White
9. Maize	Kernel form	Round	Wrinkled
	Kernel composition	starchy	sugary
10. Guinea pig	Skin colour	Black	White

Thus the phenomenon of segregation is based on the assumption that both the gametes are produced in equal frequency in F_1 generation, which was proved by Test cross.

Test cross: When F_1 hybrid is crossed with a variety or strain with recessive form of the character in question, it is called test cross. The purpose of a test cross is to present evidence that segregation for the alleles of a single gene in the F_1 hybrid produces two types of gametes in equal frequencies. This cross is also used to test whether an individual is homozygous (pure) or heterozygous (hybrid).

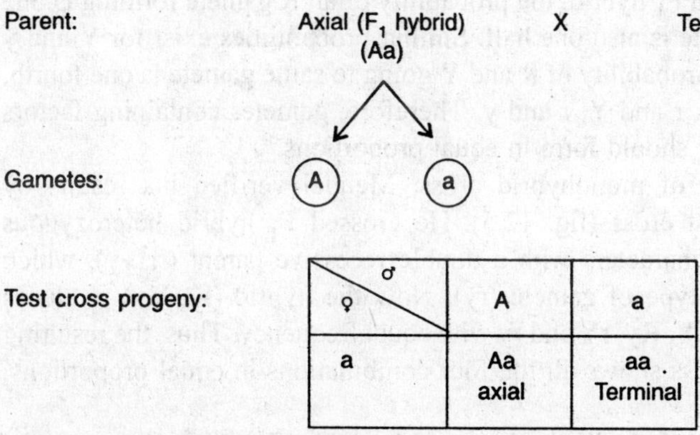

Phenotypic ratio 1 Axial : 1 Terminal

Genotypic ratio 1 Aa : 1 aa

Fig. 12.3. Test cross

According to the law of segregation, the cross between axial (AA) and terminal (aa) flowered pea varieties produces F_1 axial (Aa) plant. The F_1

hybrid (Aa) plant produces gametes carrying A and a in equal frequencies. The strain with the recessive trait (aa), used as the test cross parent, would produce only one type of gamete, all carrying a. Therefore, the union of the gametes from F_1 with those from the test cross parent is expected to produce two types of flowers axial (Aa) and terminal (aa) in the equal frequency (1 : 1) (Fig. 12.3)

Law of Independent Assortment

Mendel made crosses between pea plants differing in two characters such as texture of seed and colour of cotyledons. Such a cross in which inheritance of two characters is considered, is called a dihybrid cross. Mendel crossed homozygous round and yellow seeds pea plant with homozygous wrinkled and green seeds pea plant. As expected from the results of his monohybrid crosses, the F_1 hybrid was round and yellow seeds plant. When he selfed the F_1 hybrids, the F_2 progeny showed all the parental characters in different combinations with each other. Thus plants with round yellow seeds, round green seeds, wrinkled yellow seeds and wrinkled green seeds appeared in 9 : 3 : 3 : 1 ratio (Fig. 12.4).

Mendel applied the principles of a monohybrid cross and argued that each character is determined by two factors. Thus in a dihybrid cross there must be four factors present in each parent. Therefore the F_1 hybrid must be Rr Yy. Mendel argued that the pair of factors for roundness must be behaving independently of the pair of factors for colour of seeds. In other words, one factor for a character must be passing independently of a factor for another character. Thus in a F_1 hybrid, the probability of an R gamete forming is one half and of r gamete is also one half. Similar probabilities exist for Y and y games. Again the probability of R and Y going to same gamete is one fourth, as also of R and y, r and Y, r and y. Therefore, gametes containing factors RY, Ry, rY and ry should form in equal proportions.

As in the case of monohybrid cross, Mendel verified his results by performing the test cross (fig. 12.5). He crossed F_1 hybrid heterozygous (Rr Yy) for both characters with a double recessive parent (rr yy), which produce only one type of gamete (ry). Now the hybrid (Rr Yy) produces gametes carrying RY, Ry, rY and ry with equal frequency. Thus, the resulting progeny of test cross shows all the four combinations in equal proportions, i.e. 1:1:1:1.

From the results of his dihybrid cross, Mendel realised that at the time of gamete formation the segregation of alleles R and r into separate gametes occurs independently of the segregation of alleles Y and y. Therefore the resulting gametes contain all possible combinations of these alleles, i.e. RY, Ry, rY, ry. In this way Mendel proved that when two characters are considered in a cross, there is independent assortment of genes for each character and this become the '*Law of Independent Assortment*'.

Parent: Round Yellow RRYY X Wrinkled Green rryy

Gametes: (RY) (ry)

F_1: Rr Yr Round Yellow

F_1 Gametes: (RY) (Ry) (rY) (ry)

F_2

♀ \ ♂	RY	Ry	rY	ry
RY	RRYY	RRYy	RrYY	RrYy
Ry	RRYy	RRyy	RrYy	Rryy
rY	RrYY	RrYy	rrYY	rrYy
ry	RrYy	Rryy	rrYy	rryy

Phenotypic ratio → 9 Round Yellow: 3 Round Green: 3 Wrinkled yellow: 1 Wrinkled green

Round Yellow ⇒ RRYY, RrYY, RRYy, RrYy

Round Green ⇒ RRyy, Rryy

Wrinkled Yellow ⇒ rrYY, rrYy

Wrinkled green ⇒ rryy.

Fig. 12.4. Mendel's dihybrid cross for Round yellow pea plant with wrinkled green pea plant.

F_1 Round Yellow Rr Yy X Wrinkled Green rryy

Gametes (RY) (Ry) (rY) (ry) (ry)

F_2:

♀ \ ♂	RY	Ry	rY	ry
ry	RrYy	Rryy	rrYy	rryy

Phenotypic ratio: 1 Round Yellow: 1 Round Green: 1 Wrinkled Yellow: 1 Wrinkled Green

Fig. 12.5. Mendel's dihybrid Test cross

SUMMARY

Gregor Johann Mendel, popularly known as the father of Genetics, conducted different hybridisation experiments in garden pea plants. On the basis of his experimental results, he put forth different laws like the principle of dominance, the law of segregation and the law of independent assortment. These laws are dependent on his monohybrid crosses and dihybrid crosses.

EXERCISE

1. Write briefly on the contribution of Mendel to Genetics.
2. State briefly what is meant by 'dominant' and recessive' character.
3. State and discuss Mendel's laws. Why was Mendel successful in enunciating these laws?
4. Describe Mendel's Law of Independent Assortment with suitable examples.
5. Give an account of Mendel's Laws of Inheritance and discuss their importance in genetics.
6. When a pure breeding brown cat is mated to a pure breeding white cat, all the kittens born are brown. What result would you expect when an F_1 brown kitten is mated with
   (a) a pure breeding white cat
   (b) another F_1 brown kitten
   (c) a pure breeding brown cat.
7. In man, brown eyes (B) is dominant over blue (b). A brown eyed man married to a brown eyed woman has a blue eyed child. What would be the genotype of the parents?
8. In sweet peas yellow seed coat (Y) is dominant over green (y) and round seed shape (R) is dominant over wrinkled (r). What phenotypic ratios would be obtained in F_1 progeny of the following crosses?
   (a) YyRR × YYRr
   (b) yyRr × YrRr
   (c) Yyrr × yyrr
   (d) YyRr × yyRR.
9. A cross between yellow round and green wrinkled seeded pea plants gave the following four types of progeny in equal proportions: yellow round, yellow wrinkled, green round and green wrinkled. What are the genotypes of the parents?
10. What do you understand by dihybrid cross? State and explain Mendel's second law by an experiment.
11. What is test cross? What is its significance.
12. What is test cross? How does it help in interpreting the results of a monohybrid cross?
13. Explain the following terms
    (a) Recessive character

(b) Heterozygous
(c) Genotype
(d) Test cross
(e) Dihybrid cross
(f) Monohybrid cross

14. Write short notes on the following:
    (a) Phenomenon of dominance
    (b) Monohybrid cross
    (c) Dihybrid cross
    (d) Mendel's first law
    (e) Mendel's second law
    (f) Test cross

13

Gene Interaction

The phenotypic ratios obtained by Mendel in garden peas demonstrate that one gene controls one character and in the two alleles of a gene, one allele is completely dominant over the other. Due to this the heterozygous hybrid has a phenotype identical to the homozygous dominant parent.

After Mendel several workers performed hybridisation experiment. But these workers reported different results in some experiments where a gene was not producing an individual effect. On the contrary, genes were interacting with each other to produce novel phenotypes which did not exhibit dominance relationships observed in Mendel's experiments. *The phenomenon of two or more genes affecting the expression of each other in various ways in the development of a single character of an organism is called gene interaction.*

Some examples of gene interactions are:

1. Supplementary genes (9 : 3 : 3 : 1)
2. Complementary genes (9 : 7)
3. Epistatis (13 : 3)

Again some workers reported different results in some experiments where a single gene was producing an individual character, but which were not producing results similar to Mendel's dominance phenomenon e.g. incomplete dominance and codominance.

Incomplete Dominance

Sometimes in a heterozygote hybrid, dominant allele does not completely repress the phenotypic expression of the recessive allele and therefore an intermediate phenotype in the heterozygote hybrids occurs. This is called incomplete dominance.

Example: A monohybrid cross between a red flowered pea plant (RR) and a white flowered (rr) pea plant does not produce red or white flowered plants in F_1 as expected. But the flowers are pink, i.e. intermediate between the two parents. This is because neither red flower colour nor white flower colour is dominant, but each allele has its influence in colour development and the hybrid appears pink. In the F_1 pink flowers are self pollinated, the F_2 progeny shows red, pink and white flowered plants in the proportion 1 : 2 : 1. In this case heterozygote progeny is distinct in appearance from the homozygotes (Fig. 13.1).

This type of incomplete dominance inheritance has been found in many other plants like four-o'-clock plants (Mirabilis Jalapa), Snapdragons (Antirrhinum majus), etc.

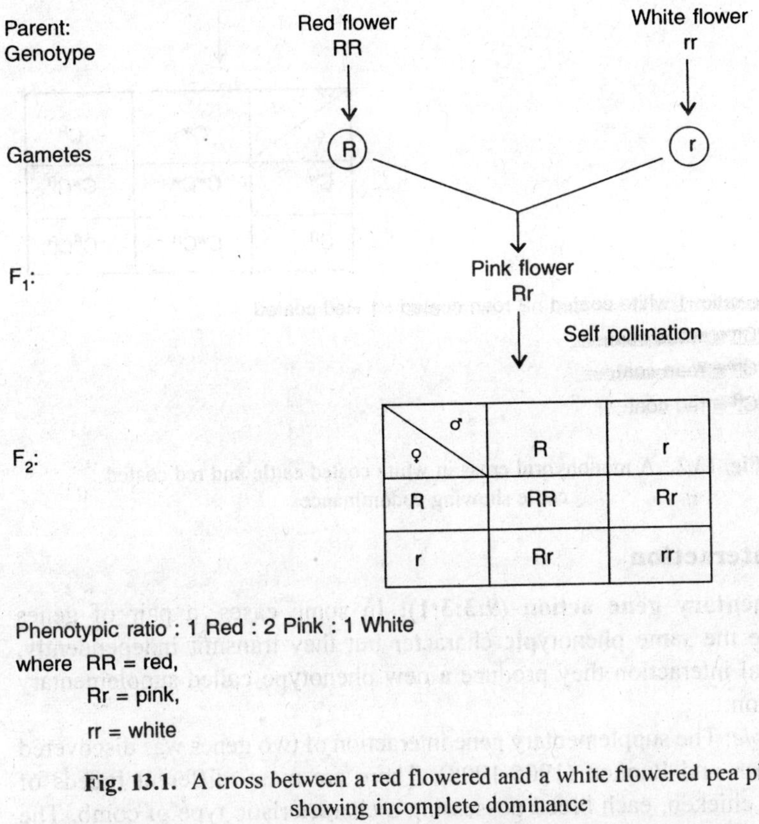

Phenotypic ratio : 1 Red : 2 Pink : 1 White
where RR = red,
Rr = pink,
rr = white

Fig. 13.1. A cross between a red flowered and a white flowered pea plant showing incomplete dominance

Codominance: Sometimes both alleles of a gene in a heterozygote hybrid have an equal phenotypic expression. Therefore, codominance is not dominance at all, and the heterozygote hybrids show phenotypic character of both parents.

Example: When a short horn breed of red coat cattle (C^RC^R) is crossed with the cattle of white coat (C^wC^w), the F_1 hybrid is found to possess *roan coat* (C^RC^w). In roan coat the red and white hairs occur in definite patches but no hair has intermediate colour of red and white. If the F_1 hybrid roan cattle are crossed, the F_2 progeny shows red coat, roan and white coat in the proportion of 1 : 2 : 1 (Fig. 13.2).

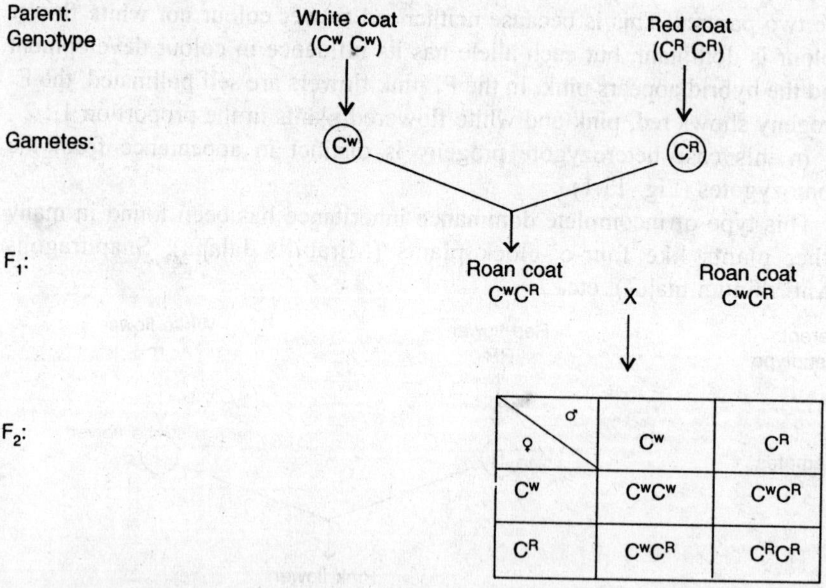

Phenotypic ratio: 1 white coated : 2 roan coated : 1 Red coated

where C^wC^w = white coat,

C^wC^R = roan coat,

C^RC^R = red coat

Fig. 13.2. A monohybrid cross in white coated cattle and red coated cattle showing codominance

Gene Interaction

Supplementary gene action (9:3:3:1): In some cases, a pair of genes determine the same phenotypic character but they transmit independently. By mutual interaction they produce a new phenotype called supplementary gene action.

Example: The supplementary gene interaction of two genes was discovered by Bateson and Punnett (1905-1908). There are many different breeds of domestic chicken, each breed possessing a characteristic type of comb. The Wyandotte breed has a comb called *rose*. The Brahmas breed has a comb called *pea* and the Leghorns breed has a comb called *single*. All these breeds are pure breed.

Bateson and Punnett crossed rose comb with a single comb. All the F_1 hens were rose comb and on inbreeding gave rise to an F_2 progeny of rose and single in the ratio 3 : 1. This cross indicates that rose is dominant over single. In another cross between pea comb chicken with single comb chicken, all the F_1 progeny had pea comb and F_2 had pea and single in the proportion 3 : 1. This cross indicates that pea is dominant over single. Bateson then crossed rose and pea varieties, and the F_1 progeny had a different comb of *Walnut* type (Fig. 13.3).

The inbreeding of walnut hybrids gives rise to F_2 progeny which consists of all the four types of chickens i.e. walnut, rose, pea and single in the ratio 9 : 3 : 3 : 1. As this ratio is typical for dihybrid inheritance it becomes clear that rose and pea combs are controlled by two pairs of genes.

Parent: Rose RRpp X Pea rrPP

Gametes: (Rp) (rP)

F_1: Walnut (Rr Pp) × Walnut (Rr Pp)

Gametes: (RP) (Rp) (rP) (rp) (RP) (Rp) (rP) (rp)

F_2:

♀ \ ♂	RP	Rp	rP	rp
RP	RRPP	RRPp	RrPP	RrPp
Rp	RRPp	RRpp	RrPp	Rrpp
rP	RrPP	RrPp	rrPP	rrPp
rp	RrPp	Rrpp	rrPp	rrpp

Phenotypic ratio: 9 Walnut : 3 Rose : 3 Pea : 1 single

Where Walnut ⇒ RRPP, RrPP, RrPp, RRPp

Rose ⇒ RRpp, Rrpp

Pea ⇒ rrPP, rrPp

Single ⇒ rrpp.

Fig. 13.3. A cross in a rose and pea comb chickens showing supplementary gene action.

The appearance of walnut comb in F_1 of cross between rose and pea shows that both of the independent dominant genes P and R are jointly responsible for the walnut comb. When present together in the zygote, P and R genes interact and produce the walnut comb. When present alone, they produce distinct comb, i.e. P produces pea comb and R produces rose comb. The presence of recessive alleles of both rose and pea combs produce the fourth type of chicken with the single comb.

In the inheritance of combs in fowls, the genes themselves do not determine the development of a character (i.e. presence or absence of comb), but simply modify a character determined by a basic gene and are therefore known as *supplementary genes*.

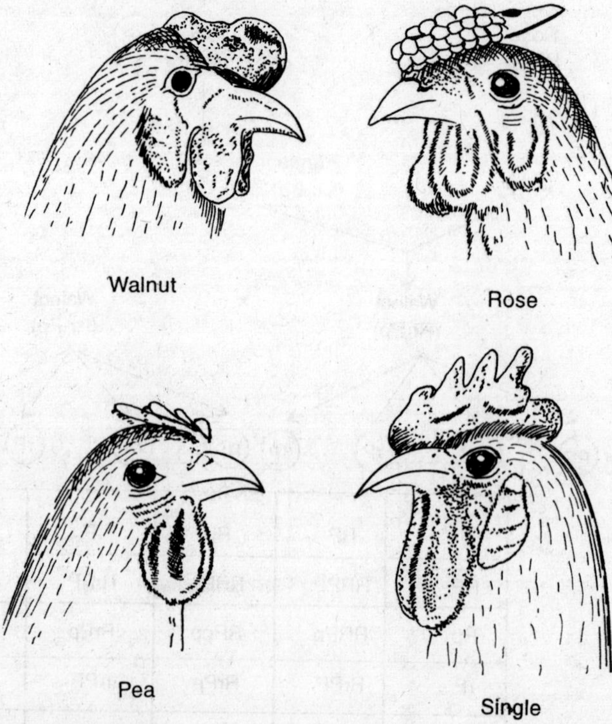

Complementary gene action (9 : 7):

In this type of gene interaction, the production of one of the two phenotypes of a trait requires the presence of dominant alleles of both the genes controlling the concerned trait. When any one of the two genes or both are present in the homozygous recessive state, the contrasting phenotype is produced. As the recessive allele of any one of the genes produces identical phenotypic character, the F_2 ratio for dihybrid cross i.e. 9 : 3 : 3 : 1 would become 9 : 7. In such a case, the genotypes aaBB, aaBb, AAbb, Aabb and aabb produce one phenotype while both dominant alleles, i.e., AaBb, AABb, AaBB when present together, produce another phenotype by interacting or complementing each other and are therefore called complementary genes.

Example: Bateson and Punnett crossed a homozygous variety of white flowered sweet pea plant with another homozygous variety of white flowered sweet pea plant. But contrary to expected result, they got all purple flowered F_1 progeny. The F_1 plants when self pollinated, produced the F_2 generation with ratio of 9 coloured and 7 white flowered sweet pea plants (Fig. 13.4).

Parent:	White	X	White
Genotype:	CCpp		ccPP
Gametes:	(Cp)		(cP)
F_1:		Purple (Cc Pp)	

Gametes: (CP) (Cp) (cP) (cp)

F_2:

♀ \ ♂	CP	Cp	cP	cp
CP	CCPP	CCPp	CcPP	CcPp
Cp	CCPp	CCpp	CcPp	Ccpp
cP	CcPP	CcPp	ccPP	ccPp
cp	CcPp	Ccpp	ccPp	ccpp

Phenotypic ratio: 9 Purple : 7 White

where Purple ⇒ CCPP, CcPP, CCPp, CcPp

White ⇒ CCpp, Ccpp, ccPP, ccPp, ccpp.

Fig. 13.4. A cross between white pea plants, showing complementary gene action.

The mechanism behind this surprising result of colour production in F_1 flower is that a given enzyme acts upon chromogen to produce the purple colour of flowers. The dominant allele of gene 'C' is responsible for the presence of chromogen while in homozygous recessive alleles (cc), chromogen is absent. Another dominant allele of gene 'P' causes the production of an enzyme, which is necessary for colour production from chromogen, while homozygous recessive (pp) condition does not produce any such enzyme. Thus both C and P are necessary for purple colour. If there is no C, the first step is blocked and chromogen is not formed and if there is no 'P', the second step is blocked and the chromogen does not get converted into pigments (anthocyanin). In this way 'C' and 'P' act as complementary to each other, for the production of colour.

The homozygous white parent does not produce purple flower because each of them have either 'C' or 'P' and not both. In the F_1 hybrid 'C' and 'P' come together to produce colour.

Epistatis (13 : 3): When two different dominant genes are present together and affect the same character in such a way that the expression of one gene inhibits or suppresses the expression of the other gene, it is called *epistatis*. The gene which suppresses the expression of another gene is called *epistatic gene* or *inhibitory gene* and the gene whose effect is covered by epistatic gene is called *hypostatic gene*. The term epistatis comes from a Greek word meaning to "*stand above*".

Example: Inhibitory gene action was well illustrated by inheritance pattern of plumage (feathers) in poultry. In poultry white birds belong to two different varieties, namely white leghorns and white wyandottes.

A cross between a white leghorn and a white wyandotte gives F_1 hybrid white birds with small dark flecks. When such birds are inbred, the F_2 progeny segregates in the ratio of 13 white to 3 coloured birds (Fig. 13.5).

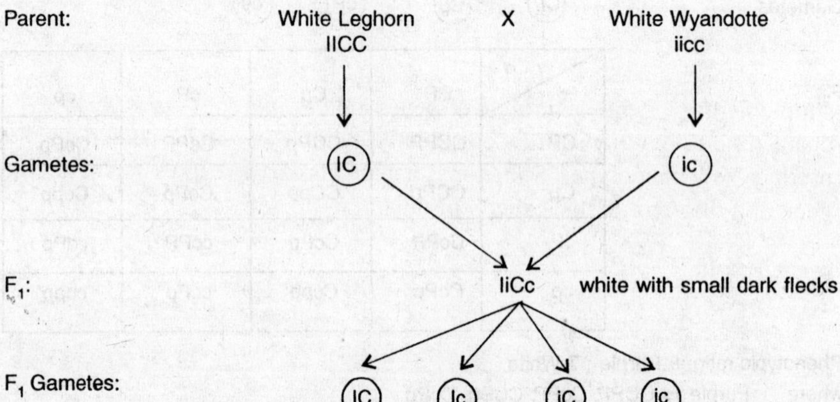

F2: ♀ \ ♂	IC	Ic	iC	ic
IC	IICC	IICc	IiCC	IiCc
Ic	IICc	IIcc	IiCc	Iicc
iC	IiCC	IiCc	iiCC	iiCc
ic	IiCc	Iicc	iiCc	iicc

Phenotypic ratio ⇒ 13 white : 3 coloured

where white ⇒ IICC, IiCC, IICc, Iicc, IiCc, IIcc, iicc.

Coloured ⇒ iiCC, iiCc.

Fig. 13.5. A cross between white leghorn and white wyandotte showing dominant epistatis.

It is assumed that the white leghorns contains a colour gene 'C', but it also contains an inhibitor gene 'I', which suppresses the expression of gene C. Thus leghorns are genetically coloured birds. On the other hand, the white wyandottes are genetically white because they do not have dominant colour gene. Therefore, in the absence of coloured dominant gene, presence or absence of inhibitory gene will not have any effect because there is no gene to be suppressed by it. But the recessive allele of the inhibitor gene (ii) produces coloured birds due to expression of colour gene 'C'.

SUMMARY

After Mendel, several workers reported different results in some hybridisation experiments, where a gene was not producing an individual effect. In the gene interaction, two or more genes affect the expression of each other in various ways in the development of single character of an organism. There are many types of gene interactions reported. In incomplete dominance, the heterozygote shows a phenotype intermediate between the parental phenotypes, while in codominance, the heterozygote expresses both the parental phenotypes equally. In complementary gene action, some genes act together to produce an effect that neither gene can produce separately. In supplementary gene action, two genes determine the same phenotypic character but they transmit independently, while by mutual interactions they produce new phenotypes. Sometimes a gene masks the expression of another gene at a different locus, called epistatis.

EXERCISE

1. A rose combed chicken is crossed with a walnut. The progeny consists of 3 walnut, 3 rose, one pea and one single. What are the genotypes of the parents?
2. In sweet peas the gene 'C' and 'P', when present together, produce purple flowers. But when either 'C' or 'P' is present alone it produces white flowers. What flower colour would the progeny of the following crosses have
   (a) Ccpp × ccPP
   (b) ccPp × CcPp
   (c) CCPp × ccPP
3. Four hens with walnut combs are each mated with a single combed cock. One mating produces only walnut comb; the second produces both walnut and pea comb; the third produces rose and walnuts and the fourth produces walnut, rose, pea and single combs. What are the genotypes of the four walnut hens used as parents?
4. Write short notes on:
   (a) Codominance
   (b) Incomplete dominance
   (c) Supplementary genes
   (d) Epistatis
5. What is epistatis? Explain with suitable example.

14

Linkage and Crossing Over

The tendency of genes to stay together during inheritance and to retain their parental combination even in the offspring, because of the genes being located relatively close to each other, in the same chromosome, is called *linkage*. The genes located on the same chromosome and inherited together are known as *linked genes*. Linked genes are separated from each other by the process of crossing over and cause genetical variations due to recombination of genes.

Linkage can be defined as close physical association between two loci on chromosome, preventing independent assortment.

DEVELOPMENT OF CONCEPT OF LINKAGE

1. Chromosome theory of heredity

In 1902, Sutton and Boveri concluded that genes are located on chromosomes. This theory, known as the chromosome theory of inheritance (heredity), provides a physical basis of heredity.

2. Coupling and Repulsion hypothesis

Bateson and Punnett (1905-1908) proposed the *Coupling and repulsion hypothesis* to explain the unexpected F_2 result of dihybrid cross between a homozygous sweet pea plant (Lathyrus odoratus) having dominant alleles for blue flower (RR) and long pollen grains (R_0R_0) with another homozygous double recessive plant with red flowers (rr) and round pollen grains (r_0r_0). When the test crossed a heterozygous blue long pollen plant with double recessive parent, instead of getting the 1:1:1:1 test cross ratio, they received 7:1:1:7 ratio as illustrated in Fig. 14.1.

Parent:	Blue Long		×		Red Round		
	($RR\ R_0R_0$)		↓		($rr\ r_0\ r_0$)		
F_1:			All Blue Long				
			($Rr\ R_0\ r_0$)				
Test Cross:	F_1 Blue long		×		Red Round		
	($Rr\ R_0\ r_0$)		↓		($rr\ r_0\ r_0$)		
Test cross progeny:	Blue long	:	Blue Round	:	Red long	:	Red Round
	7		1		1		7

Fig. 14.1. A cross between Blue Long and Red Round

It clearly indicated that, there was a tendency in the dominant alleles (RR_0) to pass together to the same gamete and also similar with the recessive alleles ($r\ r_0$). This tendency of dominant and recessive alleles to inherit together was explained as *'gamete coupling'* by Bateson and Punnett.

Further, when they had crossed blue round ($RR\ r_0r_0$) with red long ($rr\ R_0R_0$), the F_1 hybrids were all blue long ($Rr\ R_0r_0$). The F_1 hybrid, when test crossed with recessive (rrr_0r_0) parent, got a test cross ratio of 1:7:7:1, as illustrated in Fig. 14.2.

Parent:	Blue Round		×		Red Long		
	($RR\ r_0r_0$)		↓		($rrR_0\ R_0$)		
F_1:			All Blue Long				
			($Rr\ R_0\ r_0$)				
Test cross:	$F_1\ Rr\ R_0\ r_0$		×		($rr\ r_0\ r_0$)		
	(Blue long)		↓		(red round)		
	1	:	7	:	7	:	1
	Red Round		Red long		Blue Round		Blue long

Fig. 14.2. A cross between Blue Round and Red Long

It clearly indicates that the two dominant pairs of alleles repelled each other and same with or both recessive alleles to repel each other was termed *repulsion.* Bateson and Punnett could not explain the exact reasons of coupling and repulsion.

3. T. H. Morgan (1911) suggested that the two genes are found in coupling phase or in repulsion phase, because they are present on the same chromosome (couplings) or on two different homologous chromosomes (repulsion). Such genes are called *linked genes* and the phenomenon of inheritance of linked genes is called *linkage.*

4. Chromosome theory of linkage: Morgan, along with Castle, proposed the chromosome theory of linkage. According to this theory

(a) The genes which show the phenomenon of linkage are situated in the same chromosomes and these linked genes usually cannot be separated during the process of inheritance.

(b) The distance between the linked gene determines the strength of linkage, i.e. closely located genes show strong linkage while distantly located genes show weak linkage.

(c) In the chromosomes, genes are arranged in linear fashion.

T.H. Morgan and his co-workers found two types of linkage in Drosophila and other organisms.

1. Complete linkage: In complete linkage, the parental combinations of characters appear together for 2 or more generations in a continuous and regular fashion. In this type of linkage, genes are closely associated and tend to transmit together. *e.g.* Drosophila male and silkworm female show complete linkage.

In Drosophila gray body colour is normal and long wings are normal, while black body colour and short or vestigial wings are mutant-characters. In 1919 T.H. Morgan mated gray bodied and vestigial winged fruit fly. All F_1 progeny had gray bodies and normal long wings (b^+vg/bvg^+) which are dominant. When F_1 (B^+vg/bvg^+) were back crossed to double recessive, only two types of progeny (one gray bodied vestigial wings and other black body and long wings), instead of four types of phenotypes, were obtained.

Parents:	Gray Vestigial b^+vg / b^+vg	× ↓	Black long bvg^+ / bvg^+
F_1:		All Gray Long b^+vg / bvg^+	
Test cross:	F_1 0→ Gray Long b^+vg / bvg^+	× ↓	Female Black Vestigial bvg / bvg
Test Cross progeny:	b^+vg / bvg Gray vestigial		bvg^+ / bvg Black Long
Test Cross Ratio:		1 : 1	

Fig. 14.3. A cross showing complete linkage in male Drosophila

2. Incomplete linkage: The linked genes do not always stay together because homologous non-sister chromatids may exchange segments of varying length with each other during meiotic prophase. This type of exchange of chromosomal segments in between homologous chromosome is known as *crossing over*. The linked genes which are widely located in chromosomes and have chances of separation by crossing over are called *incompletely linked genes* and the phenomenon of their inheritance is called *incomplete linkage* e.g. (1) Drosophila ♀ shows incomplete linkage.

When Morgan test crossed F_1 hybrid female with double recessive male, all four types of progeny were obtained in the ratio of 41.5% Gray vestigial, 41.5% black long, 8.5% Gray long and 8.5% Black vestigial which shows occurrence of crossing over.

Parents:	Gray Vestigial	×	Black long
	b^+vg / b^+vg	↓	bvg^+ / bvg^+
F_1:		Gray Long	
		b^+vg / bvg^+	
Test cross:	F_1 ♀ Gray Long	×	0→ Black Vestigial
	b^+vg / bvg^+	↓	bvg / bvg
Test Cross ratio:	1. Gray vestigial b^+vg / bvg = 41.5%		83% Parental
	2. Black Long bvg^+/ bvg = 41.5%		Combination
	3. Gray Long b^+vg^+/ bvg = 8.5%		New
	4. Black vestigial bvg / bvg = 8.5%		Combination 17%

Fig. 14.4. A cross showing incomplete linkage in Drosophila female

Example 2. Incomplete linkage in Maize: Hutchinson observed incomplete linkage in *Zea mays* (Maize). When a maize plant with seeds having coloured and full endosperm (CS/CS) is crossed with another maize plant having recessive alleles for colourless seed and shrunken endosperm (cs). When F_1 hybrid is test crossed four types of progeny were obtained.

Parents:	Coloured Full	×	Colourless Shrunken
	CS / CS	↓	cs / cs
F_1:		CS / cs	All Coloured Full
Test cross:	CS / cs	×	cs / cs
	Coloured Full	↓	Colourless Shrunken
Test Cross ratio:	Coloured full	CS / cs	83%
	Colourless shrunken	cs / cs	Combination
	Coloured shrunken	Cs / cs	New
	Colourless full	cS / cs	Combination 4%

Fig. 14.5. A cross showing incomplete linkage in maize plant

Crossing over

The incomplete linkage takes place due to the occurrence of *new combination* or *recombination* of linked genes. The recombination is the result of *crossing over* in which the non-sister chromatids of homologous chromosome exchange the chromosomal parts or segments.

Definition: The crossing over is a process that produces new combinations of genes by interchanging of corresponding segments between non-sister chromatids of homologous chromosomes.

The chromatids resulting of crossing over are known as *cross overs.*

Characteristics of crossing over: (1) It occurs at two levels (a) at gross chromosomal level called *chromosomal crossing over* and (b) at DNA level called *genetic recombination.*

(2) Basically it results from exchange of genetic material between non-sister chromatids by break and exchange following replication.

(3) Frequency of crossing over is closely related to physical distance between genes on chromosomes and serves as a tool in constructing genetic maps of chromosomes.

Mechanism of Crossing Over (meiotic)

The process of crossing over includes following stages – synapsis, duplication of chromosomes, crossing over and terminalisation. Before undergoing recombination the chromosomes necessarily complete (1) 99.7% replication of DNA and 75% synthesis of histones which takes place prior to onset of prophase-I and (2) attachment of each chromosome by its both ends (telomers) to the nuclear envelope via the specialised structure called *attachment plaques.* This event occurs during *leptotene* stage of prophase-I.

1. *Synapsis:* During *Zygotene* stage of (meiosis-I) prophase I, the synapsis or pairing between the two homologous chromosomes is initiated. Synapsis often starts when the homologous ends of the two chromosomes are brought together on the nuclear envelope and it continues inward in a zipper like manner from both ends and aligning the two homologous chromosomes side by side (e.g. mammals). In other cases, synapsis may begin in internal regions of the chromosomes and proceed towards the ends. Thus by synapsis each gene is brought into *Juxtaposition* (being side by side) with its homologous gene on the opposite chromosome. The resultant pairs of homologous chromosomes are called *bivalents.*

2. *Duplication of chromosomes:* The synapsis is followed by duplication of chromosomes in *Pachytene* stage of prophase-I. During duplication, each homologous chromosome of bivalents splits longitudinally and form two identical sister chromatids which are held together by an unsplitted centromere. At this stage each bivalent contains four chromatids. So it is known as tetrad. The longitudinal splitting of chromosomes is achieved by the separation of already duplicated DNA molecules along with certain chromosomal proteins.

3. *Crossing over:* During pachytene stage of prophase-I, chromosomal crossing over occurs due to exchange of chromosomal material between non-sister chromatids of each tetrad. During the process of crossing over, two non-sister chromatids first break at the corresponding points due to an enzyme known as *endonuclease.* Then a segment on one side of the break connects with a segment on the opposite side of the break, so that the two non-sister chromatids cross each other. At this stage 0.3% synthesis of DNA occurs to fill the gap. The fusion of chromosal segments takes place due to an enzyme known as *ligase.* The point of crossing over is known as *chiasma.* The number of crossing over or chiasmata formation depends on the length

of the chromosomes because the longer the chromosome, the greater the number of chiasmata.

4. *Terminalisation:* After the crossing over, the non-sister chromatids start to repel each other because the force of synapsis attraction between them decreases. During *diplotene* desynapsis begins and during *diakinesis*, chromosomes detaches from the nuclear envelope and each bivalent is clearly seen to contain four separate chromatids with each sister chromatid linked at its centromere. Due to terminalisation the homologous chromosomes are separated completely.

Kinds of Crossing Over

(A) According to its occurrence in the somatic or germ cells, there are two types of crossing over:

1. *Somatic/mitotic crossing over:* When the process of crossing over occurs in the chromosomes of somatic cells of an organism during the mitotic cell division, it is known as somatic or mitotic crossing over. It is rare in occurrence and has no genetical significance.
2. *Germinal or Meiotic crossing over:* When the process of crossing over occurs in germinal cells during the gametogenesis in which the meiotic cell division takes place, it is known as germinal or meiotic crossing over. It is universal in its occurrence and is of great genetic significance.

(B) According to the number of chiasma following types of crossing over:

1. *Single crossing over:* When the chiasma occurs only at one point of the chromosome pair then the crossing over is known as single crossing over. It produces two cross over chromatids and two non-cross over chromatids.
2. *Double crossing over:* When the chiasmata occur at two points in the same chromosome, the crossing over is known as double crossing over. In it, the formation of each chiasma is independent of the other and in it four possible classes of recombination occur. In the double crossing over, following two types of chiasma may be formed:

(i) Reciprocal chiasma: In it, the same two chromatids are involved in the second chiasma formation as in the first chiasma. Thus, the second chiasma restores the order which was changed by the first chiasma and it produces two non-cross over chromatids. In reciprocal chiasma, out of four chromatids only two are involved in the double crossing over.

(ii) Complimentary chiasma: When both the chromatids taking part in the second chiasma are different from those chromatids involved in the first chiasma, it is known as complimentary chiasma. It produces four single cross overs and no non-cross over. It occurs when three or four chromatids of the tetrad undergo the crossing over.

(3) *Multiple crossing over:* When crossing over takes place at more than

two places in the same chromosome pair, then it is known as multiple crossing over. It occurs rarely.

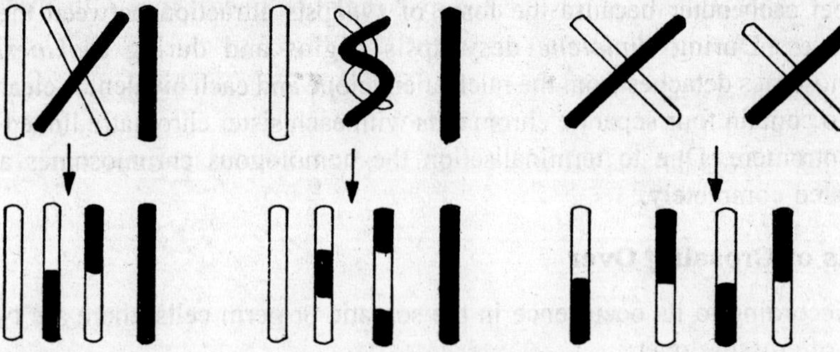

A
Single crossing over

B
Reciprocal double crossing over

C
Complimentary double crossing over

Fig. 14.6. Types of crossing over

Factors Controlling Frequency of Crossing Over

Primarily, the frequency of crossing over depends on the distance between linked genes. But a number of genetic, environmental and physiological factors also affect the frequency of crossing over. They are as follows:

1. *Temperature:* High and low temperature increases the frequency of crossing over.
2. *Radiation:* Exposure to the X-ray and other radiations increases the frequency of crossing over.
3. *Age:* With increase in age, the frequency of crossing over decreases.
4. *Distance:* The frequency of crossing over increases with increase in the distance between the genes.
5. *Chemicals:* While some chemicals increase the frequency of crossing over, other chemicals decrease it.
6. *Inversions:* Inversions suppress the crossing over.
7. *Chiasmata formation:* The formation of one chiasmata discourages the formation of other chiasmata in the near vicinity.
8. *Centromere:* Generally genes present close to the centromere show reduced crossing over.

Chromosomal Maps

Gene mapping is the assignment of a gene to a locus on a chromosome. By mapping the relative positions of many genes and other markers it is possible to generate a *chromosome map* or a map of entire genome. Gene maps are of following types:

1. Genetic/linkage/chromosome map: It assigns genes to linkage groups.
2. Cytogenetic map: It is created by correlating phenotypes with observable chromosomal rearrangements and deficiencies.
3. Physical or molecular map: It is created by ordering cloned fragments of genomic DNA and is calibrated in real units, i.e. bp, kbp, etc. The physical map has the highest resolution and is the ultimate aim of the mapping projects.

Chromosome maps

The method of construction of maps of different chromosomes is called *genetic mapping or chromosome* mapping. It includes the following processes:

(1) *Determination of linkage groups:* Before starting the genetic mapping of the chromosomes of a species, one has to know the exact number of chromosomes of that species and then determine the total number of genes of that species by carrying out hybridisation experiment. By the hybridisation technique, it is also possible to determine different linkage groups of a species. All the linked genes of a chromosome form a linked group. The number of linkage groups of a species corresponds with haploid chromosome number of that species.

(2) *Determination of Map distance:* The intergene distance on the chromosomes cannot be measured by units employed in light microscopy but by *map units*. A map unit is equal to 1% of cross overs (recombination) for e.g. if a F_1 hybrid having the genotype Ab/aB produces 8% of cross over gametes AB and ab, then the distance between A and B is estimated at 16 map units. As each chiasma produces 50% crossover products, it is equivalent to 50 map units. If the mean number of chiasmata is known for a chromosome pair, the total length of the map for the linkage group may be predicted as

Total length = mean no. of chiasmata × 50

The % of crossing over between two linked genes is calculated by test crosses in which a F_1 dihybrid is crossed with a double recessive parent e.g.

Ac / aC × ac / ac

↓

F_2: Ac / ac ; aC / ac ; AC / ac ; ac / ac

37% 37% (Parental Combination) 13% 13% (New Combination)

The last two groups were produced by crossover gametes (13+13) from the dihybrid parent. Thus 26% of all gametes were of crossover types and the distance between the loci A and C is estimated at 26 map units.

A three point test cross or trihybrid test cross gives us information regarding relative distance between genes and also shows linear order in which these genes should be present on chromosome:

For example, we test cross trihybrid individuals of genotype ABC/abc and get the following progeny:

36% ABC / abc 36% abc / abc	9% Abc / abc 9% aBC / abc	4% ABc / abc 4% abC / abc	1% Abc / abc 1% aBC / abc
72% Parental combination	18% Single crossovers between A and B	8% Single crossover between B and C	2% double crossover

To find the distance between A and B count all single and double cross overs i.e. 18% + 2% = 20% or 20 map units distance between A and B. To find distance between B and C, 8% + 2% = 10% or 10 map units distance between B and C. Therefore, the distance between A and C is 30 map units. When double crossovers are detected in a 3 point linkage experiment it is 26 map units.

(3) *Determination of Gene order:* After determining the relative distance between genes of a linkage group, it becomes easy to place genes in their proper linear order. For example, distance between the genes A – B = 12, B – C = 7 and A – C = 5 and we can determine the order of genes in the following manner.

(1) Assume that gene A is middle

B 12 A | A 5 C

B 7 C

(2) Assume that gene B is middle

A 12 B | B 7 C

A 5 C

(3) Assume that gene C is middle

A 5 C | C 7 B

A 12 B

In the third case, the distance between A – B is equitable, so gene C must be in middle.

(4) *Combining Map segments:* Finally, the different segments of maps of a complete chromosome are combined to form a complete genetic map for a chromosome.

Significance of Chromosome Maps: (1) It shows the number of chromosomes and number of linkage group of organism. (2) It shows exact location, arrangement and sequence of genes on chromosome.

Interference

In 1911, Muller discovered the phenomenon of *interference*. In most higher organisms one chiasma formation reduces the probability of another chiasma formation in an immediately adjacent region of the chromosome, probably because of physical inability of the chromatids to bend back upon themselves within certain minimum distance. This tendency of one crossover to interfere with the other crossover is called *interference*. Interference is inversely proportional to the distance between the points of crossing over, i.e. interference is maximum over short distance and decreases as the distance increases.

Types of interference: two types

1. *Positive interference:* Where one crossover inhibits another crossover. Thus when the double crossovers are absent, we can say that interference is cent per cent.

   e.g. Most Eukaryotes

2. *Negative interference:* Where one crossover stimulates another. Recombination between very close markers shows evidence of negative interference, i.e. several crossovers appear to be clustered.

Interference varies from chromosome to chromosome and within the chromosome from one point to the other. Generally, it is maximum near the centromere and at the two ends of the chromosome (telomeres).

Coincidence

Coincidence or coefficient of coincidences is inverse measure of interference and is expressed as the ratio between the actual number of double crossovers and the expected number of such double cross overs, i.e.

$$\text{Coefficient of coincidence} = \frac{\%\ \text{observed double crossovers}}{\%\ \text{expected double crossovers}}$$

The coincidence is the complement of interference, so coincidence + interference = 1.0

i.e. when interference is complete (1.0), no double crossovers will be observed and coincidence becomes zero. When interference decreases, coincidence increases. Coincidence is generally quite small for short map distance.

SUMMARY

Linkage is a close physical association between two loci on chromosome which prevents its independent assortment. The genes which are located on the same chromosome and inherited together are called linked genes. There are two types of linkage – complete linkage and incomplete linkage. The incomplete linkage takes place as a result of crossing over. Frequency of crossing over is directly related to physical distance between genes on chromosome, which serves as a tool in constructing genetic maps of chromosomes.

The mechanism of crossing over includes – synapsis, duplication of chromosomes, crossing over, terminalisation. Crossing over may be somatic or germinal. It may be single, double or multiple crossing over.

The methods of construction of maps of different chromosomes is called chromosome mapping. It includes determination of linkage groups, determination of map distance and determination of gene order.

The tendency of one crossover to interfere with other crossover is called interference. It may be positive or negative. Coincidence is inverse measure of interference.

EXERCISE

1. What is linkage? Explain in detail as to how it is correlated with Mendel's laws of independent assortment.
2. What is linkage? Describe the mechanism of linkage with suitable example.
3. What is linkage? Describe in detail its types.
4. Give an account of linkage in Drosophila melanogaster.
5. What is coupling and repulsion hypothesis?
6. Drosophila melanogaster has 8 chromosomes and man has 46. How many linkage groups are there in each case?
7. Discuss the mechanism and significance of crossing over.
8. Discuss the different factors influencing the crossing over.
9. What do you understand by crossing over? Describe the different types of crossing over.
10. What gametes are produced by an individual whose genotype is AaBb, if crossing over can occur between the genes.
11. What do you mean by linkage and crossing over? Add a note on their significance.
12. Write short notes on the following:

(i) Chiasma	(vi) Linkage groups
(ii) Crossing over	(vii) Synapsis
(iii) Incomplete linkage	(viii) Chromosomal map
(iv) Coincidence	(ix) Complete linkage
(v) Interference	

15

Sex Determination

The word sex has been illustrated from the Latin word, *Sexus,* which means section or separation. According to the kind of gamete or sex cell produced by the organism, members of almost all species are often divided into two sections, i.e. male sex and female sex. An organism possessing both male and female reproductive organs is usually called a *hermaphrodite,* e.g. earthworm, flukes, snails and numerous species of plants. The organisms where the two sexes occur in separate individuals are called unisexual or dioecious organisms e.g. human being, Drosophila, butterflies, reptiles, birds, etc.

Historically, up to the last quarter of the nineteenth century, it was still debatable as to whether or not sex arose as a response of the organism to external stimuli such as temperature and nutrition. In 1876, Hertwig observed the fusion of two nuclei during fertilisation, one derived from the egg and another from Spermatozoa in the Sea urchin (*Toxopneustes lividus*). These observations are confirmed by Van Beneden (1883) in *Para scaris equorum*, a parasite on the horse and by Boveri (1890) in the sea urchin *Echinus microtuberculatus*. In 1891, Henking observed a chromatin body in the dividing spermatocytes of the plant bug *Pyrrhocorus apterus* which stained intensely and lagged behind other chromosomes during anaphasic movement. As its function was unknown, Henking called it the 'X' chromosome. In 1901, McClung was the first to suggest that this chromosome was concerned with the determination of sex. In 1905, Wilson found that in some insects the males had the heteromorphic pair of chromosomes, while in the other insects the female had this heteromorphic pair of chromosomes. In 1904, Montgomery confirmed that there was one pair of chromosomes which was different in the two sexes and was given the name *heterochromosomes*, while the rest of the chromosomes which were same in both sexes were called *autosomes*. In 1910, T.H. Morgan for the first time described the sex linkage in Drosophila where he showed that the genes for white eye colour was carried on the X chromosome.

Most of the mechanisms of the determination of the sex are under genetic control and they are of the following types:

1. Sex chromosomal mechanism or Heterogamesis
2. Genic balance mechanism
3. Male haploidy or haplodiploidy mechanism
4. Single gene effects

Sex chromosomal mechanism (Heterogamesis)

Sex chromosomal mechanism or chromosomal basis of sex determination was proposed by McClung in 1902. The chromosomes of an individual are of two types:

Autosomes: The chromosomes which have no relation with sex, which are same in both sexes and contain genes which determine the somatic characteristics.

Sex chromosomes: The chromosomes which are responsible for the determination of the sex of an organism is called sex chromosome. e.g. X and Y chromosome.

In dioecious diploidic organisms, the following two types of sex determination mechanisms are recognised –

– Heterogametic male and

– Heterogametic female

1. *Heterogametic male:* This is the most widespread type of sex determining mechanism. In this type of sex determination, the female has two X-chromosomes, while the male organism has only one X-chromosome. As the male has only one X-chromosome, during gametogenesis it produces two types of gametes – 50% gametes carry X-chromosome while the remaining 50% gametes are without X-chromosomes. Such a sex which produces two different types of gametes is called *heterogametic sex*. In this type female sex produces similar type of gametes and is called *homogametic sex*. The heterogametic male may be of following two types.

(i) XX - X0 type: In this type of organism, the female sex has two X-chromosomes (XX) and are thus homogametic, while the male sex have only one X-chromosome (XO). The male lacking one X-chromosome produces two types of sperms - 50% with X-chromosome and

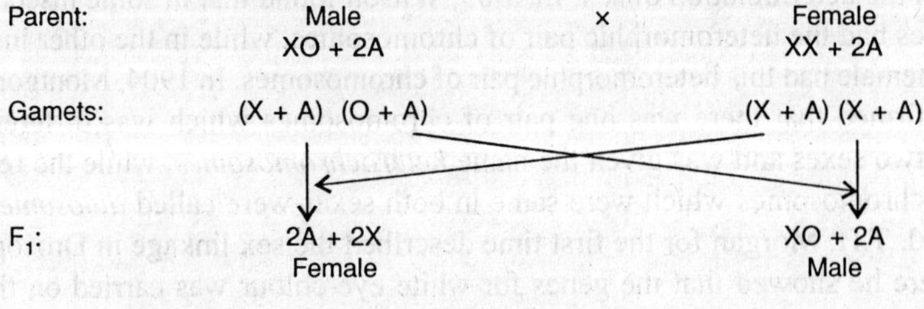

Fig. 15.1. Heterogamatic male showing XX – XO type

remaining 50% without X-chromosome. The sex of the offspring depends upon the type of the sperm that fertilizes the egg. The presence of an unpaired X-chromosome determines the male sex.

e.g. Few plants like *Vallisneria spiralis, Dioscorea sinuata*, etc. and insects specially those of the order Hemiptera and Orthoptera, etc.

(ii) XX - XY type: In few organisms, the female contains two homomorphic X chromosomes in their body (XX) and producing only one type of eggs which contains one X-chromosome. The male sex of this type of organism possesses one X-chromosome and Y-chromosome (XY). The male are heterogametic i.e. produces two types of sperms – 50% with Y-chromosome and 50% with X-chromosome. The sex of the offspring depends on the kind of sperm that fertilises the egg. An egg fertilised by an X-bearing sperm produces a female offspring while if egg is fertilized by Y-bearing sperm a male offspring is produced.

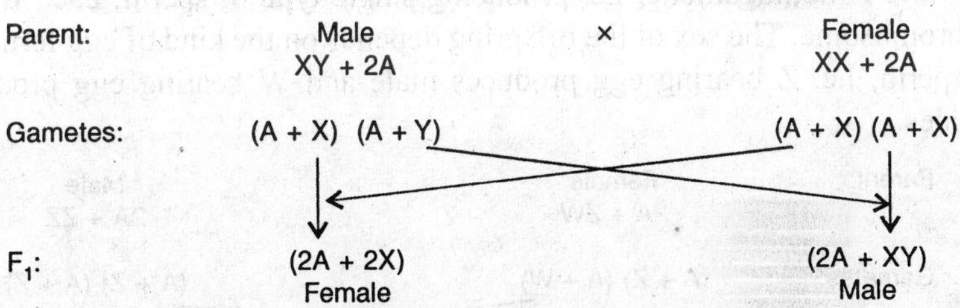

Fig. 15.2. Heterogametic male showing XX – XY type

e.g. Man and other mammals, certain insects like Drosophila, Lygaeus turicicus and in some angiospermic plants like *Melandrium album*, Humulus lupulus, etc.

(B) *Heterogametic females:* In this type of sex determination, the male sex possesses two homomorphic X-chromosomes and is therefore homogametic and produces single type of sperms, each carrying a single X-chromosome. The female sex consists of either single X-chromosome or one X-chromosome and one Y-chromosome. Thus female sex is heterogametic and produces 2 types of eggs, half with X-chromosome and half without X-chromosome (with or without Y-chromosome). To avoid confusion with that of XX-XO and XX-XY types of sex determination, instead of X and Y alphabets, Z and W are generally used respectively.

The heterogametic female may be of following two types –

(1) *ZO-ZZ type:* In few moths and butterflies, the female possesses single Z chromosome in its body cell (ZO) and is heterogametic, i.e. producing two kinds of eggs - half with a Z-chromosome and half without Z-chromosome. The male possesses two Z chromosomes (ZZ), is homogametic and is producing single type of sperms, each carrying a single Z-chromosome. The sex of the offspring depends on the kind of egg fertilised by sperm.

Parent:	Female 2A + ZO	×	Male 2A + ZZ
Gametes:	(A + Z) (A + O)		(A + Z) (A + Z)
F_1:	2A + ZZ male		2A + ZO female

Fig. 15.3. Heterogametic female showing ZO-ZZ type

(2) *ZW-ZZ type:* In some insects (gypsy moth) and vertebrates – fishes, reptiles and birds and plants such as *Fragaris elation*, the female sex has one Z-chromosome and one W-chromosome (ZW). It is heterogametic and produces two types of egg—half with Z-chromosome and remaining half with W-chromosome. The male sex has two homomorphic Z-chromosomes (ZZ) and is homogametic, i.e. producing single type of sperm, each with a Z-chromosome. The sex of the offspring depends on the kind of egg fertilised by sperm, i.e. Z bearing egg produces male and W bearing egg producer females.

Parent:	Female 2A + ZW	×	Male 2A + ZZ
Gametes:	(A + Z) (A + W)		(A + Z) (A + Z)
F_1:	2A + ZZ male		2A + ZW female

Fig. 15.4. Heterogametic female showing ZW-ZZ type

Sex Linkage

The genes which occur exclusively on the X-chromosomes are called *X linked genes* or *sex-linked genes* and the genes which exclusively occur in Y-chromosome are called *Y-linked or holandric genes.*

The inheritance of X linked and Y-linked genes is called *sex-linked inheritance*. The genes occurring only in the X-chromosomes are represented twice in female (because female contains 2X-chromosome) and once in male (because male contains only one X-chromosome).

Therefore, the X-linked genes, whether dominant or recessive, show their effects in the male phenotype. The X-linked recessive genes show the *criss-cross* pattern of inheritance i.e. X-linked recessive gene is transmitted from P_1 male parent to F_2 male progeny, through its F_1 female heterozygous which are called *carrier* and different F_1 and F_2 results in the reciprocal crosses.

Inheritance of X-linked Recessive Genes in Human

(1) *Colour blindness:* In human beings, a dominant X-linked gene is necessary

for the formation of the colour sensitive cells, the *cones* in the retina of eye. There are 3 different types of cones, each with its characteristic pigment that reacts most strongly to *red, green* and *violet* light. The recessive form of this gene i.e. recessive X-linked allele for colour blindness, is incapable of producing the colour sensitive cones and the homozygous recessive female ($X^c X^c$) and hemizygous recessive male ($X^c Y$) are unable to distinguish between red and green colour. The frequency of colour blindness in a woman is much less than in a colour blind man.

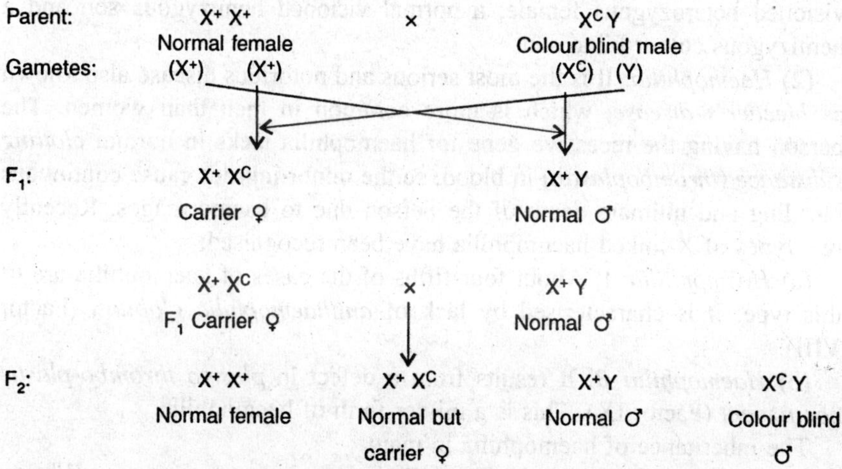

Fig. 15.5. Example of inheritance of colour blindness in human

The inheritance of colour blindness:

(a) *Marriage between colour blind man and normal vision woman:* When a colour blind man marries a normal vision woman, they will produce normal vision male and normal vision but carrier female in F_1. The marriage between F_1 normal vision carrier woman and normal vision male will produce in F_2 two normal visioned female, one normal visioned male and one colour blind male.

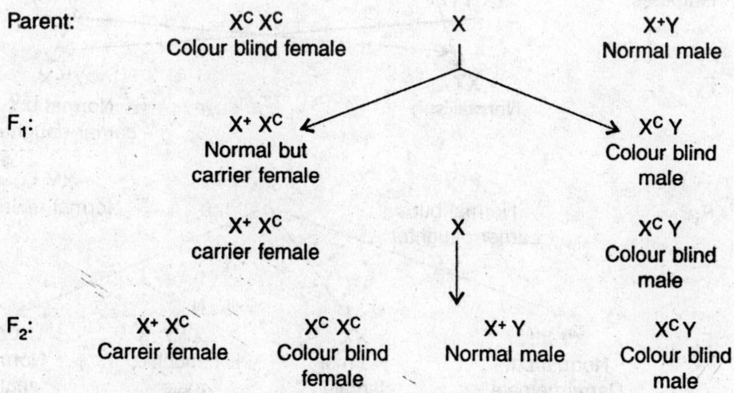

Fig. 15.6. Example of colour blindness in human

(b) *Marriage between normal visioned male and colour blind female:* When a colour blind woman marries a normal visioned man, all the F_1 sons are colour blind and daughters will be normal visioned but carriers. Because male receives one X-linked recessive gene for colour blindness from a colour blind mother and the daughter receives one X-linked dominant gene for normal vision from the father and one X-linked recessive gene for colour blindness from the mother. The marriage of F_1 carrier female with a colour blind male produces in F_2 a colour blind homozygous female, a normal visioned heterozygous female, a normal visioned hemizygous son and a hemizygous colour blind son.

(2) *Haemophilia:* It is the most serious and notorious disease also known as *bleeder's disease*, which is more common in men than women. The person having the recessive gene for haemophilia lacks in normal *clotting substance* (*thromboplastin*) in blood, so the minor injuries cause continuous bleeding and ultimate death of the person due to haemorrhages. Recently two types of X-linked haemophilia have been recognised:

(a) *Haemophilia A*: About four-fifths of the cases of haemophilia are of this type. It is characterised by lack of *antihaemophilic globulin* (Factor VIII).

(b) *Haemophilia B*: It results from a defect in *plasmo thrombo-plastic component* (Factor IX). This is a milder form of haemophilia.

The inheritance of haemophilia in man:

1. *A marriage between a haemophilic man and a normal woman:* When a haemophilic man marries a normal woman, they will produce all normal male offsprings and all normal but carrier female offsprings in F_1 generation. The marriage between F_1 normal but carrier female with normal male produces one normal female, one normal carrier female, one normal male and one haemophilic male.

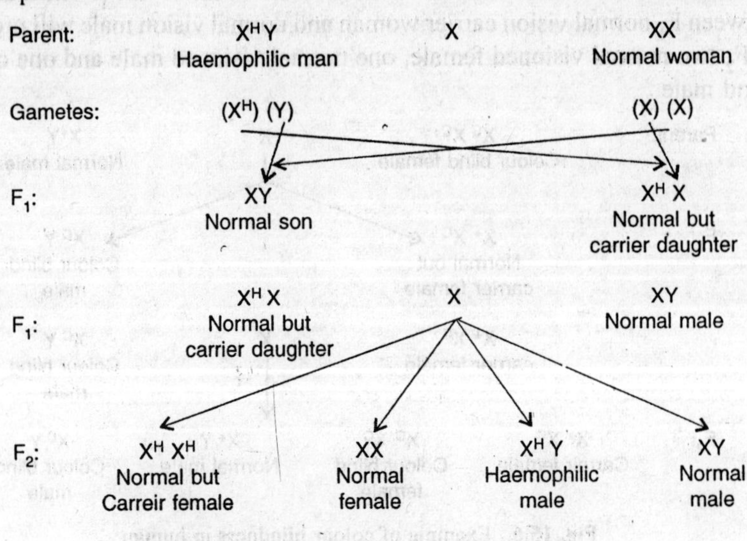

Fig. 15.7. Example of haemophila in human

2. *A marriage between a normal man and a haemophilic woman:* When the haemophilic woman marries a normal man, all the F_1 sons are haemophilic and daughters are normal but carriers. This is because the male receives one X-linked recessive gene for haemophilia from their haemophilic mother and the daughter receives one X-linked dominant gene from the father and one X-linked recessive gene for haemophilia from the mother. The marriage of F_1 normal but carrier female with a haemophilic male produces in F_2 generation 50% offspring haemophilic and remaining 50% offspring normal.

Parent:	$X^H X^H$ Haemophilic woman	X		XY Normal man
F_1:	$X^H X$ Normal but carrier daughter	↓		$X^H Y$ Haemophilic son
F_1:	$X^H X$ carrier daughter	X		$X^H Y$ Haemophilic man
		↓		
F_2:	$X^H X^H$ Haemophilic female	$X X^H$ Normal but carrier female	$X^H Y$ Haemophilic male	X Y Normal male

Fig. 15.8. Example of haemophila in human

SUMMARY

Members of most of the species are divided into two sections, i.e. male sex and female sex. The chromosomal basis of sex determination is the most common method for determination of sex. Normally, a cell contains two chromosomes – autosomes and sex chromosomes. The organism in which both the chromosomes in sex chromosome pair is identical is called homochromosomes. They produce identical gametes, while some in which both chromosomes are not identical produce different gametes. Generally, there are two types of sex determination – Heterogametic male and heterogametic female. In heterogametic male type, male produces two types of gametes while female produces identical or only one type of gametes. In heterogametic female type, male produces only one type of gametes while female produces two types of gametes.

The genes which occur exclusively on X-chromosome are called X-linked genes and the genes which occur exclusively on Y-chromosome are called Y-linked genes. The inheritance of X-linked and Y-linked genes is called sex-linked inheritance. The X-linked recessive gene shows the criss-cross pattern of inheritance. The well known examples of diseases due to X-linked recessive genes in humans are colour blindness, haemophilia, Duchenne's muscular dystrophy, etc.

EXERCISES

1. A girl of normal vision marries a boy of normal vision. The father of both the girl and the boy are colour blind. What type of vision do you expect in their offspring?
2. In a man haemophilia is sex linked and is a recessive trait. A non-haemophilic woman, whose father was haemophilic, marries a non-haemophilic man. What kind of children can they have and in what proportion?
3. Can a normal son have a colour blind mother? Explain.
4. Colour blindness is found more often in men than in women. Give reasons.
5. What is sex-linked inheritance? Explain the inheritance pattern of haemophilia in man.
6. What is sex determination? Name various mechanisms of sex determination. Give an account of genetically controlled sex-determination mechanism.
7. Give an account of the chromosomal mechanism of sex determination in man. How does it differ from the avian type of sex determination?
8. Write short notes on the following:
   (i) Sex chromosomes
   (ii) Autosomes
   (iii) Colour blindness
   (iv) Haemophilia
   (v) Sex linked genes
   (vi) Carrier daughter
   (vii) Heterogametes

16

Chromosomes

Chromosomes are thread like structures, which are typically present in the nucleus and which become visible during cell division. Waldeyer, in 1888, first coined the term chromosome for the darkly staining bodies of the nucleus. The word chromosome is derived from the Greek word '*Chromo*' meaning *coloured* and *'soma'* meaning *body*. Chromosomes are the carriers of the genes. Most of the chromosomes in a cell are called *autosomes*. In addition to *autosome*, a cell also contains one or two *sex chromosomes*, which carry the genes for determination of sex.

Table 16.1 Number of chromosome in some organisms

Common Name	Scientific Name	Chromosome number (2n)
Man	Homo sapiens	46
Gorilla	Gorilla gorilla	48
Pigeon	Columba livia	80
Mosquito	Culex pipiens	6
Cabbage	Brassica oleracea	18
Onion	Allium cepa	16
Potato	Solanum tuberosum	48
Garden pea	Pisum sativum	14
Chick pea	Cicer arietinum	16
Carrot	Daucus carrota	20
Silk worm	Bombyx mori	56
Dog	Canis familaris	78
Fruit fly	Drosophila melanogaster	8
Rat	Rattus norvegicus	42

Number of chromosomes: Generally all the individuals of a species have the constant number of chromosomes, which characterise the species (Table 16.1). The number of chromosomes varies from one species to another. But closely related species usually have similar chromosome numbers. Gametes normally contain only one set of chromosomes (haploid), while somatic cells usually contain two sets of chromosomes (diploid or 2n). Higher plants generally contain 3 to 15 chromosomes in the gametes, while higher animals generally contain 8 to 25 chromosomes in their gametes.

Chromosome size: The chromosome length varies from 1 μ (e.g. some fungi) to 30 μ (e.g. Trillium). However, in some exceptional cases (e.g. polytene chromosomes), the chromosomes may be as long as 2 mm. In some species, all the chromosomes may be of similar size, called *Symmetrical karyotype* or all the chromosomes of a species may be of different size called *asymmetrical karyotype*. Again in asymmetrical karyotype, all chromosomes may either be two distinct size groups (e.g. Yucca arakansana) or there is a gradation in size (e.g. man). In some cases, the cells at different regions like root and shoot tip show variation in chromosome size. In general, the chromosomes of plants are larger than those of animals.

Ultra structure of chromosome

Electron microscope observation shows that a chromosome includes chromonema, centromere, secondary constrictions, nucleolar organisers, telomeres and satellites (Fig. 16.1).

Light microscopic observations have revealed that each chromosome is surrounded by a membranous structure called pellicle, which encloses a jelly like substance called matrix. However, electron microscopic observations have revealed that there is nothing like a pellicle surrounding the chromosome.

1. *Chromonema*: The metaphasic chromosomes have two thread like structures called chromatids, which are twisted around each other and joined at the primary constriction or centromere. Treatment of chromosome with a trypsin reveals that each chromatid is made up of two subunits or subchromatids. These subchromatids are again made up of chromonema. The number of chromonema varies from 2 to 32. The chromonema fibrils are coiled around each other in two different ways. Paranemic coils are loosely coiled and the chromonemal fibrils are easily separable from each other. However, the plectonemic coils are very closely coiled and they are not easily separable.

2. *Centromere* or *primary constriction*: The chromosome has a constricted region which stains lightly called centromere or kinetochore or primary constriction. The position of the centromere is constant for a particular chromosome and is responsible for the various shapes of chromosomes. The chromonema are less coiled in the centromeric region and therefore stain lightly. During cell division the centromere is functional, while the remaining part of the chromosome is genetically inactive. Thus, the centromere may

contain the genes for mitotic and meiotic activity.

The position of the centromere varies in the different chromosomes. Depending on the position of the centromere, the chromosomes are categorised into following four groups (Fig. 16.2).

(a) *Metacentric*: If the centromere lies midway along the length of chromosome, the chromosome is said to be metacentric. Thus in metacentric chromosomes, the two arms are almost equal in length and during the anaphasic movement, the chromosome appears 'V' shaped. e.g. Man, Trillium, Tradescantia, etc.

(b) *Submetacentric*: If the centromere is located some distance away from the middle region of chromosome, it is called submetacentric. In submetacentric chromosome, one arm is shorter than the other. During anaphasic movement, it appears 'L' shaped.

e.g. Man

(c) *Acrocentric*: If the centromere is very near one end so that one arm is very small and the other arm is very long, then it is called acrocentric chromosome. During anaphasic movement it appears rod shaped.

e.g. Man, Grasshoppers, etc.

(d) *Telocentric*: When the centromere is situated exactly at one end of the chromosome, such chromosome will have only one long arm and it is called telocentric. Telocentric chromosomes are very rare. True telocentric chromosomes have been reported by Marks (1957) in certain species of plants. Few species of protozoan and mammals also contain telocentric chromosome.

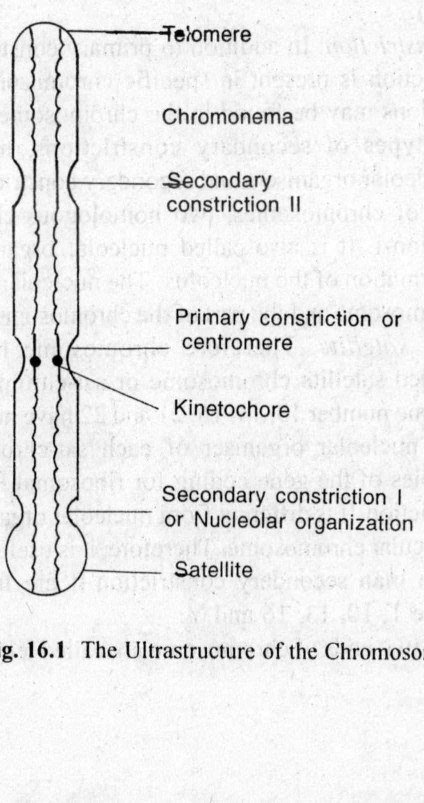

Fig. 16.1 The Ultrastructure of the Chromosome

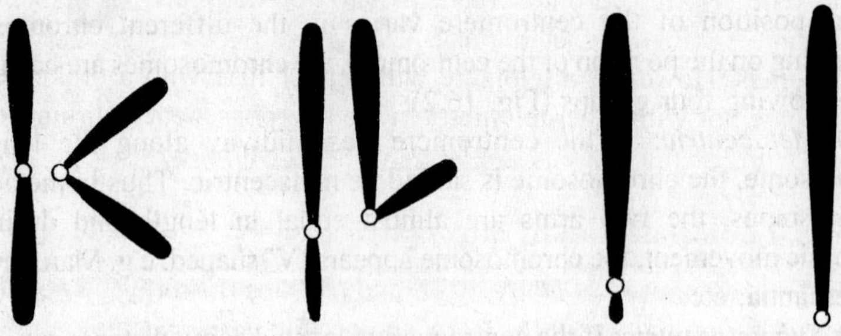

Fig. 16.2. Types of Chromosomes based on centromere position

Sometimes a chromosome may undergo breakage so that only one part has the centromere while the other fragment is without the centromere. The fragment without the centromere is called acentric. In some plants and animals, the chromosome is located more than one position along the length of the chromosome. Such chromosome is called polycentric chromosome.

Functions of centromere: A centromere provides mobility to the chromosome. Centromeres attach themselves to the spindle fibres during anaphasic movement.

Centromeric region of each chromatid appears to have a small intensely staining granule called the *spindle spherule* or *kinetochore*. The spindle fibres get attached to the kinetochore rather than to the whole region of the primary constriction.

3. *Secondary constriction*: In addition to primary constriction, sometimes a secondary constriction is present in specific chromosomes. One or more secondary constrictions may be found in the chromosomes of organisms.

There are two types of secondary constrictions, namely secondary constriction-I or nucleolar organisers and secondary constriction II. Generally, in each diploid set of chromosomes, two homologous chromosomes have secondary constriction-I. It is also called nucleolar organiser because it is necessary for the formation of the nucleolus. The nucleolar organiser appears near one end of chromosome and the part of the chromosome beyond nucleolar organiser is called *satellite*. Therefore chromosome having secondary constriction-I is called satellite chromosome or sat-chromosome. In human being, the chromosome number 13, 14, 15, 21 and 22 have nucleolar organisers and satellites. The nucleolar organiser of each sat-chromosome contains several hundred copies of the gene coding for ribosomal RNA (rRNA).

Secondary constriction-II is different from nucleolar organisers. Its location is constant for a particular chromosome. Therefore, it is useful for identification of chromosomes. In man secondary constriction-II are found on the long arms of chromosome 1, 10, 13, 16 and Y.

4. *Telomere*: The two ends of chromosome are called telomeres. Telomeres

are highly stable and they do not fuse or unite with telomeres of other chromosome. If the chromosome ends are broken, the broken ends of chromosome are highly unstable and readily fuse with broken ends of other chromosomes. However, a telomere will neither fuse with a broken end of chromosome nor with other telomere. In the interphase, each chromosome is attached with the nuclear envelope by one of its telomere.

Chemical composition of chromatin

Chromatin is composed of DNA, RNA and proteins (histone and non histone proteins). The amount of DNA, RNA and protein varies considerably due to species and different tissues of the same species. Chromatin isolated from interphase nuclei consists of about 30-40% DNA, 50-65% protein and 0.5-10% RNA. The chemical composition of metaphase chromosome differs from that of interphase chromatin. As compared to interphase chromatin, the metaphase chromosome contains less DNA and more protein and RNA. Metaphase chromosomes consist of about 15-20% DNA, 10-15% RNA and 65-75% Protein.

(a) *DNA*: Generally the amount of DNA present in normal somatic cells of a species is constant. Any variation in DNA from this normal amount is strictly correlated with a corresponding variation at the chromosome level. Gametes of a species contain only half of the amount of DNA present in its somatic cells. The DNA found in chromosomes is mainly of two types, namely unique DNA and repetitive DNA. The DNA segments whose base sequences are present only in a single copy per genome, i.e. their base sequences are not repeated in the genome, are called *unique DNA*. While DNA segments whose base sequences are present in several to a million copies per genome are called *repetitive DNA*. Repetitive DNA sequences are of two types – highly repetitive DNA and moderately repetitive DNA. The proportion of unique DNA in cells of various species may vary. Human genome has 70% unique DNA and only 30% repetitive DNA. All the structural genes, except those genes whose products are present in large quantities in cells, are made up of unique DNA.

(b) *RNA*: The RNA associated with chromosomes are ribosomal, messenger and transfer RNA. But around 5% of the total chromosome weight is due to a special class of RNA called *chromosomal RNA*. Chromosomal RNA is 40-60 nucleotides long and is always found associated with chromosomes. It is essential for the association of DNA with chromosomal protein, particularly histone proteins, in order to produce chromatin fibres. Thus, it is essential for the structural organisation of chromatin fibres and therefore of chromosomes. It also evolved in gene regulation in eukaryotes.

(c) *Protein*: Proteins associated with chromosomes are of two types – basic proteins or histone proteins and acidic proteins or non-histone proteins. About 80% of total chromosomal proteins constitute histones. Its molecular weight ranges from 10,000 to 30,000 and it is completely devoid of tryptophan.

In histone five different types of polypeptides are present. They are H_1, H_2a, H_2b, H_3 and H_4. Histone always consists of two copies of H_2a, H_2b, H_3 and H_4 and only one copy of H_1 polypeptide. Generally, H_1 fraction is lysine rich while H_2a and H_4 are arginine rich. It has been suggested that H_2a, H_2b, H_3 and H_4 are involved in the structural organisation of chromatin fibres while fraction H_1 holds together the folded chromatin fibres.

About 20% of the total chromosomal protein constitute non-histone proteins, but their amount is variable. There may be around 12 to 20 different types of non-histone proteins. Non-histone proteins show variation from one species to the other and even in different tissues of the same species. Non-histone proteins contains many important enzymes like DNA polymerase, RNA polymerase, etc. The non-histone proteins may switch on specific genes in eukaryotes.

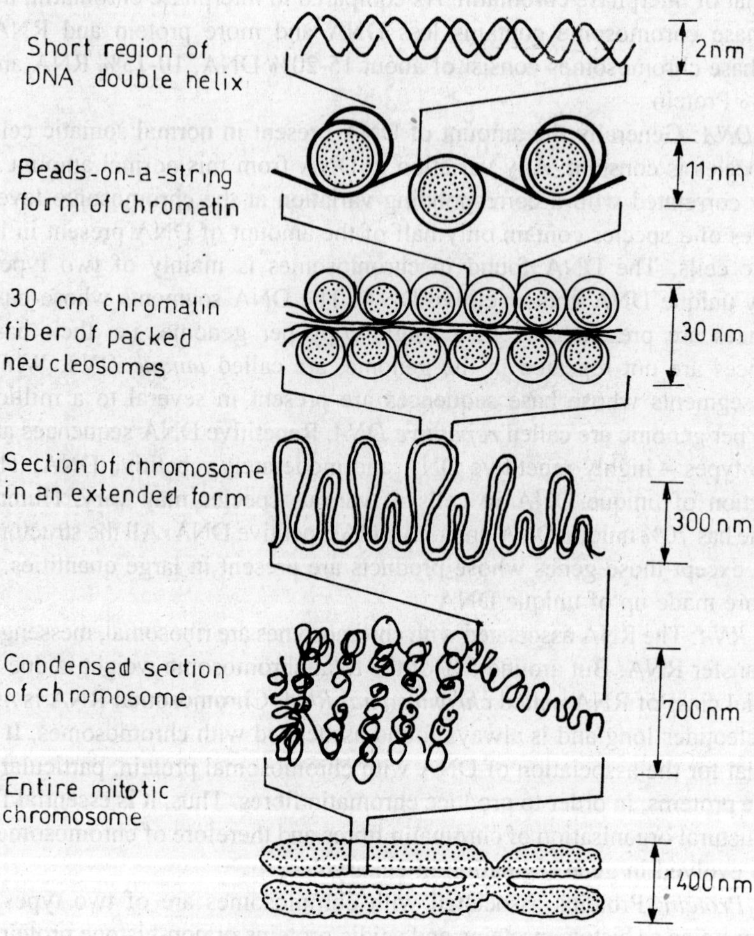

Fig. 16.3 Structural organisation of nucleosome

Structural organisation of nucleosome

In 1974, Kornberg and Thomas proposed Nucleosome model for organisation of chromatin fibres. According to this model, chromatin is composed of a repeating unit called nucleosome. One complete nucleosome consists of a nucleosome core, linker DNA, one molecule of H_1 histone and other associated chromosomal proteins.

Nucleosome core consists of a histone octamer, which is composed of two molecules of each H_2a, H_2b, H_3 and H_4 polypeptides. In addition to histone octamer, it consists of a 146 bp long DNA molecule, which turns round this histone octamer in 1¾ turns. The structure of nucleosome is essentially identical in all the eukaryotes.

The linker DNA joins two neighbouring nucleosomes. Depending on the species, the size of linker DNA varies from 8 bp to 114 bp. Each nucleosome contains at least one molecule of H_1 histone. Some studies suggest that H_1 histone stabilises nucleosome core, while some other studies suggest that it is involved in stabilising the supercoils of nucleosome chromatin fibers.

Each nucleosome contains one molecule of H_1 histone. Some studies suggest that H_1 is associated on the outside of each nucleosome core and it stabilises about 166 bp long DNA molecule. Other studies, however, suggest that the molecules of H_1 histone are involved in stabilising the supercoils of nucleosome chromatin fibers.

The nucleosome fibres are around 100 Å thick. These nucleosome fibres are supercoiled to produce a compact 1400 Å chromatine fibre, which is seen in electron micrographs of metaphase chromosomes.

Heterochromatin and Euchromatin

Chromosomes which remain condensed during interphase are called *heterochromosomes* e.g. - sex chromosomes of insects. The chromosomes which extend during interphase are called *euchromosomes*.

Chromatin material is of two types – heterochromatin and euchromatin. Chromatin material showing heteropycnosis (i.e. overcondensed part of chromosome) at any stage is called *heterochromatin*. Regions of the chromosomes which never show heteropycnosis is called euchromatin.

Table 16.2 Difference between Heterochromatin and Euchromatin

	Heterochromatin	Euchromatin
1.	Heterochromatin stains deeply	It stains less deeply.
2.	It is found in highly condensed region of chromosome	It is found in diffused or less condensed region of chromosome.
3.	It is late replicating.	It replicates during the early stage of the S phase.

(Contd.)

	Heterochromatin	Euchromatin
4.	It does not become acetylated.	It takes up acetic acid on its histone during interphase.
5.	It is more labile and is affected by temperature, sex, age of parents, proximity to the centromere and presence of additional Y-chromosome.	It is less labile.
6.	Metabolically, it is inert.	Metabolically, it is active.
7.	In it crossing over frequency is less.	In it crossing over frequency is more.

Heterochromatin is of two types – constitute heterochromatin and facultative heterochromatin. Constitute heterochromatin shows heteropycnosis in all cell types. Facultative heterochromatin, on the other hand, is heteropycnotic only in some special cell types or at some particular stages of the life cycle.

Gaint Chromosomes

Gaint chromosomes are of two types – polytene chromosome (Salivary gland chromosomes) and lampbrush chromosomes.

(A) *Polytene chromosomes/Salivary gland chromosome:* Except polytene chromosomes, normally chromosomes are not visible during interphase. They were first observed by Balbiani (1881) in the salivary glands of the midge *Chironomous* and hence are called salivary gland chromosomes. All the chromosomes of a particular type in the individual have an equal number of bands with similar distribution. In the four chromosomes of Drosophila over 5,000 bands have been found. The difference between bands and interbands are as follows.

Table 16.3 Difference between bands and interbands

Sl. No.	Bands	Interbands
1.	Stain intensely with basic stain.	Do not stain intensely.
2.	Feulgen positive.	Feulgen negative.
3.	Regions of high DNA concentration.	Regions of low DNA concentration.
4.	Chromonemata tightly packed by folding.	Chromonemata relatively extended.
5.	Absorb UV light.	Absorb little UV light.

Painter (1933) and Bridges (1936) show that the bands of polytene chromosomes are the sites of genes and that the interbands are relatively inert linker regions. But now it is proved that interbands also contain many genes.

The polytene chromosome is formed by about 1000 to 4000 unit chromatids (chromonemata). Each unit chromatid consists of a single DNA molecule associated with protein. The DNA fibres are continuous from one end of the chromosome to the other. They are tightly folded in the bands and relatively extended in the interbands.

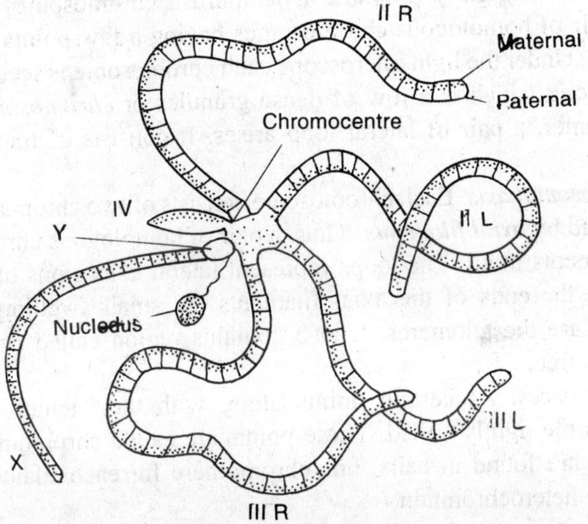

Fig. 16.4. Salivary gland chromosome of Drosophila

Chromosome puffs: The bands of polytene chromosomes become enlarged at certain times to form swelling which are called *chromosome puffs* or *Balbiani rings*. According to Beerman and Bahr (1954), chromosome puffs are regions where the tightly coiled chromosomal fibres open out to form many loops. Thus, puffing is due to unfolding or uncoiling of individual chromomeres in a band. The puffs are active genes and represent sites of RNA synthesis.

B. Lampbrush Chromosomes

Lampbrush chromosomes are so called because they appear like the brushes once used for cleaning glass chimneys of kerosene lamps. The first description of lampbrush chromosomes was by Flemming (1882) on amphibion oocytes. A detailed study of the lampbrush chromosomes was made by J. Ruckert (1892) on the oocytes of the shark.

Lampbrush chromosomes are found in animal oocytes, which grow greatly during prophase of meiosis I and which produce a lot of RNA. However, the lampbrush chromosomes of invertebrates are not as large as those of vertebrates, but they have the characteristic *hairy* appearance. The lampbrush chromosomes have also been reported in plants.

Size: In many animals lampbrush chromosomes may be more than 1000 μ in length and 20 μ in width during early prophase I. In some salamander oocytes the lampbrush chromosomes may reach a length of 5900 μ. However, towards the end of prophase I (diakinesis), the chromosomes contract and are greatly reduced in size. They may contract still further during metaphase I. Lampbrush chromosomes are very elastic and can be stretched up to about two times before they break.

Structure: During early prophase I, lampbrush chromosomes are in the form of a pair of homologous chromosomes having a few points of contact or *chiasmata*. Under the light microscope each chromosome is seen to consist of an axis along which is a row of dense granules or *chromosomes*. From each chromomer, a pair of lateral loop arises. It consists of the following parts:

1. *Chromosomal axis*: Each chromosome consists of two chromatids which are represented by *axial filaments*. Thus, a pair of homologous chromosomes has four filaments in all. The loops represent lateral extensions of the axial filaments. At the ends of the axial filaments are small swellings without loops. These are the telomeres. It also contains region called *centromere*, which is loop free.

2. *Chromomeres*: At certain points along with their length the axial filaments become tightly coiled. These points are called chromomeres. The chromomeres are found in pairs, one chromomere for each filament. They correspond to heterochromatin.

3. *Loops*: Loops are of two main types – typical and special. Most of the loops are typical. Each typical loop consists of a central axis from which are given off RNA fibrils of progressively increasing lengths. This makes the loop markedly thicker on one side. The special loops have a marked asymmetry and have granules at the end of the fibrils.

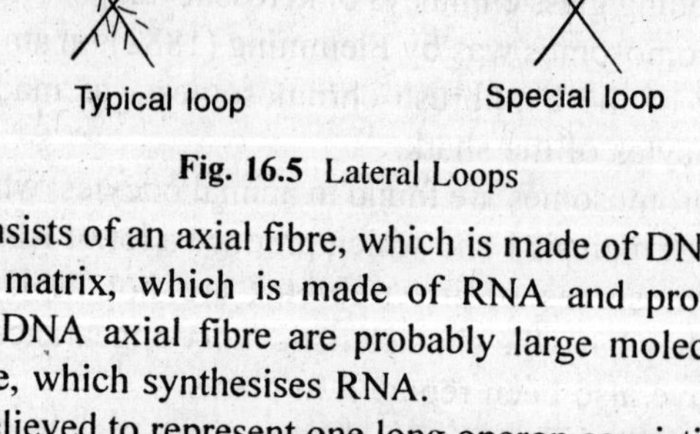

Fig. 16.5 Lateral Loops

Each loop consists of an axial fibre, which is made of DNA. Axial fibre is covered with a matrix, which is made of RNA and protein. The dense granules on the DNA axial fibre are probably large molecules of enzyme RNA polymerase, which synthesises RNA.

The loop is believed to represent one long operon consisting of a series of identical copies of the same structural genes. Each gene locus codes for 45 s rRNA in the Triturus oocytes. Between the gene loci are matrix free regions of *spacer DNA* which apparently do not produce RNA. At one end of the spacer DNA region, dense granules of RNA polymerase, attached to the promoter region can be seen. The other end is the chain termination region.

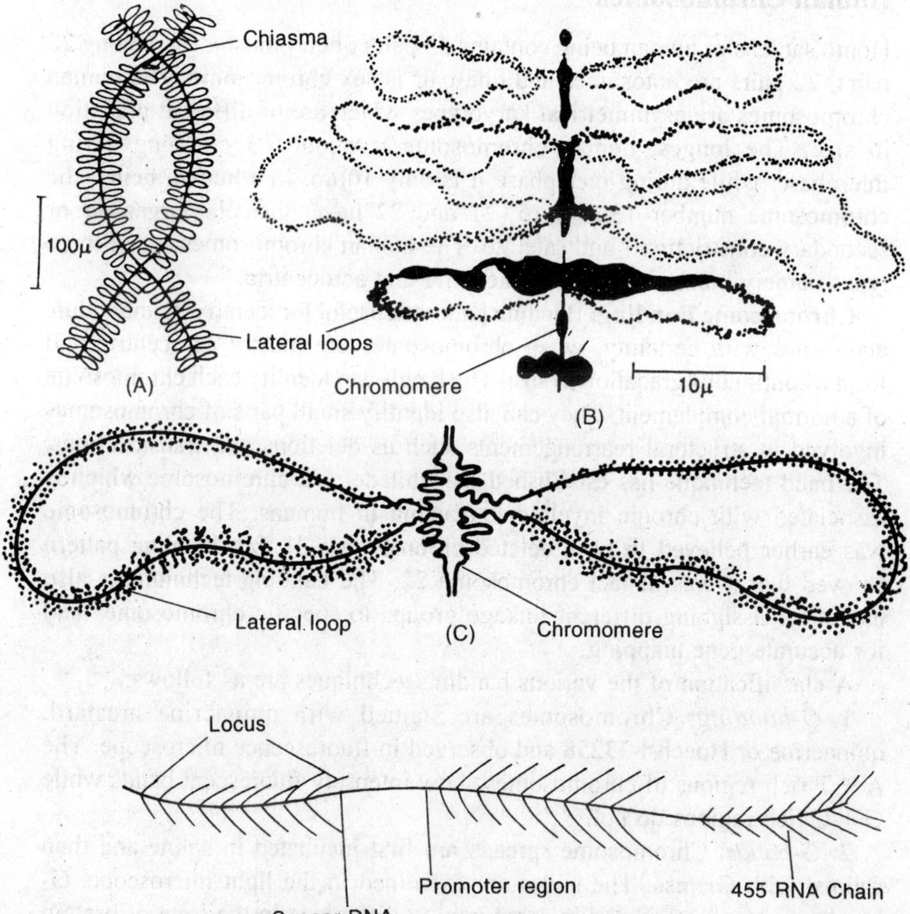

Fig: 16.6 Lampbrush chromosome

(a) A pair of meiotic chromosomes joined by two chiasmata.
(b) A part of the chromosome enlarged showing the lateral loops.
(c) A part of axis and a pair of loops.
(d) Diagram of a tandem series of genes.

Callen and Liyod (1960) have suggested that each pair of loops is associated with the activity of a specific gene. Each loop is supposed to contain repeating DNA sequence (genes) arranged in a series. In other words there are several identical copies of genes in each loop. At each chromomere is supposed to be a *master gene* copy which transfers information to several *slave gene* copies on the same loop. Only the slave genes take part in RNA synthesis and not the master gene. A high rate of RNA synthesis is possible because of the presence of repeating similar genes.

Human Chromosomes

Homo sapiens or human being contains 23 pairs of chromosomes. Among 23 pairs, 22 pairs are autosomes and one pair is sex chromosome. The human chromosomes are asymmetrical karyotypes which are of different gradation in size. The longest human chromosome is about 7.3 cm long during interphase, while during metaphase it is only 10μm. In a human being, the chromosome number 13, 14, 15, 21 and 22 have nucleolar organiser or secondary constriction I and satellite. The human chromosomes are of three types namely metacentric, submetacentric and acrocentric.

Chromosome Banding: Banding pattern is useful for identifying individual autosomes with certainty, when chromosomes are mostly acrocentric and form a continuous gradation in size. The bands can identify each chromosome of a normal complement. They can also identify small parts of chromosomes involved in structural rearrangements such as deletions and translocations. The band technique has established the Philadelphia chromosome which is associated with chronic myeloid leukaemia in humans. The chromosome was earlier believed to be a deleted chromosome 21, but banding pattern showed that it was in fact chromosome 22. The banding technique is also useful for assigning different linkage groups to specific chromosomes and for accurate gene mapping.

A classification of the various banding techniques are as follows:

1. *Q-banding*: Chromosomes are stained with quinacrine mustard, quinacrine or Hoechst 33258 and observed in fluorescence microscope. The A + T rich regions of chromosomes show intensely fluorescent bands while G + C rich regions do not.

2. *G-bands*: Chromosome spreads are first incubated in saline and then stained with Giemsa. The slides are examined in the light microscope. G-bands appear to be related in some way to difference in the state of protein sulfur along the chromosome.

3. *R-bands*: Chromosome preparations are incubated in a buffer at high temperature and stained with *Giemsa*. The banding pattern is reverse of the one observed with G-bands.

4. *C-bands*: Chromosome preparations are treated with a moderately strong alkali followed by warm saline and staining with Giemsa. The satellite DNA present around the centromere becomes deeply strained to form C-bands.

SUMMARY

Chromosomes are thread like structures visible during cell division. The number of chromosome differs from one species to another. The size of chromosomes varies from one type of cell to another, one genera to another, etc.

Chromosomes have different structures like chromonema, primary constriction, secondary constriction I and II, satellite and telomere. Depending on the position of primary constriction or centromere, the chromosomes are of different types, namely metacentric, submetacentric, acrocentric and telocentric.

Chemically chromatin fibres are made up of DNA, RNA and nucleo-proteins. These proteins are of two types, namely histone and non-histone proteins. Depending on the condensation during interphase, chromatins are called heterochromatin and euchromatin.

Gaint chromosomes are of two types, namely polytene chromosome and lampbrush chromosome.

EXERCISE

1. What is a chromosome? Describe the structure of a chromosome.
2. What is Gaint chromosome? Add a note on lampbrush chromosome.
3. Describe the types of chromosomes based on position of centromere.
4. Explain Nucleosome.
5. Differentiate between heterochromatin and euchromatin.
6. Write a note on salivary gland chromosome.
7. Write a short notes on the following:
   (i) Euchromatin
   (ii) Human chromosome
   (iii) Chromosome puffs
   (iv) Telomere
   (v) Centromere
   (vi) Heterochromatin
   (vii) Secondary constriction I

17

Chromosomal Variation/ Aberrations

The chromosomes of each species has a characteristic morphology and number. But sometimes due to certain accidents or irregularities at the time of cell division, crossing over or fertilisation some alternations in the morphology and/or number of chromosomes take place. The slightest variation in the organisation of chromosome is visualised phenotypically and is of great genetical interest. Chromosomal variation is of two types:

(A) Variation in chromosome number

(B) Variation in chromosome morphology

A. Variation in Chromosome Number

Each species has a characteristic number of chromosomes in the nuclei of its gametes and somatic cell. The gametes contains a basic set of chromosomes called genome, while the somatic cell of a diploid organism contains two copies of same genome (2n). When haploid gametes of both sexes unite in the process of fertilisation a diploid zygote with 2 genomes is formed. The diploid zygote develops in an adult organism, which upon attaining sexual maturity, produces haploid gametes. Thus alternation of haploidic and diploidic generation continues in the organisms. But sometime the size of haploidic or diploidic genome or number of chromosomes in a haploid or diploid cell does not remain constant but alters variously. Variation in number of sets of chromosomes (ploidy) is commonly found in nature. It may be of the following two types – Euploidy and Aneuploidy.

I. EUPLOIDY

The word euploidy has originated from the Greek words 'eu' which means even or true and 'ploid' which means unit. Euploidy indicates genomes containing chromosomes that are multiple of some basic number (n). The euploids are those organisms which contain balanced set or sets of chromosomes in any number. The euploids may be of following types:

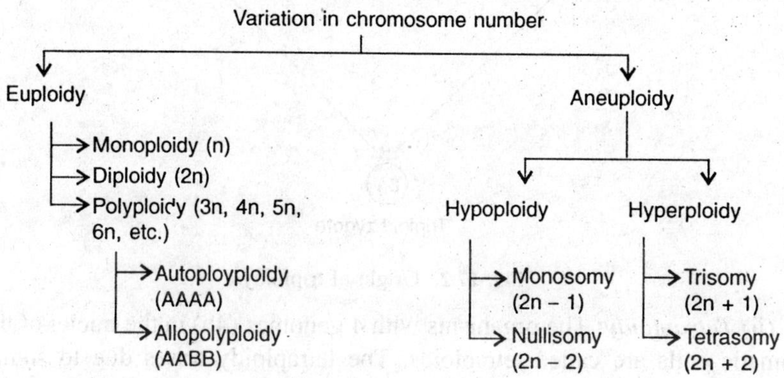

Fig. 17.1. Schematic representation of variation in chromosome number.

1. *Monoploidy:* The monoploid organisms have only single set of chromosomes or only one genome (n) in the nuclei of their somatic cells. Monoploids in some cases are found normally and are produced due to parthenogenesis, e.g. in male of the insects like bees, wasps, ants, etc. The monoploid plants are often weak or less vigorous and sterile. Some examples of monoploid organisms are – some algae, fungi, bryophytes, Sorghum, Triticum, Hordeum, Datura, hymen-opteran insects like wasps, bees, etc.

As monoploid organism contains only one copy of each chromosome and only one allele of each gene, in it each gene is expressed whether it is dominant or recessive. This facilitates genetic experiments which can be helpful in genetic studies.

2. *Polyploidy:* An organism with nore than two sets of genomes is called polyploidy. Ploidy levels higher than tetraploid are not common in natural populations, but our most important crop plants and ornamental flowering plants are polyploids. e.g. wheat (Hexaploid, 6n), Strawberries (Octaploid 8n) liver cells of man, etc. Generally, polyploidy is common in plants but rare in animals.

Depending upon number of sets of genomes present, polyploidy is divided into following types:

(i) *Triploidy:* The triploid organisms have 3 sets of chromosomes or genomes (3n) in the nuclei of their somatic cells. A triploid may originate by the union of a monoploid gamete (n) with a diploid gamete (2n). The extra set of chromosomes of the triploid is distributed in various combinations to the germ cells, resulting in the genetically unbalanced gametes. Triploids are usually sterile organisms for example, some cells of birds, reptiles, man and other vertebrates.

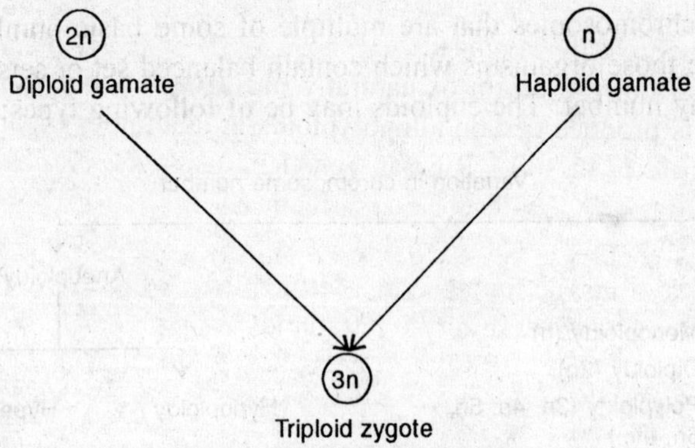

Fig. 17.2. Origin of triploidy

(ii) *Tetraploidy:* The organisms with 4 genomes (4n) in the nuclei of their somatic cells are called tetraploids. The tetraploidy arises due to somatic doubling of the chromosome number. Doubling is due either spontaneously or induced by exposure to chemicals such as colchicine, acenaphthene, veratrine or heat or cold, etc. e.g. Wheat, Chrysanthemum, Solanum, few cells of man, Salamander, etc.

Depending upon the nature of genome, the polyploid organisms are divided into following types:

(i) *Autopolyploidy:* The autopolyploids are those polyploids which consists of multiple copies of same basic set of chromosomes. For example, if diploid species has 2 similar sets of chromosomes (AA), autotriploid will have three similar sets of genomes (AAA). Similarly autotetraploid contains four sets of genome (AAAA). Since autopolyploidy, particularly autotriploid, remains sterile and cannot produce seeds, therefore it has great commercial value in producing seedless varieties of economical plants. It has greater vegetative growth. For example, in Japan, H. Kihara produced seedless watermelons which are autotriploids. Other seedless autopolyploids are grapes, sugarbeet, banana, etc.

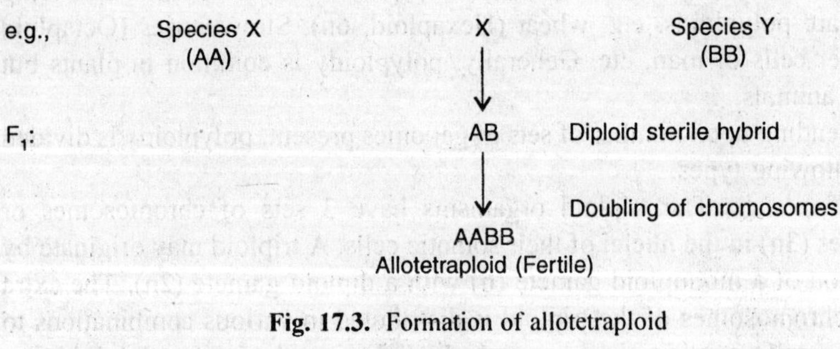

Fig. 17.3. Formation of allotetraploid

(ii) *Allopolyploidy:* (Greek 'allo' means other, different). When the polyploidy results due to the doubling of chromosome number in a F_1 hybrid, which is derived from two different species, then it is called allopolyploidy and resultant species is called allopolyploids. It is the most common natural form of polyploidy in plants.

To find out the origin of naturally occurring allopolyploids some cytogeneticists produce certain allopolyploids in laboratory by artificial means.

Example 1: In 1927, Russian geneticist *G. D Karpechenko* performed a cross between radish-Raphanus sativum (2D = 18) and cabbage Brassica oleracea (2n = 18). The F_1 hybrid was sterile diploid plant. Among these sterile F_1 hybrids, he found certain fertile plants which were found to contain 36 chromosomes. These fertile tetraploids were called *Raphanobrassica.*

Parent: Radish RR (2n = 18) X Cabbage CC (2n = 18)

↓

F_1: RC (2n = 18) Diploid sterile plant

↓ Doubling of chromosomes

RRCC (2n = 36)
Tetraploid fertile plant

Fig. 17.4. Formation of Raphanobrassica

Example 2: Triticum spelta is a hexaploid wheat which was artificially synthesised in 1946 by E.S. McFadden, E.R. Sears and H. Kihara. They crossed an emmer wheat *Triticum dicoccoides* (tetraploid 2n = 28) with goat grass *Aegilops squarrrosa* (diploid 2n = 14). This artificially synthesised hexaploid wheat was found to be similar to the primitive wheat *Triticum spelta.* When this synthesised hexaploid wheat was crossed with naturally occurring Triticum spelta, the F_1 hybrid was completely fertile. This suggested that hexaploid wheat must have originated in the past due to natural hybridisation between tetraploid wheat and goat grass followed by subsequent chromosome doubling.

Parent: Triticum dicoccides AABB (2n = 28) X Aegilops squarrosa DD (2n = 14)

↓

F_1: ABD (2n = 21)

↓ Colchicine

AA BB DD (2n = 42) Synthesized hexaploid wheat

Fig. 17.5 Synthesis of hexaploid wheat in laboratories

Example 3: *Gossypium hirsutum*, the new world cotton plant, is another interesting example of allopolyploidy. L.O. Beasley crossed old world cotton *Gossypium herbaceum* (2n = 26) and American cotton Gossypium *raimondii*

(2n = 26) and doubled the chromosome number in the F_1 hybrids. The allopolyploids produces resembled the cultivated new world cotton and when crossed with it gave fertile F_1 hybrids. This result thus suggested that tetraploid *Gossypium hirsutum* originated from two diploid species, namely, G. herbaceum and G. raimondii.

Parent: G. herbaceum AA (2n = 26) X G. raimondii BB (2n = 26)

↓

F_1: AB (2n = 26)

↓ Colchicine

AA BB (2n = 52) G. hirsutum

Fig. 17.6. Synthesis of tetraploid cotton in laboratory

Example 4: Similarly, *Nicotiana tobacum* and *Nicotiana rustica* are also allotetraploids. The evidence indicates that N. tobacum is derived from the cross between N. Sylvestris and N. tomentosa.

Parent: N. Sylvestris AA (2n = 24) X N. tomentosa BB (2n = 24)

↓

F_1: AB (2n = 24)

↓ Doubling of chromosomes

AA BB (2n = 48) N. tobacum

Fig. 17.7. Origin of N. tobacum

Phenotypic effects of Polyploidy

The increase in the genome's size beyond the diploid level is often caused following detectable phenotypic characteristics in a polyploid organism:

(i) *Morphological effect:* Generally the polyploid organisms are gigantic. The polyploid plants have large sized pollen grains, cells, leaves, stomata, xylem, etc. They are more vigorous than diploid organisms.

(ii) *Physiological effect:* In tetraploid cabbage and tomato, ascorbic acid content is higher as compared to diploid plants. Also corn meal of tetraploid maize seeds contains 40% more vitamin A than corn meal of diploid maize seeds.

(iii) *Effect on fertility:* Polyploidy reduces fertility of an organism in variable degrees.

(iv) *Speciation:* Polyploidy is a great evolutionary significance because it is a means of origin of new species.

II. ANEUPLOIDY

The word aneuploidy originated from Greek word 'aneu' meaning uneven and 'ploidy' meaning units. Thus in aneuploidy, variation in chromosome number may occur which do not involve variation in whole set of chromosomes, but variation in only parts of a chromosome set. Aneuploidy is of two types: *Hypoploidy and Hyperploidy*. Loss of one or more chromosomes is called hypoploidy, while addition of one or more chromosomes is called hyperploidy.

Hypoploidy is of two types:

1. *Monosomy:* Diploid organisms which are missing one chromosome of a single pair are called monosomic. Its genomic formula is 2n – 1. During gametogenesis monosomic organisms form two types of gametes – (n) and (n – 1). The (n – 1) gametes do not survive in plants, but in animals they may cause genetic imbalance, which result in high mortality or reduced fertility of resulting organisms.

e.g. Turner's syndrome (45, X) in woman. (Table 17.1)

Monosomy may be single monosomic (2n – 1), double monosomic (2n – 1 – 1) or triple monosomic (2n – 1 – 1 – 1). In double monosomy, two non-homologous chromosomes are missing and in triple monosomy, three non-homologous chromosomes are missing.

2. *Nullisomy:* An organism which has lost a pair of chromosome is a nullosomic with genomic formula (2n – 2). A nullsomic diploid often does not survive. However, a nullisomic polyploid may survive but has reduced vigour and fertility. Nullisomic condition for different chromosomes produces distinct morphological effects.

Hyperploids are organisms with increase in one or more chromosomes. It may be of following types:

1. *Trisomy:* When somatic cells of an organism contain three copies of any one chromosome, the condition is known as trisomy. It is denoted by 2n + 1. As in trisomy, a pair of chromosome has an extra number, a trivalent structure may be formed during meiotic prophase. Gain of a complete chromosome causes a major imbalance, but its effects are not as deleterious as those due to the loss of a full chromosome. Therefore, trisomics are viable in diploid species. In several species full set of trisomics is available, e.g. maize, tomato, pea, Datura, etc.

Trisomy can produce different phenotypes, depending upon which chromosome is present in the triplicate. Trisomy may be single (2n + 1) or double (2n + 1 + 1).

In human being, the following syndromes are observed due to trisomic condition for different chromosomes. Down's syndrome (trisomy 21), Edward's syndrome (trisomy 18), Patau's syndrome (trisomy 13), Triplo-X

(XXX females), Klinefelter's syndrome (XXY male), etc. For detail see Table 17.1.

In other animals, trisomy 22 has been reported in chimpanzees, which shows Down's syndrome like phenotypic features. Trisomy-21 has been reported in the gorilla.

2. *Tetrasomy:* The somatic cell of tetrasomic organisms contains one pair of a chromosome in excess of the normal somatic complement. It is denoted by 2n + 2. Generally, gain of a chromosome pair is more deleterious than that of one chromosome. Therefore, only few tetra-somics of diploid species can be recovered and maintained. Tetrasomics are viable in polyploid species, e.g. wheat.

Aneuploidy is caused due to chromosomal non-disjunction during meiosis. The failure of homologs (at meiosis) or sister chromatids (at mitosis) to separate properly to opposite poles is called non-disjunction. Non-disjunction at meiosis I produces all abnormal gametes while non-disjunction during meiosis II produces 50% normal and 50% abnormal gametes.

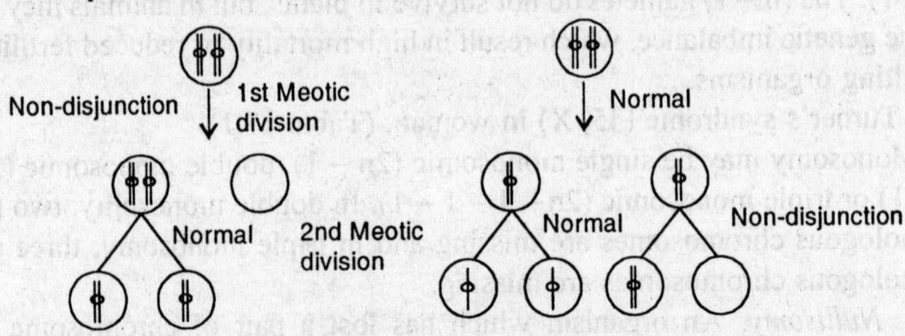

Fig. 17.8. Chromosomal non-disjunction during meiosis

Aneuploidy in man: It is estimated that about 4% of human zygotes are chromosomally abnormal, but only 10% of them survive, to be born and remaining 90% abnormal embryos either fail to implant in the uterus or abort in the early stages of embryonic development after successful implantation. The following syndromes associated to aneuploidy conditions are common in human

(a) *Down's Syndrome or Trisomy 21:* It is the best known and most common syndrome due to the trisomy for chromosome 21 in human. This syndrome was first described by a British Physician, Langdon Down, in 1866. Down Syndrome is the most common chromosomal abnormality in live births (1/650 births). Persons with Down Syndrome have short stature, loose jointedness particularly in the ankles, broad and short skull, epicanthal fold, wide nostrils, large tongue with characteristic furrowing, stubby hands, and impaired mental abilities. About 17% of the Down syndrome babies born alive die within the first year, but the average life expectation is around 16-20 years.

Trisomy 21 or Down syndrome can be caused by chromosome non-disjunction in one of the meiotic cell division. The non-disjunction event can occur in either parents, but it seems to be more likely in female parents. In addition, the frequency of non-disjunction increase with maternal age. Among mothers younger than 25 years old, the risk of having a child with Down's syndrome is about 1 in 1500 births, whereas among mothers older than 40 years, the risk of having a child with Down's syndrome is about 1 in 100 births.

(b) *Patau's syndrome or trisomy 13:* Trisomy 13 was first described by K. Patau in 1960. Its incidence is about one in 20000 births. Persons with Patau syndrome appear to be markedly mentally retarded, having sloping forehead, harelip and cleft palate. Polydactyly of both hands and feet is almost present, the hands and feet are deformed, cardiac and various internal defects are common. Death usually occurs within hours or days, but generally the foetus may aborted spontaneously.

(c) *Edward's syndrome or Trisomy 18:* It was first described in 1960 by *John H. Edwards*. Its incidence is about one among 8000 births. It is characterised by multiple malformations, primarily low-set ears, small receding lower jaw, small mouth and nose with general elfil appearance, severe mental retardation, horse shoe or double kidney, cardiac malformations, Harelip and cleft palate, short stature. After birth, 90% trisomy 18 die within first six months. Like the Down's syndrome, occurrence of Edward's syndrome is also related with maternal age.

(d) *Triplo-X females:* Such females are called super females because they possess an extra X chromosome. All Triplo-X females are slightly mentally retarded with limited fertility showing abnormal sexual development. Its incidence is about one among 700 births.

(e) *Klinefelter's syndrome:* It was first described by H.F. Klinefelter in 1942. Individuals of Klinefelter's syndrome are phenotypically males but with some tendency towards femaleness, particularly in secondary sex characteristics. They may have small testicles, slight mental retardation, enlarged breasts, less body hair, small prostrate glands, longer limbs, knock knees and higher pitched voice.

The XXY karyotype can originate by fertilisation of an exceptional XX-egg with a normal Y bearing sperm or fertilisation of an X-bearing egg with an exceptional XY- sperm. The XXY karyotypes account for about 75% of all cases of Klinefelter syndrome. Other cases involves more complex karyotypes such as XXXY, XXXYY, XXYY, XXXXY, XXXXYY, XXXXXY. Persons with more than two X chromosomes usually have some degree of mental retardation. Its incidence is about one in 500 male births.

(f) *XYY males:* Presence of an extra Y chromosome in males results in extra height, mental retardation, severe facial acne during adolescence and criminal bent of mind.

(g) *Turner's syndrome (XO female):* It was first described by H.H. Turner in 1938. It is associated with the monosomy for X chromosome and has a frequency of about one per 2500 female births. More than 90% of the XO zygotes abort spontaneously. Adults with Turner syndrome are female with retarded sexual development, usually sterile, short stature, webbed neck, low set ears, cardiovascular abnormalities, hearing impairment, shield like chest, etc.

B. Variation in chromosome structure

Chromosomes have a structure with definite organisation. They carry genes, which remain arranged on the chromosome in linear order. In ordinary conditions, the chromosomes remain unchangeable but under certain natural or artificial conditions, some structural changes may occur in them. These

Table: 17.1 Aneuploidy in Human beings

Karyotype	Chromosome Formula	Clinical Syndrome	Estimated frequency at birth	Phenotype
47, +21	2n + 1	Down	1/700	Short, broad hands with palm crease, short stature, hyperflexibility of joints, mental retardation, broad head with round face, open mouth with large tongue
47, +13	2n + 1	Patau	1/20,000	Mental retardation, deafness, minor muscle seizures, cleft lip and/or palate, cardiac anomalies, posterior heel prominence.
47, +18	2n + 1	Edward	1/8,000	Congenital malformation of many organs, low-set malformed ears, small mouth and nose with general elfil appearance, mental retardation, horseshoe or double kidney, short stature, 90% die within first six months after birth.
47, XXX	2n + 1	Triplo-X	1/700 female births	Female with usually abnormal genitalia and limited fertility, slight mental retardation
47, XXY 48 , XXXY 48, XXYY 49, XXXXY 50, XXXXXY	2n + 1 2n + 2 2n + 2 2n + 3 2n + 4	Klinefelter	1/500 male births	Male, subfertile, with small testes, developed breasts, feminine pitched voice, long limbs, knock knees.
47, XYY	2n + 1	Super male		Extra height, slight mental retardation, severe facial acne during adolescence, criminal bent of mind.
45, X	2n - 1	Turner	1/2500 female births	Female with retarded sexual development, usually sterile, short stature, webbing neck, cardiovascular abnormalities, etc.

changes are collectively called variations in chromosome structure. The induced chromosomal variation may be caused by radiation and chemicals such as Lysergic acid diethylamide (LSD).

The chromosomal variation may remain confined to single chromosome or may extend to both chromosomes of a pair or may involve two or more pairs of chromosomes. It may be of two kinds— (1) Intrachromosomal variation, (2) Interchromosomal variation

1. *Intrachromosomal variation:* The intrachromosomal or homosomal aberrations occur in one and the same chromosome. These are of three types – Deficiency or deletion, Duplication and Inversion.

(a) *Deficiency or deletion:* A deficiency means deletion of a small portion of a chromosome resulting in loss of one or more genes. A deficiency originates from breakage occurring at random in both chromatids of a chromosome or only in one chromatid. The breakage may be caused by various agents such as radiation, chemicals, drugs or viruses at any time during the cell cycle, either in somatic or in germ cells. Depending upon its location, a deletion may be *terminal,* when a single break occurs near the end of the chromosome; or *inter-stitial* when two breaks occur in the middle portion of the chromosome.

Each break produces two raw ends which may behave in one of the following ways: – (i) there might be reunion of the broken ends called *restitution* so that the original chromosome structure is restored; (ii) the broken ends may not unite giving rise to a chromosomal segment without a centromere which is eventually lost during cell division; (iii) if two single breaks occur in two different chromosomes in a cell, the deleted segment of one chromosome may unite with the raw broken end on the other chromosome, this is called *exchange union.*

A deletion of any considerable portion of chromosome is usually lethal to a diploid organism because of its genetic imbalance.
e.g. Notch-wing mutation in Drosophilla, waltzing mice, cri du chat syndrome in human.

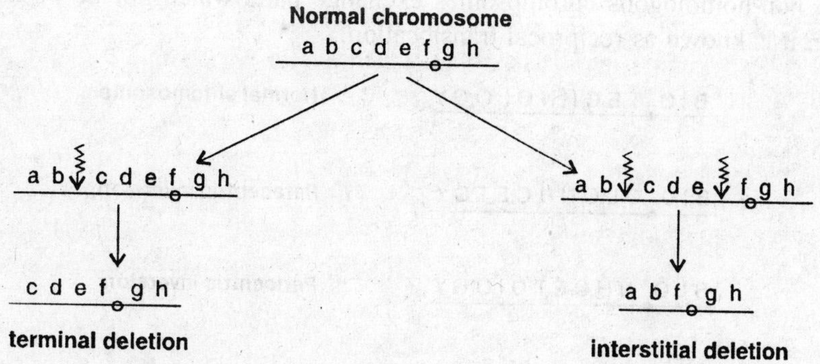

Fig. 17.9. Production of terminal and interstitial deletion.

Cri du chat syndrome: The word 'cri du chat' is originated from French word, meaning 'cry of the cat'. It was first described by *Lejeune* et al in 1963. This condition is caused by a deletion in *the short arm of chromosome 5*; about half portion of the short arm appears to be missing. Individuals heterozygous for this deletion and normal chromosome have the karyotype 46 (5P –), where the term in parentheses indicates that part of the short arm (P) of one of the chromosome 5 is missing.

These individuals are severely impaired mentally as well as physically, have malformation in the larynx, moon faces, saddle nose, small mandible, malformed low-set ears and small head. Their plaintive cat like crying gives the syndrome its name.

(b) *Duplication:* The presence of a port of a chromosome in excess of the normal complement is known as duplication. The duplication of chromosome is not deleterious to the organism like that of deficiency and it usually protects the organism from effect of deleterious recessive genes or from lethal deletions. The duplication is useful in the evolution of new genetic material and in altering the phenotype due to reallocation of chromosomal material. Example: Bar eye in Drosophilla.

A B C D E F G H	⟶	A B C D E E F G H
Normal chromosome		**Duplication**

(c) *Inversion:* Inversions result when there are two breaks in a chromosome and the detached segment becomes reinserted in the reverse order. Depending upon the inclusion or absence of the centromere within the inverted segment, it is of two types – pericentric and paracentric inversion. When both breaks occur in one arm of the chromosome it is paracentric inversion and when a break occurs in each of the two arms, the centromere is included in the detached segment and leads to a pericentric inversion.

(2) *Interchromosomal variation:* Sometimes, when a segment of a chromosome becomes detached and unites with another nonhomologous chromosome, the interchromosomal variation occurs. It is also called *translocation.* If the two non-homologous chromosomes exchange parts which not be of same size, it is known as reciprocal translocation.

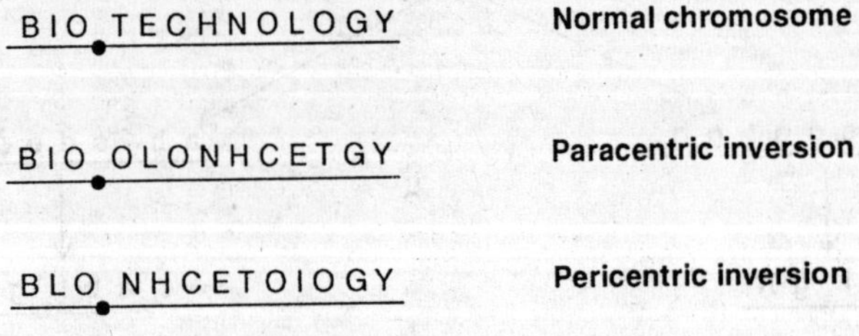

Fig. 17.10. Types of inversion

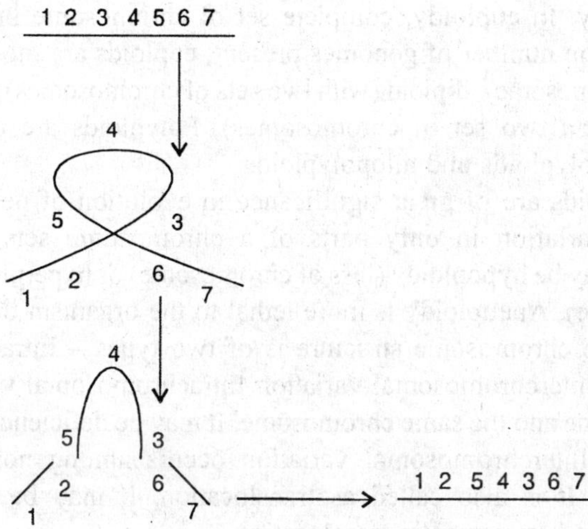

Fig. 17.11. Origin of an inversion in a chromosome

The translocation may be of two types:

(a) *Homozygotic translocation:* In this type normal meiosis occurs and that cannot be detected cytologically. Genetically, they are marked by altered linkage groups and by the fact that a gene with *new neighbours* may produce a somewhat different effect in its new location.

(b) *Heterozygotic translocation:* In this considerable degree of meiotic irregularity occurs. The heterozygous translocation produces semisterile organisms because majority of gametes fail to receive the full complement of genes required for normal development. The heterozygotic translocation is common in maize plant, evening primrose (Oenothera), etc.

Fig. 17.12. Schematic representation of homozygotic and heterozygotic reciprocal translocation.

SUMMARY

Chromosome of each species has a characteristic morphology and number. The slightest variation in chromosome number and morphology is visualised phenotypically. Variation in chromosome number is of two types – euploidy and aneuploidy. In euploidy, complete set of chromosome number varies. Depending upon number of genomes present, euploids are monoploid (with one set of chromosome), diploid (with two sets of chromosomes) and polyploid (with more than two set of chromosomes). Polyploids are of two types, namely, autopolyploids and allopolyploids.

Allopolyploids are of great significance in evolution of new species. In aneuploidy, variation in only parts of a chromosome sets takes place. Aneuploidy may be hypoploidy (loss of chromosome) or hyperploidy (addition of chromosome). Aneuploidy is more lethal to the organism than euploidy.

Variation in chromosome structure is of two types – Intrachromosomal variation and Interchromosomal variation Intrachromosomal variation takes place only in one and the same chromosome. It may be deficiency, duplication or inversion. Interchromosomal variation occurs among nonhomologous chromosomes. It is also called as translocation. It may be homozygotic translocation or heterozygotic translocation.

EXERCISE

1. Write a short note on
   (i) Down's syndrome
   (ii) Edward's syndrome
   (iii) Klinefelter syndrome
   (iv) Allopolyploidy
   (v) Triplo-X
   (vi) Cri du chat syndrome
2. What is chromosomal aberration? Write an account of structural changes in chromosomes.
3. What is a genome? Write a note on polyploidy.
4. With suitable example write numerical variation in chromosome.
5. With suitable example explain deficiency.
6. What is translocation? Explain its types.
7. What is chromosomal non-disjunction? Explain its effects on organisms.
8. What are syndromes? Describe defects associated with Down Syndromes and explain how does Down Syndrome develop?
9. Describe different types and causes of chromosomal diseases in man.
10. Briefly write the genetic basis of following disorders in man.
    (i) Down's Syndrome
    (ii) Cri du chat Syndrome
    (iii) Turner's Syndrome

18

Mutation

The replication and distribution of genetic material is extremely precise so that the genetic material or information is usually passed on from one generation to the next generation without change. But occasionally, errors do occur both during replication and distribution of the genetic material, giving rise to sudden heritable changes in the characters of organisms. Such alterations are called *Mutation.* The individual organisms showing these changes are known as *Mutants.* The allele producing the changed phenotype is called *mutant allele.*

Definition: Mutation is a random, unidirected, heritable variation caused by an alteration in the nucleotide sequence at some point of the DNA of the cell. It may be due to addition, deletion or substitution of one or more bases. An alteration in quality or quantity of genetic material of an organism is called Mutation.

A mutation arising due to a change in the base sequence of a gene is called *gene mutation* or *point mutation.* But changes in chromosome number or structure also produce heritable changes in phenotype, which are termed as *chromosomal mutations* or aberrations.

In gene mutation, there is a change in the normal base sequence of the DNA molecule. This results in a change in the structural characteristics or enzymatic capacity of the individuals. The unit of gene mutation is muton.

Mutations which occur under natural conditions are called spontaneous mutations. Mutation may also be brought about in the laboratory by exposure to a variety of agents. Such agents which bring mutation are called *mutagens* and mutations are called *induced mutation.* The mutagens are broadly classified into two types –

1. Physical mutagens
2. Chemical mutagens

There is no difference in the nature of spontaneous and induced mutations.

The main characteristics of mutations may be as follows:

1. Generally mutant alleles are recessive to their wild type or normal alleles.
2. Mutations are random events in terms of the time of their occurrence and the gene in which they occur.
3. Most mutations have harmful effects, but very less mutations are beneficial.
4. Mutations are recurrent, i.e. the same mutation may be expected to occur in different individuals of a given generation or those belonging to different generations.
5. The rate of spontaneous mutations for most of the genes is very low, ranging between 10^{-7} to 10^{-4}.
6. The rate of spontaneous mutations varies considerably from one gene to another. Some genes are very stable, e.g. Waxy Locus of maize, while some genes show a relatively high mutation rate, e.g. yellow body locus of Drosophilla.
7. Some genes show exceptionally high rates of spontaneous mutations and are called *mutable genes.*
8. Some genes increase the spontaneous mutation rates of some other genes of the genome, called *Mutator genes.*
9. Some genes suppress the mutation of other genes of the genome, and are called *antimutator genes.*
10. Some sites within genes show very high rates of mutation, while some other sites show little or no mutation. The highly mutable sites within a gene are commonly knows as *hot spots.*
11. Mutagens (physical or chemical) increases the frequency of mutations.
12. The rate of induced mutation varies considerably from one gene to other.
13. Mutations can occur in any tissue/cell (either somatic or germinal) of an organism.
14. Mutation can occur during the developmental stage of organisms.
16. The frequency of induced mutations is markedly affected by the environmental conditions prevailing just prior to, during and after the mutagen treatment.
15. Most mutant alleles are pleiotropic.
16. Mutations occur in both forward (from wild type allele to mutant allele) and reverse (from mutant allele to wild type allele) directions.
17. Generally, the rate of forward mutations is much higher than those of reverse mutations.
18. Some mutant alleles do not mutate back, i.e. do not show reverse mutation. Such mutant alleles are believed to be produced by deletions.
19. Several different mutations may occur in a single gene giving rise to multiple allelic serise at such loci.

CLASSIFICATION OF MUTATIONS

Mutations can be classified in several ways. Some of the common ones are as follows –

A. The direction of mutation

On the basis of direction of the mutational event, it is classified into – forward and reverse mutations. A mutation from the wild type allele to a mutant allele is called *forward mutation,* while mutation from mutant allele to the normal allele is known as *reverse mutation.*

B. The cause of mutation

On the basis of cause of their origin, mutations are grouped into spontaneous and induced. Spontaneous mutations occur naturally without any apparent cause, while *induced mutations* originate in response to a treatment with some mutagen.

C. Tissue of origin

On the basis of tissues in which they occur, mutations may be divided into somatic and germinal. A mutation occurring in a somatic cell is called *somatic mutation.* On the other hand, *germinal mutations* occur in reproductive tissues, i.e. cells that would produce gametes.

D. Effect on Survival

On the basis of their effect on survival, mutations are classified into – lethal, sublethal, sub vital, vital and super vital.

(i) *Lethal mutations* result in the death of the cell or organisms in which they occur.

(ii) *Sublethal Mutations* do not lead to the death of all the individuals that carry them, but they cause the death of more than 90% of the individuals carrying them.

(iii) *Sub vital mutations* reduce the chances of survival of the organism in which they occur.

(iv) *Vital mutations* do not affect the survival of the individual in which they occur. These mutations neither enhance nor reduce the viability of the individuals carrying them.

(v) *Super vital* mutations enhance the survival of those individuals that carry them as compared to wild type individuals. It results in the improvement of biological fitness under certain conditions.

E. Cytological Basis

Depending on their cytological basis, mutations are called chromosomal, gene or cytoplasmic mutations. *Chromosomal mutations* are associated with detectable changes in either chromosome number or structure, while mutations produced by alterations in the base sequences of genes are known as *gene mutations* (*point mutations*). Mutations associated with changes in mitochondrial or chloroplast DNA are called *cytoplasmic mutations.* Cytoplasmic mutations show typical cytoplasmic inheritance.

F. Molecular Basis

On the basis of molecular changes associated with them, gene mutations are subdivided into – Base substitution and Frame shift mutations.

(i) *Base substitution:* It is the most common mutation. When a single base in a DNA molecule is replaced by another base, it is called base substitution. Generally it occurs during replication or repair of DNA. In base substitution, if a purine base is replaced by another purine base (A by G or G by A) or pyrimidine base is replaced by another pyrimidine base (T by C or C by T), the substitution is called *transition.* But, when purine base is substituted by a pyrimidine base and vice versa, the substitution is called *transversion.*

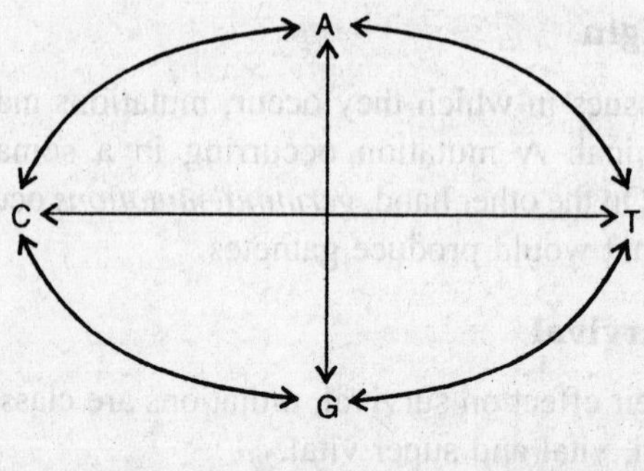

Fig. 18.1. Transition (Light lines) & transversion (Dark lines)

Base substitution usually results in a simple change of one amino acid in a polypeptide chain synthesized.

e.g. 1. Original message of reading frame

CAT ATC GAT CAT ATC GAT CAT ....

2. Substitution or replacement (A is replaced by G)

CAT ATC GAT CGA ATC GAT CAT ....

(ii) *Frame shift mutations:* Generally mutation caused by deletion or insertion of one or a few nucleotide is called *frame shift mutation.* This name is derived from the fact that there is a shift in the reading frame backward or forward by one or two nucleotides. Addition or deletion of one or a few nucleotides or bases results in a new sequence of codons, which may code for entirely different amino acid. This results in a drastic change in the protein synthesised and the resulting protein is usually *non functional.* If the addition or deletion of nucleotides or bases is three, or multiple of three, the frame of resulting protein will not shift but the resulting protein is normal, i.e. it may lack one or few amino acids or may contain one or few extra amino acids.

In frame shift mutation, the site of mutation has an important fact on whether the protein formed will be slightly or drastically altered. Because translation takes place in the 5' → 3' direction, a frameshift mutation near the 3' end of the gene results only in change in the terminal part of the polypeptide chain. This may result in a functional protein. The several variants of haemoglobin are believed to arise in this manner.

e.g. 1. *Original message of reading frame*

CAT ATC GAT CAT ATC GAT .....

2. Deletion of – G

CAT ATC **ATC ATA TCG AT** ......

3. Addition of + G

CAT ATC **GGA TCA TAT CGA** T .....

If there is simultaneous deletion and addition of a base, then the message will be changing in between deletion and addition while remaining codons remain unchanged.

4. Addition of + G and deletion of – C

CAT ATC **GGA TCA TAT** GAT ......
+G − C

G. Types of amino acid replacement in the polypeptide

Depending on the type of amino acid replacements they produce in the polypeptide chains, the mutations are called

(1) *Missense mutation:* It is generally produced by base substitution. Missense mutation affects the replacement of a single amino acid by another amino acid in a polypeptide chain synthesised.

(2) *Nonsense mutation:* If a single base substitution modifies a sense codon to termination codon, the chain may terminate at a point of mutation (e.g. CAG → UAG). Such mutation is called *nonsense mutation* or *chain terminating mutations.* It results in incomplete polypeptide chain which may not be biologically active.

(3) *Mutation in termination codon:* Mutations, which are reverse of nonsense mutation also occur. This mutation can convert a termination codon to a sense codon (codon which specify some amino acid). e.g. The α-chain of human hemoglobin is normally 141 amino acid long. A mutation (U → C) converts the termination codon UAA to CAA, the codon for glutamine. Therefore, α-chain synthesis proceeds beyond the normal termination point, producing a polypeptide chain containing 172 amino acids.

(4) *Silent mutation:* Any gene mutation which does not result in phenotypic expression is called a silent mutation. Silent mutation are due to (i) The

genetic codon is degenerative, i.e. more than one codon may specify an amino acid, e.g. AAA and AAG both specify lysine.

(ii) The change in codon may result in amino acid substitution but this is not sufficient to modify the function of the protein appreciably.

(iii) The mutation may occur in a gene, i.e. no longer functional or whose protein is not essential at the particular stage of testing.

(iv) Simultaneous presence of *suppressor mutations* may cause a mutation to become silent.

(v) *Frame shift mutation:* Frame shift mutations are produced due to deletion or addition of a number of bases which are not divisible by three.

Induced mutations

A variety of agents increases the frequency of mutation. Such agents are called *mutagens.* The different mutagens may be classified into the two broad groups (i) physical mutagens – different types of radiation and (ii) chemical mutagens – a variety of chemical compounds.

(A) *Physical mutagens:* Physical mutagens are different types of radiations having mutagenic properties. The energy content of a radiation depends upon its wavelength i.e. shorter the wavelength, the greater the energy value of the radiation. High energy radiations can change the atomic structure of a substance and bring about excitation of the atom. Radiation which brings about such a state is called *ionizing radiation.* e.g. α-rays, β-rays, γ-rays, X-rays, etc. Ionizing radiation has greater penetration power than non-ionizing radiation. Ionizing radiation causes single strand breaks in DNA and produces deletions.

Ultraviolet rays are the only non-ionizing rays with mutagenic properties. The primary mutagenic effect of uv-light is due to the production of pyrimidine dimers and pyrimidine hydrates. In pyrimidine dimers – thymine dimers are most commonly formed, while cytasine dimers and cytosine-thymine dimers are also produced. Formation of pyrimidine dimer involves the disruption of H-bonds of DNA double helix and causes distortion of the molecule. If the damage is not repaired, replication is blocked, leading to lethal effects.

(B) *Chemical mutagens:* The first chemical mutagen discovered was mustard gas. A large number of chemicals is reported to be mutagenic. Many agricultural chemicals, e.g. several insecticides, fungicides, herbicides, etc. have been found to induce mutations. Chemical mutagens can be divided according to the way in which they bring about mutations – (1) Base analogues, (2) Alkylating agents, (3) acridine dyes, (4) deamination agents and (5) other mutagenic chemicals.

(1) *Base analogues:* Base analogues are very similar to the bases normally found in DNA and, as a result, are incorporated into DNA during replication instead of normal base. e.g. 5-bromouracil, 5-chlorouracil, 5-iodouracil, 2-aminopurine and 2,6-diaminopurine.

Generally uracil can base pairs with adenine, but 5-bromouracil can also make bonds with Guanine and thus normal A = U base pair changes into G = C base pairs.

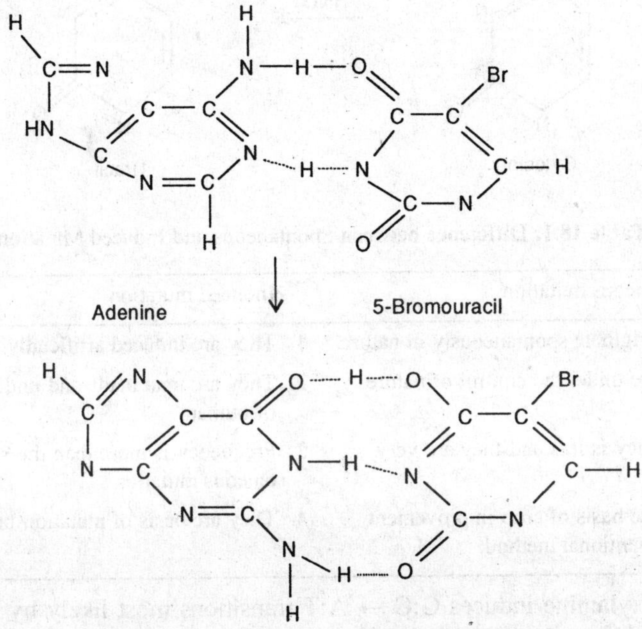

(2) *Alkylating agents:* e.g. Mustard gas, sulphur mustard, nitrogen mustard, ethyl-methane sulphonate, methylmethane sulphonate, N-methyl-N′-nitro-N′-nitrosoguanidine.

Alkylating agents have reactive alkyl groups which they transfer to DNA bases and to phosphate groups of DNA. Alkylating agents produce a variety of genetic effects, namely transition, transversions, frame shift mutations and chromosome breakage and rearrangements.

(3) *Acridine dyes:* These are also known as agents producing distortions in DNA. Acridine dyes such as acriflavin, proflavin and acridine orange cause mutations by insertion or deletion of one to few base pairs, which produce frame shift mutations.

(4) *Deamination Agents:* It is a potent mutagen which acts both on replicating and nonreplicating DNA. e.g. Nitrous oxide, hydroxylamine and alkylating agents.

Nitrous oxide reacts with bases containing amino groups and change the structure of such base by deamination (removal of amino group). For example, deamination of cytosine forms uracil. Therefore, instead of C ≡ G pairing, there is U = A pairing.

(5) *Other chemical mutagens:* A number of other chemicals posses mutagenic properties e.g. Hydroxylamine, sodium azide, etc.

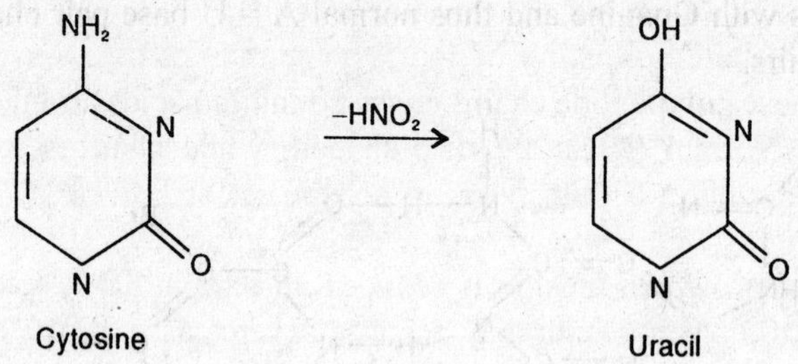

Table 18.1: Difference between Spontaneous and Induced Mutation

Spontaneous mutation	Induced mutation
1. They originate spontaneously in nature.	1. They are induced artificially.
2. They are under the control of nature.	2. They are man made and under control of human.
3. Frequency is less and they are very slow.	3. Frequency is more than the spontaneous mutation.
4. They are basis of crop improvement by conventional method.	4. They are basis of mutation breeding.

Hydroxylamine induces G:C → A:T transitions most likely by inducing hydroxylation of the amino group of cytosine. This produces hydroxylaminocytosine which base-pairs with adenine in the place of guanine.

Some examples of Mutation for the benefit of human being

Mutation is the ultimate source of all genetic variation; it provides the raw material for evolution. Without mutation, all genes would exist in only one form. Alleles would not exist and thus genetic analysis would not be possible. Most important, organisms would not be able to evolve and adapt to environmental conditions. Mutation thus is an important phenomenon in which some level of mutation is essential to provide new genetic variability to allow organisms to adapt to new environments. At the same time, if mutations occurred too frequently, they would totally disrupt the transmission of genetic information from generation to generation. Also, some mutations have negative effects on the organisms, like they produce different disease conditions sometimes lethal to that organism. The examples of some harmful mutations in human beings are as follows:

e.g. 1. There are some human genetic disorders in which the cells are highly sensitive to DNA damaging agents and the patients are prone to cancer. Many of these disorders follow the autosomal recessive mode of inheritance like *Xeroderma pigmentosum* which shows deficiency in cells to repair DNA damage by uv-rays.

e.g. 2. A change in only one amino acid can have a drastic effect on the phenotype. For example, Haemoglobin which is found in the R.B.Cs is a protein molecule consisting of four polypeptide chains – two-α and two-β chains. These polypeptide chains consist of amino acids arranged in a definite sequence. Normal RBCs are disc shaped, while changes in haemoglobin structure results in certain types of anaemia called *sickle cell anaemia* and *Haemoglobin-D anaemia.* In sickle cell anaemia, the RBCs become sickle shaped when oxygen tension is reduced and are much less effective in the transport of oxygen. Death may occur in severe cases of sickle cell anaemia. Normal haemoglobin has the amino acid *glutamic* acid in the 6th position, while in the haemoglobin of sickle cell anaemia it is substituted by valine and in hemoglobin-D – anaemia it is substituted by lysine.

	1	2	3	4	5	6	7	8
Normal Hb	valine	histidine	leucine	thronine	proline	glutamic acid	glutamic acid	lysine
Sickle cell anaemia	valine	histidine	leucine	thromine	proline	valine	glutamic acid	lysine
Hb-D anaemia	valine	histidine	leucine	thromine	proline	lysine	glutamic acid	lysine

(A) *Beneficial mutations in plants:* Even though most mutations makes the organisms less efficient and are thus harmful, the possibility of developing new desirable traits through induced mutations has important for many plant breeders. Plant breeders have reported induced mutants in barley wheat, oats, soyabeans, tomatoes and many fruit trees, that may improve presently cultivated strains.

e.g. 1. Barley plant mutants have been obtained that provide increased yield, resistant to smut, stiff straw, increased protein content and hull less seed.

e.g. 2. Wheat mutant – Sharbati Sonora – have been providing increased yield and dwarf plant.

e.g. 3. Rice mutant – P-500-23 – have been providing high yield. Other examples are Navel orange (from Africa), Golden delicious apples, Emperor Seedless grapes, etc.

(B) *Beneficial mutations in microorganisms:* One application of induced mutations come from concentrated efforts to improve the yield of *penicillin* by the mould *penicillium*. When penicillin was first discovered, the yield was low and production was seriously limited. The millions of spores are irradicated and a few of the surviving colonies produced considerably more amount of penicillin than the average. Such mutant overproducers of penicillin have proven invaluable in the commercial production of this important antibiotic.

Bendle and Tatum used the nutritional mutants of *Neurospora crossa* to discover conjugation in microbes.

In 1946 Luria and Delbrueck noticed that when E. coli cells were infected with the bacteriophage T_1, most of the cells lysed releasing progenies of newly synthesized viruses. However, some bacteria cells did not lyse but multiplied and formed colonies. It is observed that these bacterial cells have become mutant forms which are resistant to phage.

(C) *High Resolution Dissection of Biological process via mutation analysis:* Mutation has been used extensively to elucidate the pathways by which biological processes occur. Metabolism occurs via sequence of enzyme catalysed reactions. By isolating and studying mutation in the genes coding for that particular enzyme involved, the sequence of steps in a pathway can often be determined. Because an appropriate mutation will eliminate the activity of a single polypeptide, mutations provide an extremely powerful method with which dissection of biological processes is possible.

For e.g.

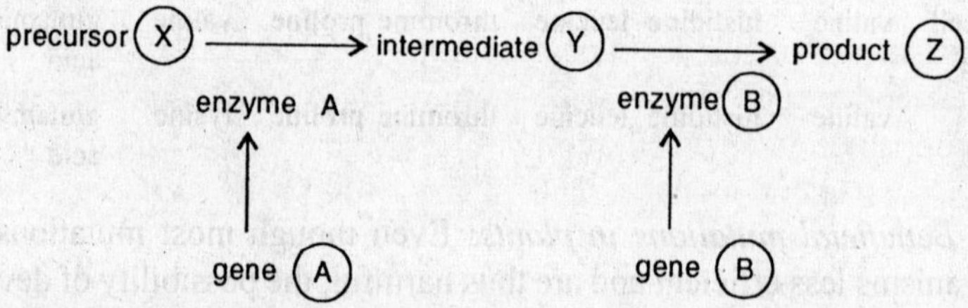

Intermediate **Y** is produced from precursor **X**, by the action of enzyme **A** [product of gene **A**] Intermediate **Y** is rapidly converted to product **Z** by enzyme **B** [product of gene **B**].

So intermediate **Y** may be present in very low quantities and be very difficult to identify biochemically. However mutant organisms that have mutation in gene **B** will synthesise either inactive form of enzyme **B** or no enzyme **B**; therefore intermediate **Y** will often accumulate to much higher concentration, facilitating its isolation and identification. Similarly, a mutant in gene **A** may be useful in the identification of precursor **X**.

Furthermore, the functions and mechanisms of action of individual gene products can often be identified by comparative biochemical and biophysical analysis of mutant and wild type organisms.

By using mutant organisms one can elucidate the pathway of morphogenesis, e.g. Bacteriophage T_4 pathway of morphogenesis.

Other biological processes that have been successfully dissected by mutational analysis are:

(i) The photosynthetic electron transport chain in *chlamydomonas reinbardi and maize.*

(ii) Nitrogen fixation in bacteria

(iii) S. Benzer and colleagues used mutant varieties to dissect behaviour and learning in Drosophila.

(iv) It is also used for understanding differentiation and development in higher plants and animals.

Theoretically, mutational dissection should be applicable to any process, i.e. under genetic control.

(D) *Beneficial mutations in animals:* A classical example of beneficial mutations in animals is that of *short legged sheep.* In 1791 a farmer named Seth Wright in New England, noticed a short legged lamb which could not jump over the fence and run away like all other sheep in his small flock. Thinking this may be an advantage, he started breeding work on the short legged lamb.

The first short legged lamb arose as a result of a recessive mutation in one of its recent ancestors. The breeding work results indicate that this lamb was homozygous for the mutant gene.

SUMMARY

Mutation is a random, unidirectional, heritable variation caused by an alteration in the nucleotide sequence at some point of DNA of the cell. It may be due to deletion, addition or substitution of one or more bases. Mutation is of two types – spontaneous mutation and induced mutation. Induced mutation is caused by some mutagens, which may be physical mutagens or chemical mutagens. The mutations show different effects on the survival. It may be lethal, sublethal, sub vital, vital or super vital. At molecular level, the mutation is base substitution mutation or frame shift mutations.

Some mutations have beneficial effects, while others have negative effects on the organisms. Mutation is the conventional method of crop improvement.

EXERCISE

1. In how many different ways can a base in DNA be altered to produce a mutation. Give one example of each.
2. Classify the mutations on the basis of its effect on survival.
3. Describe the molecular basis of mutations.
4. What is induced mutation? Write a note on physical mutagens.
5. Differentiate between
   (i) Spontaneous mutation and induced mutation
   (ii) Physical and chemical mutagens.
   (iii) Transition and transversion.
6. Write a note on significance of mutations. Give some example of mutations for the benefit of human.
7. What is mutation? Classify the mutations on the basis of types of amino acid replacement in the polypeptide.
8. Write a short note on the following:
   (i) Nonsense mutation

(ii) Mutagens
(iii) Hot spots
(iv) Spontaneous mutations
(v) Germinal mutations
(vi) Base mutations
(vii) Missense mutations
(viii) Base analogues
(ix) Deamination Agents
(x) Point mutations.

19

Microbial Genetics

Bacteria is a prokaryotic type of cell in which the nuclear membrane is absent around the genome. Bacterial chromosome is a double stranded, circular, large single DNA molecule.

For many years it was believed that bacteria lacked sexuality and even the genetic material was not reported. The work of Bendle (1945) on Neurospora, Lederberg (1946) on bacteria, Lindergren (1953) on Yeast, Hastman and Demerce on Bacteriophage have clearly demonstrated the sexuality in bacteria and other microbes. Bacterial cells reproduce sexually during which transfer of DNA from donor to recipient cell occurs. There are three different ways of transfer of genetic material (DNA) in bacteria

- Transformation
- Transduction
- Conjugation

A. TRANSFORMATION

Griffith discovered bacterial transformation first time in 1928. He was working on *Diplococcus pneumonae*, a causative agent for pneumonia. He found two strains of bacteria which differ morphologically from each other. (i) Wild type of strain with smooth cell wall surface due to presence of capsule and are virulent, i.e. pathogenic strain. (ii) The other mutant strain was rough due to lack of capsule and a virulent, i.e. non-pathogenic. When Griffith:

1. Injected virulent strain into mice, all mice died due to severe pneumonia.
2. Injected avirulent strain which was harmless, the mice were not affected, i.e. they survived.
3. Injected heat killed virulent strain into mice, it was found that it failed to kill the mice, i.e. all mice survived.
4. When mixture of heat killed virulent strain and avirulent strain was injected into mice, all mice died due to pneumonia.

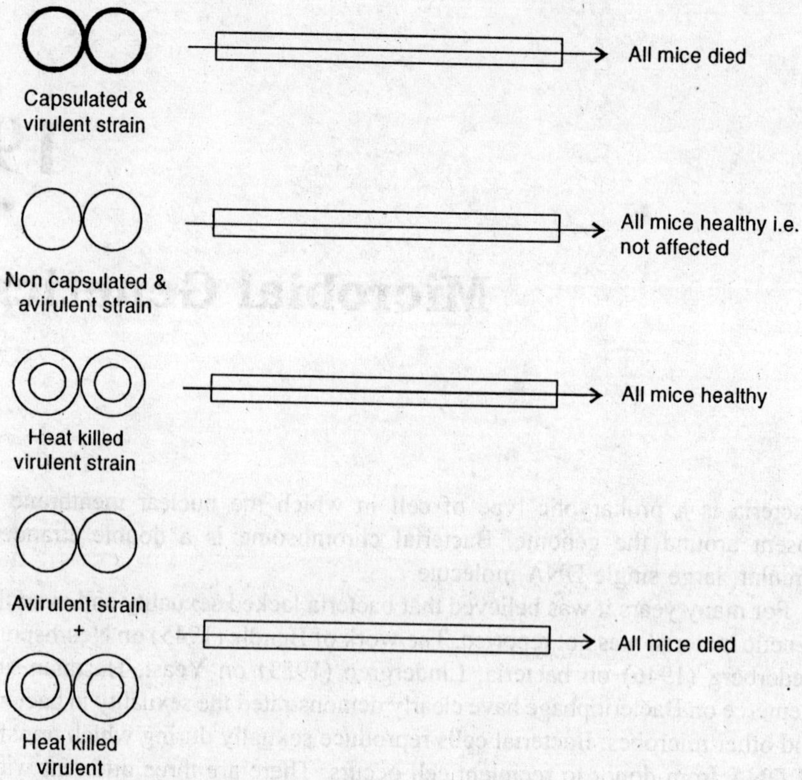

Fig. 19.1. Griffith's experiment

From the above experiment, Griffith concluded that in the mixture avirulent strain received something from heat killed virulent strain and acquired ability to produce capsule and became virulent. He considered that the transforming substance was protein. Griffith called this phenomenon as *transformation*.

In 1944 Avery Macleoid and McCarty extracted DNA, Poly-saccharide and proteins from heat killed virulent bacterial cell in pure form. They added each biomolecular separately to the culture media in which avirulent bacteria were growing. It was found that only DNA brought about transformation and not protein or poly-saccharides. After receiving DNA from heat killed virulent strain, avirulent strain started synthesizing capsule and became virulent. Polysaccharides and proteins did not affect avirulent strain.

Protein + avirulent strain $\longrightarrow$ avirulent strain

Polysaccharide + avirulent strain $\longrightarrow$ avirulent strain

DNA + avirulent strain $\longrightarrow$ virulent strain

B. TRANSDUCTION

This is one of the methods of transfer of genetic material (DNA). In 1952 it was discovered by *Zinder and Lederberg* in *Salmonella* species. Later on it was observed in many bacterial species.

Transduction is the method of DNA transfer from one bacterial cell (Donor) to other bacterial cell (recipient) by bacterial virus, i.e. bacteriophage. Transduction is of two types – generalised transduction and specialised transduction. In generalised transduction bacteriophages can transfer any bacterial gene, while in specialised transduction, the particular bacteriophage transduces only specific genes.

Important steps during transduction

1. Bacteriophage injects phage DNA into bacterium (Donor cell).
2. Phage DNA replicates rapidly within bacterial cell.
3. Viral DNA directs the synthesis of new phage protein coats/capsids.
4. Bacterial DNA also breaks into small pieces and is incorporated into phage DNA.

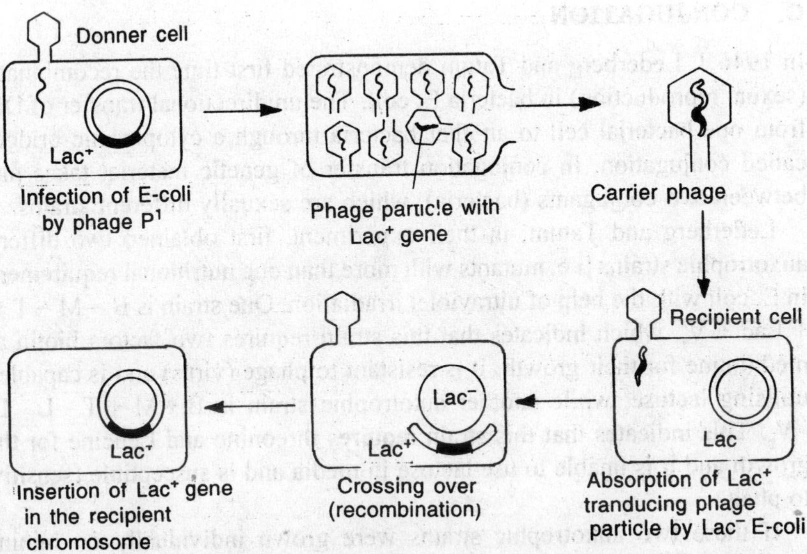

Fig. 19.2. Schematic representation of transduction

5. Within 10 to 20 minutes new DNA and new protein coat combine and new viral particles synthesise, i.e. in single bacterial cell, number of viral particles are synthesized.
6. These phage particles are released by destruction of bacterial cell (Lysis of bacterial cell).
7. Then the particles infect the other bacterial cell (recipient). The DNA fragment of donor enters into recipient bacterial cell and joins with recipient cell DNA.

Difference between Transduction and Transformation

Transduction	Transformation
1. Transfer of genetic material takes place between two living cells.	1. Transfer of genetic material takes place from dead cell to living cell.
2. DNA from one bacterial cell is carried to other bacterial cell by bacteriophage.	2. DNA transfer does not require any transferring agent. It takes place due to cell to cell contact, i.e. bacteriophage are not involved.
3. Donor bacterial cell provides a site for replication of phage DNA; therefore, hundreds of phages are synthesised within a single donor cell.	3. There is no transferring agent and therefore no site is made available for replication of DNA.
4. Newly synthesised phages are released due to destruction of cell.	4. Lysis of donor cell is absent because donor cell is already dead.
5. Transduction is of two types – generalised and specialised transduction.	5. Transformation is of only one type.

C. CONJUGATION

In 1946 J. Lederberg and Tatum demonstrated first time the recombination (sexual reproduction) in bacteria E. coli. The unidirectional transfer of DNA from one bacterial cell to another bacteria through a cytoplasmic bridge is called conjugation. In conjugation transfer of genetic material takes place between two conjugants (bacteria), which are sexually different strains.

Lederberg and Tatum, in their experiment, first obtained two different auxotrophic strains [i.e. mutants with more than one nutritional requirements] in E. coli with the help of ultraviolet irradiation. One strain is B – M – T + L + Lac + V_r, which indicates that this strain requires two factors biotin and methionine for their growth. It is resistant to phage (virus) and is capable of utilising lactose, while another autotrophic strain is B + M + T – L – Lac –V_s. This indicates that this strain requires threonine and Leucine for their growth and it is unable to use lactose in media and is susceptible (sensitive) to phage.

If these two auxotrophic strains were grown individually on minimal medium [medium which lack any specific growth factor], they were unable to grow. Thus individually both strains were unable to grow in minimal medium. But if a mixture of these two auxotrophic strains was cultured on the same minimal medium, the *prototroph colonies* (organisms which grow on minimal medium) appeared in absence of growth factors biotin, methionine, threonine and Leucine. This suggests that development of recombinant occurs with genotype B + M + T + L + When these colonies were tested for lac operon and phage resistance, a new combination appeared.

In bacteria, the genetic material is called nucleoid (not enclosed within the nuclear membrane). Nucleoid consists of single double stranded large circular

DNA molecule (chromosome). In addition to main chromosome, the bacterial cell contains one or more small circular DNA molecules. These extra chromosomal DNA molecules are called *plasmids*. Only a few plasmids are involved in the conjugation, called *conjugative plasmid* or *sex factors* or *fertility factor* or *F-factor* or *episome*.

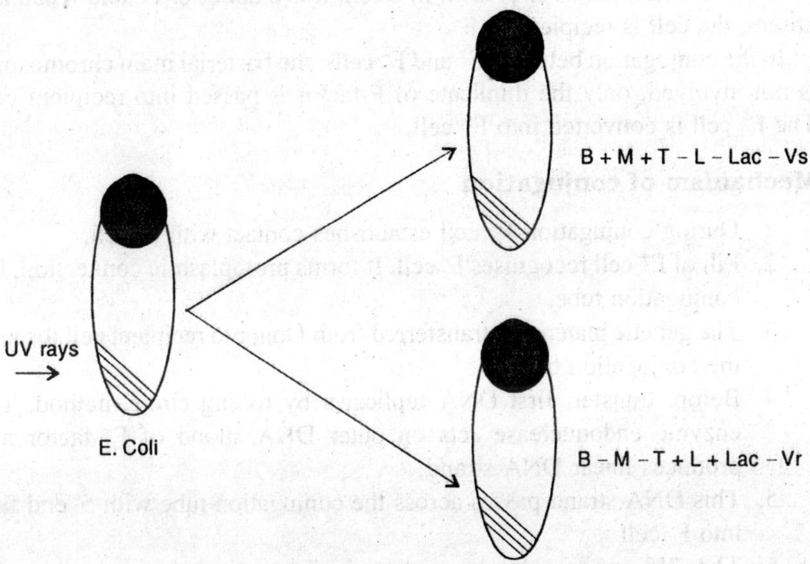

Fig. 19.3. Isolation of auxotrophic strains

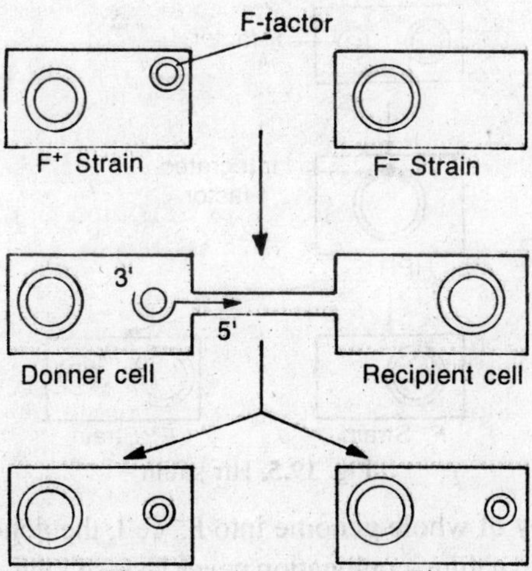

Fig. 19.4. Bacterial conjugation

F-factors: F-factors are circular pieces of double stranded DNA (in addition to main chromosome, i.e. extrachromosomal). They contain about 2% of the total DNA and are attached to cell membrane. F-factor of E. coli contains 15 genes, in which 8 genes control the formation of *Sex pili* or *F-pili*. F-factor may be free in cytoplasm or it is sometimes integrated with bacterial chromosome.

The sex difference between two conjugants is determined by presence of F-factor. When F-factor is present in a cell, it is a donor or F^+ and when it is absent, the cell is recipient or F^-.

In the conjugation between F^+ and F^- cells, the bacterial main chromosome is not involved, only the duplicate of F-factor is passed into recipient cell. The F^- cell is converted into F^+ cell.

Mechanism of conjugation

1. During conjugation F^+ cell establishes contact with F^- cell.
2. Pili of F^+ cell recognises F^- cell. It forms protoplasmic connection, i.e. conjugation tube.
3. The genetic material is transferred from Donor to recipient cell through the conjugation tube.
4. Before transfer, first DNA replicates by rolling circle method. The enzyme endonuclease acts on outer DNA strand of F^+ factor and produces linear DNA strand.
5. This DNA strand passes across the conjugation tube with 5' end first into F^- cell
6. The 2nd inner strand remains in F^+ cell and synthesises its complementary strand
7. The two conjugants separate after mating.

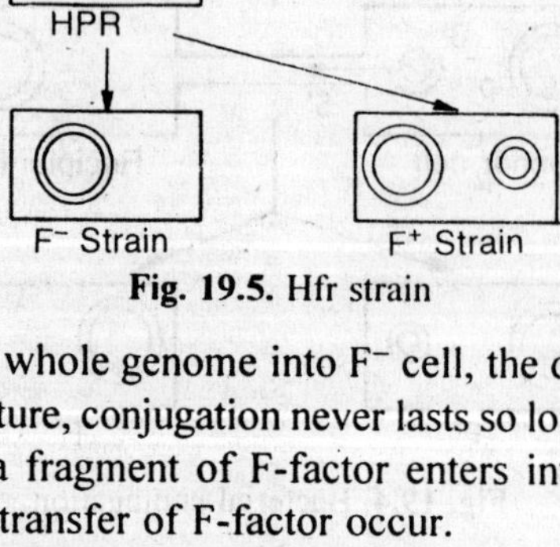

Fig. 19.5. Hfr strain

To inject a copy of whole genome into F^- cell, the donor cell requires 90-100 minutes. But in nature, conjugation never lasts so long. Due to interruption in conjugation, only a fragment of F-factor enters into recipient cell; very rarely does the entire transfer of F-factor occur.

HFr or High frequency recombination: When F-factor is integrated with bacterial chromosome, the bacterial cell is called as Hfr. This Hfr strain undergoes sexual recombination (conjugation) with F^- cells at very high speed around 10^{+3} faster than $F^+ \times F^-$ conjugation. However, it losses ability to convert F^- cells to F^+ cells. Hfr can be converted into F^+ by detachment of F-factor during recombination.

Isolation of auxotrophs

The efficiency of isolation of auxotrophs will depend largely on the availability of techniques for an easy and rapid scoring of the mutant phenotypes in very large populations. In 1946 Bendle and Tatum discovered a method of detecting nutritional mutants (auxotrophs) in the bread mold *Neurospora crassa*. Wild type Neurospora grows on a minimal medium containing inorganic acids, sugar, nitrogen, a few salt and a vitamin biotin. Bendle and Tatum found nutritional mutants of Neurospora which failed to grow on minimal medium in the absence of a particular nutrient.

The exact requirements of nutritional mutants can be determined by adding a different growth substance to each of a series of different cultures of minimal media. For example a mutant strain of Neurospora deficient in the enzyme tryptophan synthetase would be able to grow on the minimal medium only if tryptophan is added to the medium. This technique has turned out to be applicable to many micro-organisms.

In 1949 B.D. Davis found an elegant way of detecting auxotrophs in E. coli by a penicillin selection method. Certain antibiotics like penicillin inhibit cell division only in growing, actively dividing bacteria. Thus when penicillin is added to a minimal medium containing bacteria, the wild type cells which are growing are killed. The auxotrophs which are not growing are not affected by penicillin and survive. After a certain time period the auxotrophs are separated from the medium by centrifugation, washed free of penicillin and plated on fresh minimal medium supplemented by the necessary growth substance. The auxotroph will divide and form pure colony.

Auxotrophs belong to a class of mutants broadly known as conditional mutants; their growth is dependent on the presence of a particular environmental condition. The temperature sensitive mutants fail to grow normally in a certain temperature range. Usually higher temperature affects growth because of defects in thermolability of specific proteins. Temperature sensitive mutants in bacteria can be detected by the technique of replica planting. The replica plate is incubated at or above the maximum temperature which permits growth of the wild type. If the replica colony fails to grow under these conditions, it indicates that it is probably a temperature sensitive mutant.

Replica Plating Technique

In 1952 J. Lederberg and E. Lederberg devised a technique called replica plating, which permits mutants to be demonstrated directly.

The replica plating technique is as follows. Suppose to isolate phage resistant E. coli cell, E. coli cells are grown on a master plate of nutrient agar until they form a continuous lawn of cells. A sample of cells is transferred by pressing on it a sterile velvet surface supported on a circular block (replicator) to a fresh plate of agar medium already coated with phage particles. The method produces an identical pattern of colonies on the second plate. Some mutant phage resistant colonies also grow on this plate.

If the mutant colonies have originated from the master plate, then it would be possible to obtain a culture of such cells by inoculating from locations on the master plate corresponding to position of resistant colonies on the second phage containing plate. Lederberg, therefore, repeated the procedure by preparing a second master plate and replica plating onto a phage coated plate. This was done 3 to 4 times, each time plating cells thinly. In this way, single resistance colonies could be identified and a true breeding phage resistant strain would be developed. The experiment proves that resistant cells arose by mutation on the first master plate and were not induced by the phage to become resistant.

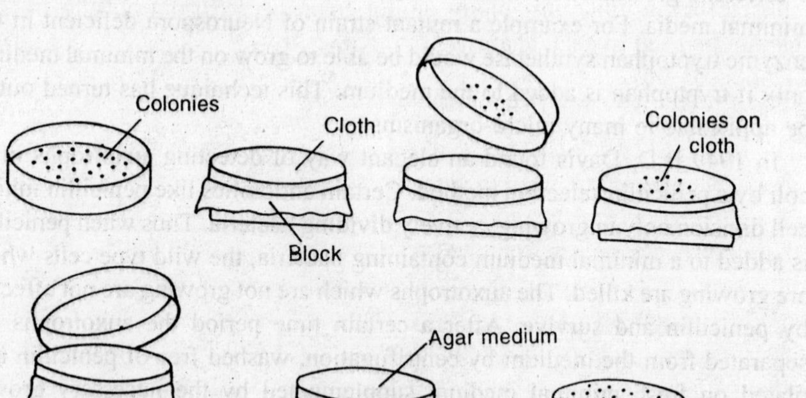

Fig. 19.6. Replica plating technique

Analysis of mutations in biochemical pathways: The reactions involved in the biosynthesis of a compound occur in a precise sequence so that the product of previous reaction serves as substrate for the next reaction. In a biosynthetic pathway, if a particular reaction is blocked, the substrate for that reaction will accumulate, while the products of that reaction and those of all the subsequent reactions will be absent.

For example, if reaction Z is blocked, product Y will accumulate and product Z and A will be absent (Fig, 19.7). In this case, if product Z is supplied to an organism it will be able to produce A and behave as the wild type. However, supply of product Y or the substrate X will not lead to the production of A since both these substances are produced before the blocked reaction Z.

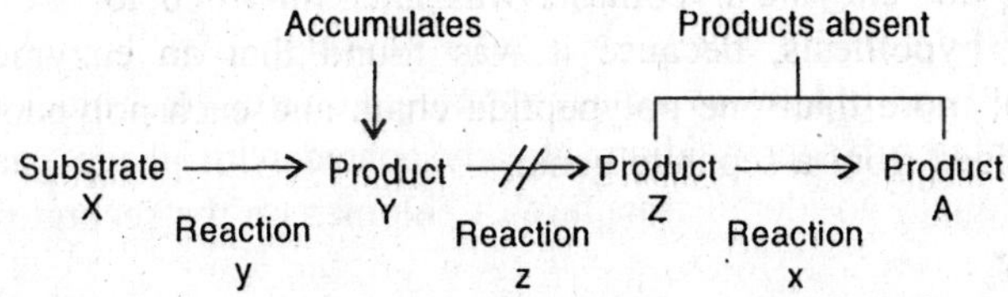

Fig. 19.7. Schematic representation of Biosynthetic pathway showing block in a reaction Z which results in accumulation of product y, while product Z and A are absent.

Thus the precise sequence of intermediate substances produced in a biosynthetic pathway can be established by comparing the substances utilised by different mutant. For example, if mutant P can utilise substances Y, Z and A, while mutant Q can utilise substances Z and A, a substance Y must be produced before Z and A. In this manner, the precise sequence of reactions in a biochemical pathway can be determined. However, some intermediate substances are quite unstable or do not enter the cell from outside and thus cannot be used in such analysis.

One-gene-one-enzyme hypothesis

The relationship between genes and enzymes was first precisely emphasised by Bendle and Tatum in 1946 while working with the mutants of *Neurospora crassa* and proposed one-gene-one-enzyme hypothesis. This hypothesis states that each gene controls a single specific enzyme. One of the first proofs for this hypothesis came from arginine biosynthetic pathway. This pathway consists of eight chemical reactions, each catalysed by a specific enzyme and the presence of each enzyme is controlled by a single distinct gene as specific enzymes are absent in mutants for specific genes.

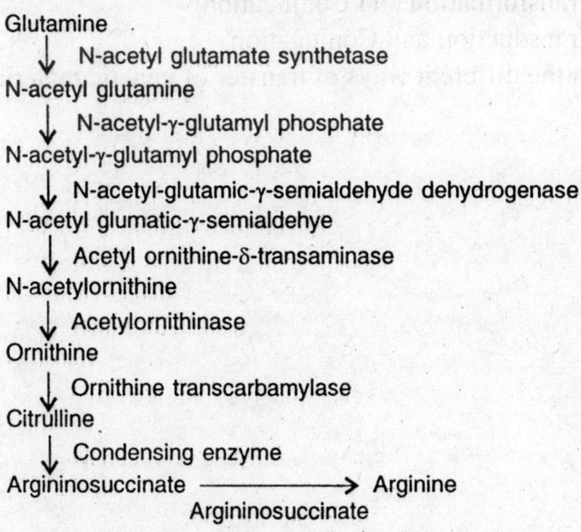

Fig. 19.8. Biosynthesis of arginine in E. coli

One-gene-one-enzyme hypothesis was later modified to one-gene-one-polypeptide hypothesis, because it was found that an enzyme may be composed of more than one polypeptide chain and each polypeptide chain was synthesised from a separate gene.

SUMMARY

In microorganisms, the transfer of genetic material takes place by transformation, transduction or by conjugation. In transformation, the transfer of genetic material takes place from dead cell to living cell, while in transduction transfer of genetic material from one cell to other cell takes place by bacteriophage. In conjugation genetic material is transferred from donor cell to recipient through conjugation tube or sex pili.

To study microbial genetics different techniques are used, like isolation of auxotroph mutant from wild type of cells. In modern days, Replica plating technique is successfully used to isolate mutants/auxotrophs from vast wild cells. Analysis of mutations in biochemical pathways is very useful to emphasise the different intermediate products of biosynthetic pathway.

EXERCISE

1. Write short notes on
   (i) Transduction
   (ii) Transformation
   (iii) Conjugation
   (iv) One-gene-one-enzyme hypothesis
   (v) Replica plating technique
2. Differentiate between
   (i) Transduction and Transformation
   (ii) Transformation and Conjugation
   (iii) Transduction and Conjugation
3. Explain the different ways of transfer of genetic material in bacteria.

20

Extrachromosomal Inheritance

The genes of nuclear chromosomes have different roles in inheritance, cellular metabolism, development and mutation of the organisms in which they occur. Though the genes of nuclear chromosomes have a significant and key role in the inheritance of almost all characters from one generation to other generation, they alone cannot be considered as the sole vehicles of inheritance because certain experimental evidence suggest transmission of some genetic information from one generation to the next generation through the cytoplasm and is called extranuclear or cytoplasmic inheritance. The role of cytoplasm in heredity is determined from results of reciprocal crosses in which the sources of male and female gametes are reversed. In chromosomal inheritance it makes no difference in the transmission of a gene whether it comes from which parent and identical phenotypes are obtained in reciprocal crosses. But if there is extranuclear inheritance, the resulting phenotypes in a reciprocal crosses are nonidentical and indicate uniparental transmission, in most cases from the maternal parent. The bigger role of the maternal parent results from the unequal cytoplasmic contributions of the male and female parents as most of the cytoplasm of the zygote comes from the egg. However, a few cases of paternal transmission are also known. Extrachromosomal inheritance is also known as extranuclear inheritance, non-mendelial inheritance, uniparental inheritance, maternal inheritance or cytoplasmic inheritance.

The traits with extranuclear inheritance patterns are identified by the accumulated evidence from a number of criteria as mentioned below:

(1) Reciprocal crosses show marked difference for the characters governed by cytoplasmic genes (cytogenes or plasmagens).

(2) Whenever a trait fails to demonstrate the classical segregation patterns and deviates from standard ratios.

(3) When the trait fails to show linkage to any known nuclear linkage groups and assorts independently from nuclear genes.
(4) Plasma genes generally show somatic segregation during mitosis, a feature of rare occurrence in the case of nuclear genes.
(5) Several plasma genes have been shown to be associated with chloroplast DNA or mitochondrial DNA, which is evidence for cytoplasmic inheritance.

Extrachromosomal inheritance in eukaryotes

Many geneticists have studied various cases of extrachromosomal inheritance in different eukaryotes. Some of the most important examples of extranuclear inheritance in eukaryotes are as follows.

(a) *Plastid Inheritance:* The inheritance pattern due to plasma genes located in plastids is known as *plastid inheritance.* Plastid inheritance was first discovered by Correns and Baur in 1908. Al! the knowledge about plastid inheritance in higher plants is based on studies on chlorophyll variegation in leaves. Variegation refers to the presence of white or yellow spots of variable size on the green background of leaves.

Correns and Baur found that the four o'clock plant, Mirabilis Jalapa, in contrast to other higher plants, contains 3 types of leaves and parts as follows (i) Full green leaves or branches having chloroplasts (ii) White leaves or branches having no chloroplast and (iii) variegated or mixed leaves or branches with green and white patches. These leaves have leukoplast in white area and chloroplast in green area. As the leucoplasts are incapable of producing food by photosynthesis, the white branches of this plant survive by receiving nourishment from green parts.

Table 20.1 Chloroplast inheritance in four o'clock plants

Leaf phenotype of the branch used as the female parent	Leaf phenotype of the branch used as the male parent	Leaf phenotype of the progeny
Green	Green	Green
	White	Green
	Variegated	Green
White	Green	White
	White	White
	Variegated	White
Variegated	Green	Green, white and variegated
	White	Green, white and Variegated
	Variegated	Green, white and variegated

Corren reported that the flowers on green branches produced only green offspring regardless of the genotype and phenotype of pollen parent. Likewise, flowers from the white branches produced only white offspring regardless of genotype and phenotype of pollen parents. However these offspring die soon because they lack chlorophyll and cannot carry on photosynthesis. The flowers from the variegated branches yielded mixed progeny of green, white and variegated plants in various ratios.

The above experiment clearly shows that the phenotype of progeny is the same as that of the female parent. Except when the female parent is variegated, all three combinations occurs.

(b) *Mitochondrial inheritance:* In fungi Neurospora crassa a number of mutations of mitochondria are inherited via the female parents. But the *poky strain* of Neurospora crassa was the best studied mutant, which was first isolated by Mitchell and Mitchell in 1952. Poky strain grows very slow, forms small colony and exhibits poorly differentiated mitochondria. The mitochondria are deficient in the membrane bound cytochrome a and b, which is essential for respiratory electron transport chain. It also possesses a greatly reduced number of small ribosomal units.

This trait seems to be inherited in certain strains in non-mendelial fashion, indicating its extrachromosomal nature. In a cross between normal female and poky male the resultant ascospores all produce normal colonies. And in reciprocal cross between poky female and normal male parents, all the resultant colonies are poky. This indicates that the poky trait is found to be inherited in a fashion of maternal inheritance.

Mitochonderial mutants are also known in Paramecium, Trypanosoma and yeast (Saccharomyces cerevisiae).

(c) *Maternal effect:* Where the development of some characters in several organisms is either governed or influenced by the genotype of the female parent, it is known as *maternal effect.* Such phenotypic expression of maternal genes in the offspring may be short lived or may persist throughout the life span of the individual. The substances which produce the maternal effects in the progeny are found to be transcriptional products of maternal nuclear genes like mRNA, tRNA, rRNA, which have been produced during oogenesis and which exist in the ooplasm of fertilised eggs in the form of inactive protein coated mRNA molecules (informosomes) or inactive rRNA and tRNA. These maternal genes transcriptional products produce their phenotypic effects during early cleavage and blastulation when there is very little or no transcription because maternal and paternal genes of zygote remain engaged in mitotic replication or duplication of DNA. There may be other reasons of maternal effect which are still little understood.

Some examples of maternal effects are as follows.

Example: The flour moth *Ephestia Kuhniella* has dark brown eyes due to the dominant gene (AA), in which gene A directs the production of pigment precursor called *Kynurenine.* However, when the organism is homozygou for a recessive allele of this gene (aa), Kynurenine is absent, causing red eye colour.

When homozygous recessive (aa) female is crossed with heterozygous (Aa) male, their offspring will appear in the normal genotypic ratio of 1 Aa: 1 aa and phenotypically these offsprings are one brown eyed and one red eyed. But in a cross between heterozygous (Aa) female and homozygous (aa) male, all larvae are brown eyed. However when larvae mature, only half of them are dark brown eyed and other half are red eyed. The explanation for these unexpected dark eyed larvae is simple and is based on the egg cytoplasm contributed by the heterozygous mother, i.e. all eggs of the dark brown female contains Kynurenine no matter what their genotype and therefore begin development as pigmented larvae. However, half the larvae are homozygous recessive (aa) and unable to synthesise further Kynurenine. Such larvae gradually develop into red eyed adults. *This is the short lived maternal effect.*

However, in some cases the maternal effect does not diminish during development but lasts throughout adult life. Such instance occurs when development is started in a particular direction that cannot be reversed. Example: In the snail *Limnaea,* the shell is spirally coiled. The direction of coiling is determined by a single pair of genes. The clockwise coiling or dextral coiling is due to dominant gene 'R', while anticlockwise or sinistral coiling is due to recessive gene 'r'.

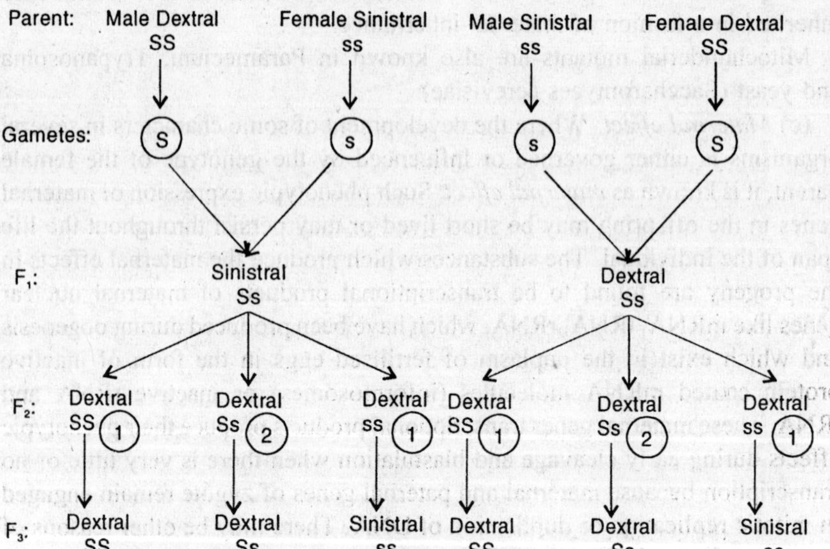

Fig. 20.1. Maternal effect in direction of coiling of shell in Limnaea.

In the reciprocal crosses between these snails, the direction of coiling of the offspring shell is always determined by the genotype of the maternal snail, even when it is completely different from genotype of zygote. Thus self fertilised dextrally coiled heterozygotes (Rr) will produce offspring that always develop into dextrally coiled, even though some fertilized eggs are homozygous recessive (rr) and should be sinistrally coiled (Fig. 20.1). Similarly the products of a self fertilised sinistrally coiled heterorygotes (Rr) will be dextrally coiled even though some eggs are homozygous recessive (rr). These maternal effects last for only one generation.

Among mammals, the maternal effects can be more broadly defined because mammalian mother may affect the development of their offspring not only through the egg cytoplasm but also through the uterine environment. Some examples of maternal effects among mammals are as follows.

(1) Maternal blood group genotype may be the cause of serious defects in fetal development because of Rh incompatibility.

(2) Embryonic defects caused by maternal diabetes and maternal phenyl Ketonuria.

(3) A maternal influence on left and right handness.

SUMMARY

Beside chromosomal inheritance, few characters are inherited extra-chromosomally in many organisms. The extrachromosomal inheritance may be due to extranuclear DNA or due to cytoplasmic factors (maternal effects). Other than nuclear DNA, eukaryotic cell also contains DNA in mitochondria and chloroplast, which play an important role in extrachromosomal inheritance.

EXERCISE

1. What is meant by cytoplasmic inheritance? Explain with suitable examples.
2. Give an account of extranuclear inheritance.
3. How will you distinguish between maternal effect and extranuclear inheritance? Illustrate with suitable examples.
4. Differentiate between nuclear inheritance and cytoplasmic inheritance.
5. Describe one example of a mutation in mitochondria that produces a visible phenotype in a Neurospora.
6. "Extrachromosomal traits are mostly transmitted through the maternal parent." Justify.
7. Write short notes on:
   (a) Plastid inheritance
   (b) Cytoplasmic inheritance
   (c) Mitochondrial inheritance
   (d) Maternal effects
   (e) Poky strain of neurospora

Part C

Microbiology

21

Introduction to Microbiology

Microbiology is the study of living organisms of microscopic size. The word 'Microbiology' is derived from three Greek words – mikros (small), bios (life) and logos (study). Therefore, microbiology deals with study of microorganisms. Microorganisms are very tiny and invisible to the naked eye. They can be looked into and studied only with the help of a microscope.

Microbiology is concerned with the form, structure, reproduction, physiology, metabolism and classification of microorganisms. Microbiology includes study of distribution of microorganisms in nature, their relationship with each other and with other living organisms, their effect on human beings, on animals and on plants. It also includes study of abilities of microorganisms to make physical and chemical changes in our environment and their reaction to physical and chemical reagents.

Since microbiology is not a mere study of the structural diversity and classification of microbes, the various aspects of microbiological studies can be divided basically into the following branches:

1. *Phycology:* It deals with the study of autotrophic eukaryotic microorganisms. e.g. - algae.
2. *Mycology:* It deals with the study of Fungi e.g. - yeasts, molds, mushrooms, etc.
3. *Virology:* It deals with the study of viruses. Viruses are on the border line between living and non-living matters. Viruses cause disease to plants and animals including human beings.
4. *Protozoology:* It deals with the study of protozoans in all their aspects. Protozoans are known to cause many diseases like malaria, amoebic dysentery, sleeping sickness, etc.
5. *Bacteriology:* It deals with the study of bacteria. Bacteria are the largest group among microbes not only in number but also in importance. Bacteria have profound influence on human being, including health, industry, agriculture, etc.

6. *Medical microbiology:* This branch deals with the pathogenic microbes, their life cycle, physiology, genetics, reproduction, etc. Diagnostic procedure for the identification of causative agents and preventive measures are studied in this branch.
7. *Agricultural microbiology:* It deals with the study of role of microbes in agriculture, both harmful and useful. It also includes study of role of microbes in soil fertility, plant and animal diseases, etc.
8. *Industrial microbiology:* In this branch the role of microbes in industrial production is studied. Many microorganisms produce medicinal products like antibiotics and vaccines, fermented beverages, industrial chemicals, production of protein and hormones by genetically engineered microorganisms.
9. *Food microbiology:* In this branch various aspects such as food processing, food preservation, canning, pasteurisation of milk, study of food born microbial diseases and their prevention is studied.
10. *Water microbiology:* Microbiological examination of water, water purification, biological degradation of waste are studied in this branch.
11. *Aero microbiology:* In this branch dispersal of disease causing microbes through air, microbial population in air, their quality and quantity in air are studied.
12. *Environmental microbiology:* In this branch, the role of microbes in maintaining the quality of the environment, microbial influence in degradation and decay of natural, domestic and industrial waste, their role in biochemical cycles are studied.
13. *Geochemical microbiology:* In this branch the role of microbes in coal, mineral and gas formation, prospecting for deposits of coal, oil and gas and recovery of minerals from low grade ores using microbes are studied.
14. *Exomicrobiology:* It includes study of life in outer space.

Microorganisms are closely associated with the health and welfare of human beings. Some microorganisms are beneficial while others are harmful to human being. For example, beneficial activities of microorganisms involve the making of yogurt, cheese and wine, production of penicillin, interferon and alcohol. Microorganisms are also involved in the processing of domestic and industrial waste. The harmful effect of microorganisms on human being includes – some microorganisms can cause diseases, spoil food and deteriorate materials like iron pipes, glass lenses and wood pillar. Many microorganisms are useful to the human being.

Microorganisms are very popular models for studying funda-mental life processes because:

(i) They are smaller in size. e.g. E. coli is 500 times smaller than average plant and animal cell. Therefore, within a small space (petriplates, culture tubes, etc.), a large number of microbes can be grown.

(ii) The rate of cell division of microorganisms is very high under normal laboratory conditions, e.g. some bacteria double their number in every 20 minutes, while mammalian cells divide only once in 24 hours. Therefore within a short time, interval study of many generations is possible.

(iii) Generally microbes can grow as a single cell and can therefore be easily grown in laboratory conditions.

(iv) They are highly adaptable to certain organic substances, which are used as carbon or energy source.

HISTORY OF MICROBIOLOGY

Microbiology began when people learnt to grind lenses from pieces of glass and combine them to produce magnification great enough to enable microbes to be seen.

In 13th century, Roger Bacon (1220-1292) suggested that diseases are produced by invisible creatures. This opinion was supported by Girolamo Fracastoro Verona (1483-1553) and Anton von Plenciz in 1762, but these people had no proof. In 1658 a monk, Athanasius Kicher, referred to *worms* invisible to naked eyes in decaying bodies, meat, milk and diarrhoeal secretions. Kicher was the first person to recognise the significance of bacteria and other microbes in diseases, but his description lacked accuracy.

The most important discovery in the field of microbiology is by Antony Van Leeuwenhoek. He observed tiny microbes in a drop of pond water which he called *animalcules.* Since then started the glorious era of microbiology.

1. Antony Van Leeuwenhoek (1632-1723)

Antony Van Leeuwenhoek, a Dutch cloth merchant, is credited as the discoverer of the microbial world. Leeuwenhoek had a hobby of lens grinding and microscope making. During his lifetime he made more than 250 microscopes, consisting of home ground lenses mounted in brass and silver. The microscopes of Leeuwenhoek could magnify objects about 200-300 times. He was so successful in grinding lenses that he gave up textile business and took to full time lens making.

Leeuwenhoek had little formal education but his keen interest in nature made him examine a variety of materials like hair fibres, plant structures, crystals, insect's eye, a variety of fluids such as pond water, blood, scraping from his own teeth, etc.

For some 50 years Leeuwenhoek kept meticulous records and made accurate drawings of his observations. In a series of letters to the Royal Society of London, he described a variety of microorganisms such as protozoa, algae, yeast and bacteria. His description was so precise that it is now possible to assign to them specific genera without any additional description.

In 1683 he described and sketched different forms of animalcules, i.e. rod, spheres and spiral shapes. This was the first recorded observation of bacterial morphology and even today it holds good for bacteria.

2. Louis Pasteur (1822-1895)

Louis Pasteur, a Professor of organic chemistry in France, may be rightly called the "father of microbiology" due to his enormous contribution to the development of microbiology. In recognition of his path breaking discoveries, a special Institute – The Pasteur Institute was built for him in Paris by public support. His main contributions are:

(i) Tartaric acid, an organic compound, is formed of two types of crystals, which could be separated microscopically.

(ii) *Discovery of microbial fermentation:* Pasteur intensively studied methods and process of fermentation and found that fermentation of fruits and grains resulting in alcohol was brought by microbes. By examining many batches of ferment, he found microbes of different kinds. He observed that in good lots of wine one type of micro-organisms were predominant, while in the poor products another kind of microorganisms were predominant. He suggested that the undesirable types of microbes can be destroyed by heating at 62.8°C for half an hour. This process was called *Pasteurization.* Today pasteurization is widely used in Fermentation Industry, but we are most familiar with it in the Dairy Industry.

(iii) *Disproving the theory of spontaneous generation:* The discovery of microorganisms immediately raised the questioh regarding the origin of microorganisms. It was by then well recognised that living organisms such as plants and animals do not originate spontaneously. But still some people believed that microbes, such minute creatures, needed no parents, i.e. microbes could originate spontaneously. This theory was known as the theory of spontaneous origin.

In order to disprove the spontaneous origin theory, Pasteur invented a flask with long, narrow Swan neck opening. The nutrient solution was heated in the flask and air untreated, unfiltered could pass in or out through this swan neck opening. The result was that the dust particles and microbes in the air settled at the bend and did not reach the nutrient solutions. Thus microbes did not appear in the solution. If the nutrient solution was tipped into the neck and flown back, microbes appeared again. This experiment clearly proved that the microbes cannot generate spontaneously.

(iv) Pasteur also showed that air contained microbes and the number of microorganisms in the atmosphere varied from place to place. To prove this, he took a large number of sealed flasks containing sterile nutrient broth and opened a few flasks at a time for a short period at various places and resealed them. Out of the 20 flasks, which he opened and resealed on a dusty road, 8 showed spoilage, while out of the 20 flasks that he opened on the top of a mountain, only five showed spoilage and out of 20 flasks that he opened near a glacier, only one showed spoilage.

(v) *Discovery of aerobic and anaerobic microbes:* Since animals require oxygen for growth, scientists in the middle of 19th century assumed that microbes also required oxygen. However, during the studies on butyric acid fermentation, Pasteur discovered that microbes can survive in the absence of oxygen. Microscopic observation of the motile butyric acid bacteria showed that the motility of the bacteria stopped when they came in contact with the air. To confirm this, he passed a current of air through the butyric fermentation broth and found that this resulted in complete arrest of the fermentation. Thus he concluded that some microbes could live only in the absence of oxygen. He designated life in the presence of oxygen as *aerobic* and in the absence of oxygen as *anaerobic.* He also found that many other microorganisms, including yeasts could grow either in the presence or in the absence of oxygen. These microbes, he termed as *facultative anaerobes.*

(vi) *Work on silkworm disease:* In 1965, Pasteur was asked to find the cause of *pebrine*, a disease affecting silkworms. He demonstrated that the disease was caused by a protozoa and by choosing microbes free egg for breeding, the disease would be controlled.

(vii) *Protection against infection:* In 1880 Pasteur isolated bacterium responsible for *chicken cholera* and grew it in pure culture. When he inoculated a group of chickens with several week old culture of chickens cholera, he observed that all the chickens remained healthy. After some weeks he repeated the experiment by using two groups of chicken, group A had been inoculated at the previous experiment with old culture, while group B had not previously inoculated. This time he inoculated both A and B group with fresh culture of chicken cholera and found that all the chickens in B group got sick and died but those in A group remained healthy. Pasteur thus concluded that bacteria lose their ability to produce disease (virulence) after standing and growing old. But these attenuated (having decreased virulence) bacteria still retained their capacity for stimulating the host to produce substances, i.e. *antibiotics* that protect host against subsequent exposure to the virulent organisms.

Pasteur also demonstrated that the anthrax disease in sheep was caused by rod shaped bacteria (bacillus) and cultured them in sterilised broth and showed that the inoculation of culture would cause the disease in healthy animals. He also showed that inoculation of weakened microbes to animals would provide them with immunity and they would not contract the disease. Thus he had discovered the anthrax vaccine.

In this principle, Pasteur also developed vaccine for rabies, a dangerous human disease, caused due to rabies dog bite.

3. Joseph Lister (1827-1912)

Joseph Lister, an English surgeon, had made several contributions to the improvement of the microbiology.

(i) *Prevention of wound infection:* Lister's contribution to surgery are many, but his major contribution has been in preventing wound infection after surgery by the following antiseptic methods. As a surgeon, Lister was attracted towards post operative complications in patients due to wound infection and started studying the reasons for it. He observed a similarity between the wound inflammation and fermentation described by Pasteur, which caused due to microorganisms. He believed that the microbes entered the wound through air, hands, surgical instruments or band aids and caused sepsis. When he protected the wound with dressing that killed microbes, inflammation would never occur. This was the basis of antiseptic technique.

(ii) *Pure culture technique:* Microorganisms occur in nature in extremely large populations made up of many different species. In order to study the characteristics of a particular species it is necessary to separate it from all other species. In 1878, Lister for the first time obtained pure culture of bacteria by using serial dilution technique in liquid media. With a specially developed syringe, he diluted a curdled milk containing a mixture of bacteria until a single organism was delivered into a container of sterile milk. After incubation, bacteria in a container were of a single kind, rod shaped bacteria which he called as *Bacterium lactic.*

4. Robert Koch (1843-1910)

Robert Koch, a German medical practitioner switched over to microscopic studies which engaged him full time. His main contributions are:

(i) *Germ theory of disease:* He showed that microbes cause disease in animals. While studying the anthrax disease in cattle, Koch had found rod shaped organisms in the blood of the diseased animals. To test if these organisms were the causative agents of the disease, he injected healthy mice with the blood from a diseased animal. He found that disease was transmitted to the mice which died. He took the blood of the dead mice and injected it into a healthy mouse and again the mouse died. Koch repeated this experiment 20 times and each time the healthy mice died.

Koch cultured these bacilli in the aqueous humour from the eyes of a cow. By injecting these organisms into healthy animals, he was able to reproduce the disease. From these studies he was able to lay down certain basic criteria for the identification of microorganisms as causative agents of disease. These criteria are known as '*Koch's postulates*' and are—

(a) The causative agent of the disease must be present in every diseased animal.
(b) The causative agent must be isolated from the diseased animal and grown in pure culture.
(c) The isolated infectious organism must reproduce the disease when injected into a healthy animal, and
(d) The same infectious agent must be isolated from the experimentally diseased animal.

These postulates of Koch continue to be useful even today with some modification.

(ii) *Discovery of tuberculosis bacilli:* Koch showed that all forms of tuberculosis were caused by the bacillus - *Mycobacterium tuberculosis.*

(iii) Koch introduced the method of making smears of bacteria on glass slides and staining them with aniline dyes to observe them more clearly under the microscope.

(iv) *Plate Method for isolating pure culture:* Koch accidentally observed that a slice of potato had colonies of bacteria distinct from one another. However the potato had certain disadvantages like – the moist surface of potato allowed motile bacteria to spread. As the surface of potato is opaque it is difficult to observe bacterial colonies and it is not a good nutrient medium for many bacteria. Because of these disadvantages, Koch thought that if a liquid nutrient medium could be solidified by the addition of a solidifying agent such as gelatin, then it could be used for the isolation of bacteria. He poured the molten gelatin medium on glass plates and allowed it to solidify. After the gel had solidified, the surface was streaked with a platinum wire dipped into a bacterial suspension. The plates were then covered and incubated. Following incubation, growth was found along the streak and single colonies were found at the end of the streak. These single colonies contained only one type of bacteria and thus pure cultures. Thus, the use of gelatin for the preparation of a solid medium and the streak plate method for the isolation of pure bacterial cultures came into existence.

5. Alexander Fleming

In 1927, Alexander Fleming, a microbiologist working at St. Mary's Hospital in London, discovered *Penicillin,* the first wonder drug.

Fleming in his laboratory was conducting experiment in search of new antibacterial agents, particularly against wound infections. In the course of experiment, he observed a plate culture of *Staphylococcus aureus* was contaminated by a blue green coloured mold. The area around the edges of the mold colony was clear i.e. the growth of the Staphylococcus cells near the mold was inhibited or was killed by the mold. Flemming concluded that something that is excreted by the intruding mold proved deadly to the staphylococcus cells. The intruding mold was then identified as a common blue green mold *Penicillium notatum*.

Fleming cultivated this mold and obtained from it little extraction. In order to test his hypothesis that the mold secretions could kill Staphylococcus, he injected this extraction into laboratory mice infected with Staphylococci, Streptococci and Pneumococci. The mysterious compound present in this extraction killed all the germs in the body of the mice as it had done on the Streptococci on the petri dish. Fleming named this compound *Penicillin.*

SUMMARY

Microbiology is the branch of biology which deals with the study of microorganisms. Microbiology is not a mere study of structure and classification of microbes. The various aspects of microbiology can be studied under the headings like phycology, mycology, virology, protozoology, Bacteriology, medical microbiology, Agriculture microbiology, Industrial microbiology, Food microbiology, water microbiology, environmental microbiology and Geochemical microbiology. Many microbiologists contributed to the development of microbiology like Antony Van Leeuwenhoek, Louis Pasteur, Joseph Lister, Robert Koch, Alexander Fleming.

EXERCISE

1. Write short notes on:
   (i) Aeromicrobiology
   (ii) Water microbiology
   (iii) Pasteurisation
   (iv) Pure culture technique
   (v) Protection against infection
2. Write the contribution of the following in the field of microbiology:
   (i) Robert Koch
   (ii) Louis Pasteur
   (iii) Joseph Lister

22

Microscopy

Microscopy is a technique of using the microscope. Microscopes are used to observe structures below the range of magnified lens. The study and development of microbiology had not been possible without microscope because visual observation of microbes was made possible by the use of microscopes.

The invention of light microscope in the Sixteenth Century is the single most important contribution ever made to the advancement of biology. Ancient Romans, Greeks and Indians had known lenses of glass and quartz. In the 14th Century spectacles and single lenses were used as magnifiers. Campini, an Italian Scientist, found that lens can be ground to the desired curvature. Since the curvature of the lens increases its magnifying power, thereafter lens became a powerful tool in the hands of biologists. In 1600 AD, Leeuwenhoek constructed a simple, single lens microscope. In 1665, Robert Hooke, an Englishman, constructed the first compound microscope, that could be truly referred to as the forerunner of the modern day microscope. A microscope is an optical instrument consisting of a lens or a combination of lenses for obtaining magnified images. The main function of a microscope is to increase the magnification as well as to show contrast and resolution. Magnification may be defined as the ability of lenses to enlarge the image of objects. Magnification by a microscope is the product of the ocular and objective lenses. For example, if the ocular is 10X and the objective is 10X, the specimen is magnified 100 times.

Theoretically, if the magnifying power of the ocular and objective lenses are increased, it should be possible to get higher and higher magnification. But magnification beyond resolving power is of no use because the image will be blurred and details are not clear. This is due to the fact that in a microscope beside lenses, wavelength of the illumination is also important and it decides the resolving power of microscope.

Table 2.1: Units used for Microscopic Measurements

Name of Unit	Fraction of a meter	Abbreviation
Centimeter	10^{-2}	cm
Milimeter	10^{-3}	mm
Micron	10^{-6}	μ
Milimicron/nanometer	10^{-9}	mμ/nm
Angstrom	10^{-10}	°A

Resolving power of microscope is defined as the ability of lenses to distinguish between two smaller objects which are situated very close. The resolving power of a microscope depends on two factors – the wavelength of the light used for illumination and the numerical aperture (NA) of the objective.

Depending on the principle on which magnification is based, modern microscopes are of two types, namely light microscope and electron microscope. In a light microscope, magnification is obtained by a system of optical lenses using light waves, while in an electron microscope the magnification is obtained by a system of magnetic field using a beam of electrons. Light microscopes are of following types:

1. Bright field microscope
2. Dark field microscope
3. Phase contrast microscope
4. Fluorescence microscope
5. Ultraviolet microscope
6. Interference microscope

Electron microscopes are of two types:

1. Transmission electron microscope (TEM)
2. Scanning electron microscope (SEM)

Table 2.2: Characteristic features of different types of microscopes

Type of Microscope	Maximum Magnification	Resolution	Applications
Bright field	1000X to 2000X	100-200 nm	Used for study of gross morphology of yeasts, molds, algae, protozoa and bacteria. Staining is necessary.
Dark field	1000X to 2000X	100-200 nm	Used for study of gross morphology of living specimens. Staining is not necessary. Specimen appears bright in a dark background.
Phase contrast	1000X to 2000X	100-200 nm	For studying cellular structures in living cells. Staining is not necessary.

(Contd.)

Type of Microscope	Maximum Magnification	Resolution	Applications
Ultraviolet	2500X	100 nm	Better resolution due to the usage of UV light.
Fluorescence	1000X to 2000X	100-200 nm	Uses fluorescent staining. Useful in diagnostic and identification of microbes.
Interference	1000X to 2000X	100-200 nm	Used for observation of structural details of microbes. It produces sharp multicoloured image with three dimensional view.
TEM	500000X to 1000000X		Used for study of ultra structure of eukaryotic and prokaryotic cells and viruses. Anatomical details of cells can be viewed. Highest magnification and resolution obtained due to usage of short wavelength electron beam.
SEM	10,000 X 1000000X	1-10 nm	Used for surface study of object. Three dimensional image is produced.

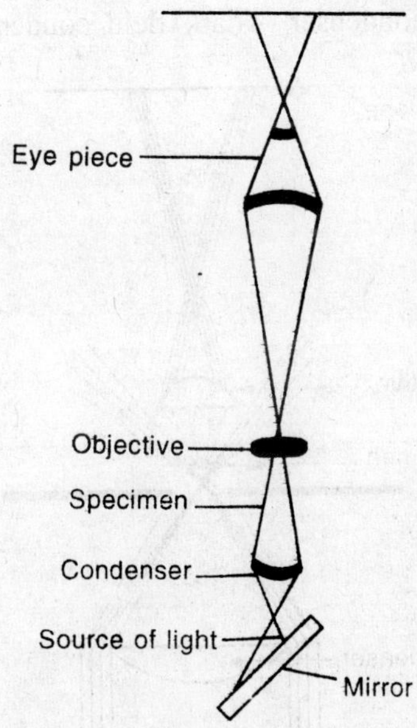

Fig. 22.1. Path of light in Bright Field microscope

Bright Field (Light) Microscope

The bright field or light microscopes are the most common instruments in the microbiology laboratory. In bright field microscopy, the microscopic field (the area observed) is brightly lighted and microorganisms appear dark because they absorb some of the light. Generally microorganisms do not absorb much light, but staining them with a stain generally increases their light absorbing ability which results in much contrast and colour differentiation.

Generally light microscopes produce a useful magnification of about 1000X to 2000X. The basic limitation of light microscope is not magnification but the resolving power.

Compound microscope, simple or dissecting microscope are the most common bright field microscopes used in biology laboratories. Generally compound microscopes are equipped with three different types of objectives, each capable of different degree of magnification. These objectives are oil immersion, high dry and low power objectives. The total magnification of a microscope is determined by multiplying the magnification power of the objective and eyepiece.

Limitation of bright field microscope: Viewing living organism is not possible because staining is necessary due to which cell will die.

Dark Field Microscope

The effect produced by the dark field microscopic technique is bright object against a dark background. This is obtained by attaching the light microscope with a special kind of condenser – dark field condenser. In a dark field

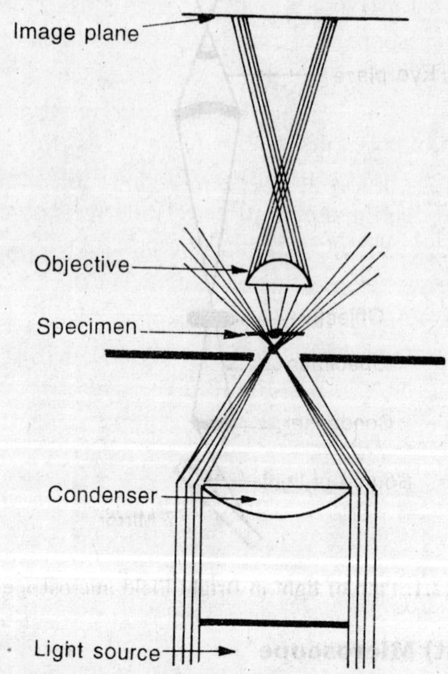

Fig. 22.2. Path of light through Dark field microscope

condenser, the centre of the top lens is opaque, so that none of the central rays of light can pass through, and the specimen on the stage is illuminated only with the peripheral oblique rays. The light rays are focussed in such a way that they do not reach the specimen directly but instead the light rays transverse across the specimen horizontally, almost at right angle to the objective lens. As a result the surrounding appears dark but the specimen appears bright refractile structure.

Uses of dark field microscope: It is useful for the examination of live, unstrained microorganisms or other specimens. It is also useful for the study of very small bacteria which are generally invisible under a light microscope. Motility of microbes can be observed because microbes are in a moist, living condition.

Phase Contrast Microscope

The main difficulty in examining cellular organelles is that most protoplasmic materials are completely transparent. For this reason stained slides are usually used in bright field cytological studies. But staining of cell is resulting in cellular death, so it is natural that there some changes occurs. Thus it results in study of cellular structure which is not an original structure.

Phase contrast microscope is a special microscope that enables the viewer to see transparent protoplasmic components without staining and thus without killing. The first phase contrast microscope was developed in 1933 by Federick Zernike and hence it is also called Zernike microscope.

Phase contrast microscope is a normal bright field microscope, attached with a phase contrast objective and phase contrast condenser. This special attachment makes it possible to distinguish unstained structures within a cell which differ only slightly in their refractive index or thickness.

The principle behind this technique is that the light passing through one material and into another material of a slight difference in thickness, will undergo a change in phase. This difference in phase is translated into variation in brightness of the structure and hence is detected by the viewer. As in the phase contrast microscope small phase changes results in interference of light waves resulting in high contrast images, some structures which are difficult to stain and observe under light microscope become visible.

Fluorescence Microscope

Fluorescence microscope involves staining of the specimens with fluorescent dyes. A fluorescence microscope differs from an ordinary bright field microscope in the following ways:

1. It utilises a mercury vapour arc lamp as a light source.
2. A dark field condenser is used instead of bright field condenser.
3. It employs three sets of filters to alter the light rays that pass through the instrument and reach the eye.

In this microscope, the object appears bright when fluorescent and the background appears light.

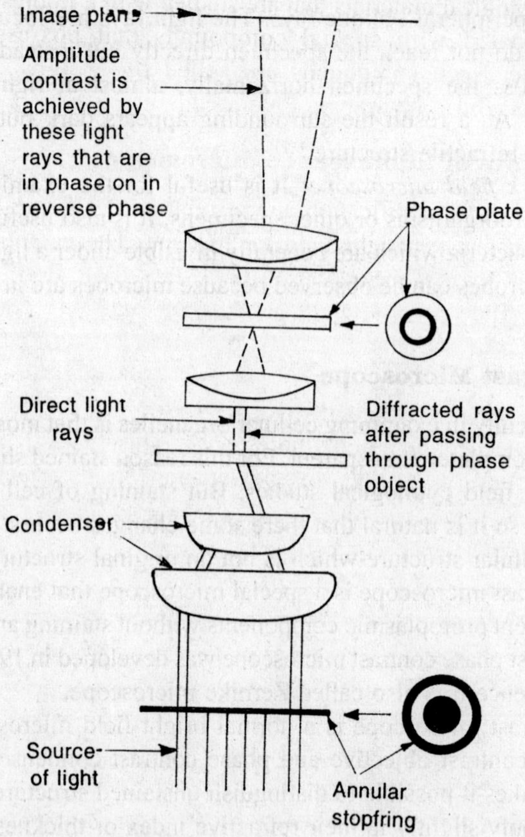

Fig. 22.3. Path of light through phase contrast microscope

The principle of fluorescence

Stokes (1852) used the word fluorescent to describe the reaction of fluorespar to uv-rays. The atom absorbs a quantum of light and an electron is boosted to a higher energy level. When this displaced electron returns to its original ground state, it emits a quantum of light. If this light is emitted only during the time of exposure or for a very small fraction of second afterwards, it is known as *fluorescence.* If the emission of light continues for some seconds after the exciting light is cut off, it is called *phosphorescence.*

In fluorescence microscopy, uv-light is used as the exciting light and resulting fluorescence rays are always of longer wavelength than exciting rays, because this follows Stokes's phenomena. According to this phenomenon, the light always loses energy when it pass through liquid, so that the emitted light has to be a longer wavelength. When the fluorescent dye flourescein isothiocyanate is illuminated with blue light, it emits green light.

This is due to the fact that light rays of shorter wavelength have more energy than those of longer wavelength.

Microbiological materials that are studied with a fluorescence microscope must be coated with special compounds that possess this quality of fluorescence. Such compounds are called *fluorochromes*, for example - Auramine O, Acridine orange, Fluorescein, etc.

Components of fluorescence microscope

The principle of magnification and resolution of a fluorescence microscope is same like that of bright field microscope. The essential components of fluorescence microscope are light source, heat filter, exciter filter, condenser and barrier filter.

Light Source: In fluorescence microscope, bright mercury vapour arc lamp is used as a light source because such bulbs produce an adequate supply of shorter wavelength light rays such as UV, violet and blue rays, which are needed for good fluorescence. This type of lamp requires high voltage, around 18000 volts, so a power supply transformer is always needed.

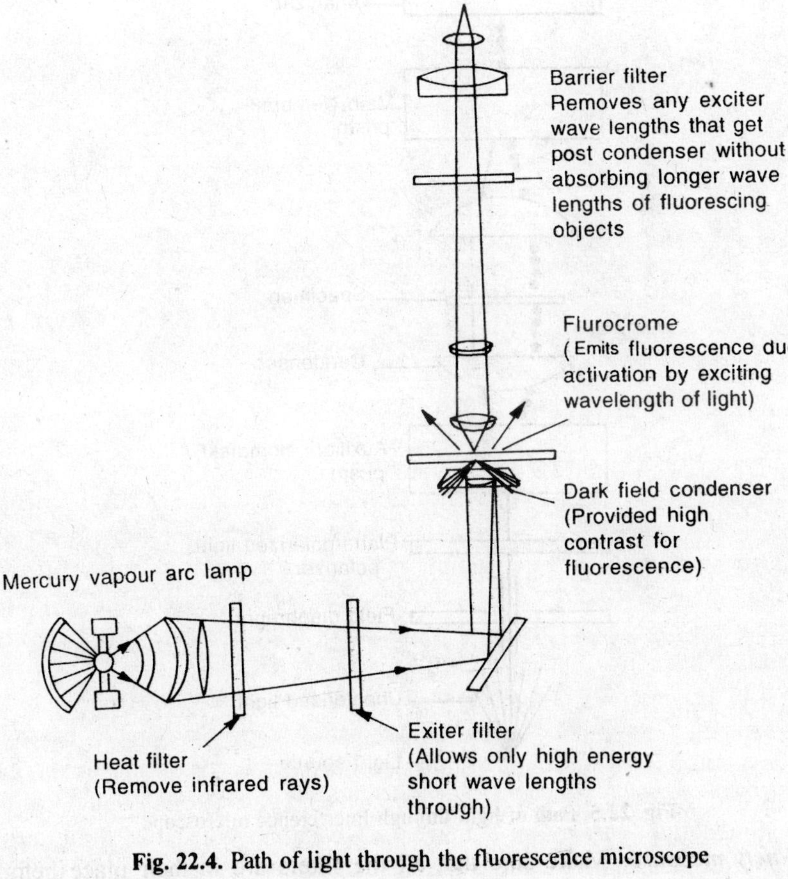

Fig. 22.4. Path of light through the fluorescence microscope

Heat filter: Infrared rays generated by the mercury vapour arc lamp produce a considerable amount of heat. Besides, these rays serve no purpose in fluorescence and place stress on the filters within the system. So to remove these rays, a heat filter is used. However, heat filter does not prevent the transmission of UV and the visible light rays.

Exciter filter: After light has been cooled down by the heat filter, it passes through the exciter filter, which absorbs all the wavelength except the short wavelength, which is needed to excite the fluorochromes on the slide. These filters are very dark and allow only the green, blue, violet or UV rays.

Condenser: Generally dark field condenser is used to achieve the best contrast of a fluorescent object in the microscopic field. Another benefit of dark field condenser is that the majority of UV rays are deflected by the condenser, thus protecting the observer's eyes.

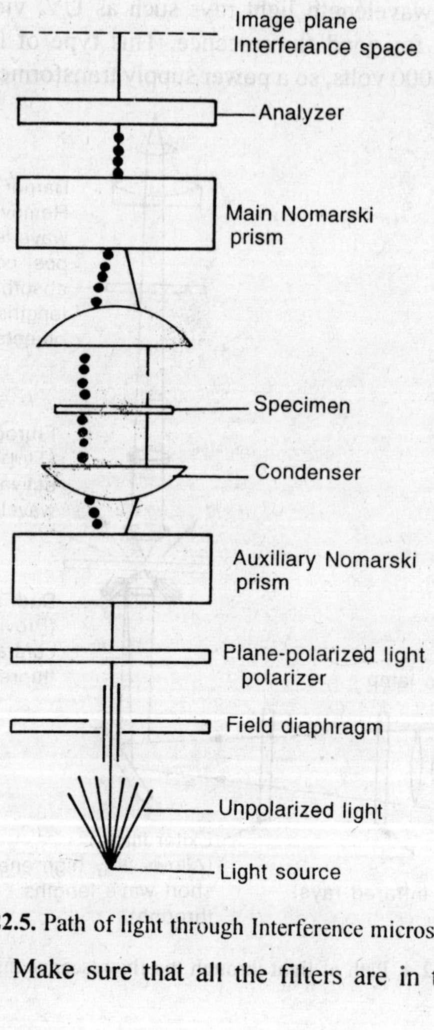

Fig. 22.5. Path of light through Interference microscope

Safety measure: Make sure that all the filters are in their place before looking into the microscope because removal of any one or all filters is harmful to the eyes.

Interference Microscope (Stereomicroscope)

Both the interference microscope and phase contrast microscope are based on the principle of change in refractive index which will result in contrast in phase. In phase contrast microscope, a single beam of light passing through the object is used, while in an interference microscope, two beams of light are used, which combine after passing through the specimen. This produces an interference pattern. Interference microscope uses colour to increase the contrast.

Nomarski Differential Interference contrast microscope is the most frequently used interference microscope. It contains polarising filters, an interference contrast condenser and a prism analyser plate. In this microscope, a beam of polarised light is split into two beams at right angles to each other and passed through the specimen and are combined and forced through an analyzer. The resultant interference pattern will produce a pseudo three dimensional image, because two beams of light passing through the specimen produce a stereoscopic effect. Images seen through this microscope are normally brilliantly coloured (non-stained) because of the phase change of light waves that pass through the various components of an object.

Electron Microscope

Electron microscope differs in many aspects from light microscope. The electron microscope provides tremendous magnification because of the much higher resolution obtained with the extremely short wavelength of electron beam used to magnify the specimen. Electron microscope uses electron beam and magnetic field to produce the image, while light microscope uses light waves and glass lenses to produce image (Fig. 22.6).

Electron microscope employs 60 to 80 kv electrons having wavelength of only 0.05 A°. By using electron microscope it is possible to resolve objects as small as 10 °A and it magnifies the image up to 400000 X to 1000000 X.

The basic components of electron microscopes are electron generator (gun), 3 sets of electromagnetic coils, screen for viewing and vacuum pump. The electron generator is a tungsten filament, which when heated by electric current emits a stream of electrons. From there, this electron beam passes through the condenser. Condenser is a magnetic coil which corrects the bending in the beam and directs it towards the object.

For electron microscope, the specimen to be examined is prepared as an extremely thin, dry film on small screen and introduced into the instrument at a point between the magnetic condenser and magnetic objective. This point is comparable to the stage of the light microscope. The electron beam passes through it and is scattered depending on the varying refractive index of the specimen. From there, the electron beam passes through the second set of magnetic coils which focus the electrons and form an intermediate image. The third set of magnetic lenses, the projection lenses, produce the final image. However, the final image cannot be directly viewed as the electron rays are harmful to the eyes. They are either projected on the fluorescent screen through the airtight window or recorded on a photographic plate by a camera built into the instrument.

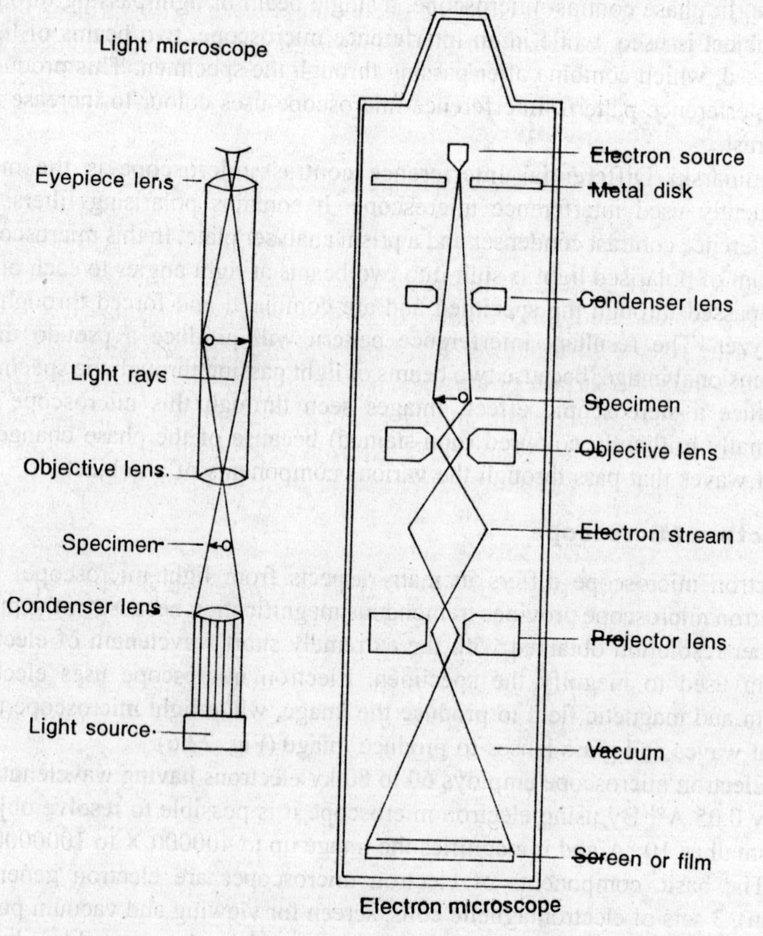

Fig. 22.6. The comparison between electronic and light microscope

The entire electron microscope must be in a vacuum or otherwise the electrons get scattered due to collisions with air molecules and thus fail to get focussed.

There are two main types of electron microscopes – Transmission electron microscope (TEM) and Scanning electron microscope (SEM).

Transmission Electron Microscope (TEM)

The general construction pattern of transmission electron microscope is same as explained above. It is used for studying the structural details of the various components present in a cell. The specimen forms a contrasting image with all the structural details.

Numerous techniques are available for use with TEM, which extends its usefulness in characterising cellular structure. Some of these are – Shadow casting, Negative staining, Ultrathin Sectioning, Freeze etching, Localization of cell constituents, Localization of enzymes in thin sections, Autoradiography, etc.

Scanning Electron Microscope (SEM)

SEM is primarily used for visualising the surface architecture of the specimen instead of internal details.

The general construction and working principle of a SEM is different from that of TEM. In SEM the specimen is subjected to a narrow beam of electrons which rapidly move over (scan) the surface of the specimen. This causes the release of a shower of secondary electrons and other types of radiation from the surface of the specimen. The intensity of these secondary electrons depends on the shape and the chemical composition of the specimen. These secondary electrons are collected by a detector, which generates an electronic signal. These signals are then scanned in the manner of a television system to produce an image on a cathode ray tube.

The SEM lacks the resolving power obtainable with TEM, but has the advantage of revealing a striking three dimensional picture. The surface topography of a specimen can be revealed with a clarity and a depth of the field not possible by any other method.

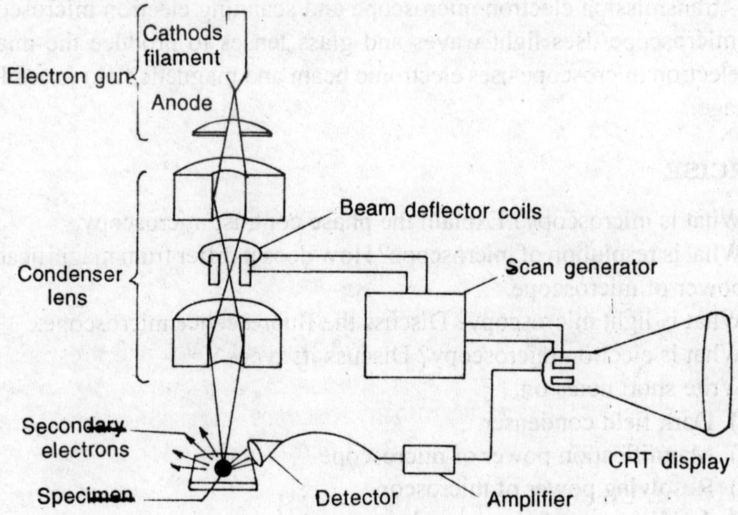

Fig. 22.7. The scanning electron microscope

Applications of Electron microscope

1. Viewing of ultra structure of cell organelles is possible.
2. Structural details of large biomolecules such as neuclic acid, protein, etc. can be studied.
3. Minute details of the topographic architecture can be studied.

Limitations of Electron Microscope

Despite the great advantage of better resolution and higher magnification, there are several limitations to electron microscopic studies. These limitations are –

1. The specimen being examined in a chamber that is under a very high vacuum, the cells cannot be examined in a living state.
2. The drying process may alter some morphological characteristics.
3. As electrons have a low penetration power, it is necessary to use thin sections of specimen cells to reveal the internal structures of the cell.

SUMMARY

Microscopy is a technique of using the microscope. Microscopes are used to observe structures below the range of magnified lens. The main function of a microscope is to increase the magnification as well as to show contrast and resolution.

Modern microscopes are of two main types – light and electron microscope. Light microscope is of different types like bright field microscope, dark field microscope, phase contrast microscope, fluorescence microscope, ultra violet microscope and interference microscope. Electron microscope is of two types – transmission electron microscope and scanning electron microscope. Light microscope uses light waves and glass lenses to produce the image, while electron microscope uses electronic beam and magnetic field to produce the image.

EXERCISE

1. What is microscopy? Explain the phase contrast microscopy.
2. What is resolution of microscope? How does it differ from magnification power of microscope.
3. What is light microscopy? Discuss the fluorescence microscope.
4. What is electron microscopy? Discuss its types.
5. Write short notes on:
   (i) Dark field condenser
   (ii) Magnification power of microscope
   (iii) Resolving power of microscope
   (iv) Applications of electron microscope

23

Sterilisation Techniques

Microorganisms are ubiquitous. Microbes grow and multiply under favourable conditions. In laboratory, for the purpose of study, they are allowed to grow and multiply. Since they cause contamination, infection and decay, it becomes necessary to remove or destroy them from materials or from areas. This is the main object of sterilisation. Control of microorganisms is essential in many situations:

1. To prevent the transmission of pathogenic microorganisms and to check the disease spread.
2. To prevent the contamination of pure cultures in scientific and microbiology laboratories.
3. To prevent the microbial contamination in industrial processes.
4. To maintain aseptic condition for surgery.
5. To prevent decomposition and spoilage of food and food products.
6. To ensure safety from contaminating organisms in pharma-ceutical industries.

Sterilisation is defined as the process by which an article, surface or medium is freed of all microorganisms either in vegetative state or in spore state.

The method of sterilisation used would depend on the purpose for which sterilisation is carried out, the material which has to be sterilised and the nature of microorganisms that are to be removed or destroyed. The various agents used in sterilisation are categorised into two main groups – physical agent and chemical agents.

A. PHYSICAL METHOD OF STERILISATION

Different physical methods employed for sterilisation are as follows:

1. *Sunlight:* Sunlight possesses appreciable bactericidal activity and plays an important role in the spontaneous sterilisation under natural

conditions. This action is mainly due to its content of ultraviolet rays. Under natural conditions, its sterilizing power varies according to circumstances.

2. *Drying:* Since moisture is essential for the growth of bacteria, therefore, drying has a deleterious effect on microorganisms. Its sterilizing power varies with different bacteria and also with conditions under which they are exposed to drying. Spores are unaffected by drying.
3. *Heat:* The most reliable method of sterilisation is heat. Heat sterilisation is the method of choice; only the materials damageable by heat are not sterilised by this method, or sterilised at lower temperature for longer periods or by repeated cycles.

Many factors influence the sterilisation by heat. They are as follows:

(i) Nature of heat – dry heat or moist heat.
(ii) Temperature and time for sterilisation
(iii) The number of microorganisms present
(iv) Nature of microorganisms, i.e. vegetative cells or spore cells.
(v) The characteristics of microorganisms like species, strain, sporing capacity, etc.
(vi) The type of material from which the microorganisms have to be eradicated.

(a) *Dry Heat:* The killing effect of dry heat is due to protein denaturation, oxidative damage and the toxic effect of elevated levels of electrolytes. In a dry or completely moisture-free atmosphere microorganisms are more resistant to heat. They are killed when oxidation of the cell constituents occurs, which requires very high temperatures.

Depending upon the material to be sterilised, various methods of dry heat sterilisation are available.

(i) *Flaming:* For sterilizing inoculating loops, points of forceps searing spatulas and are held in bunsen flame till they become red hot. Before flaming to prevent spattering, dip the loop containing infective proteinaceous materials in chemical disinfectants. To destroy bacteria from the mouths of culture tubes, glass slides, cover slips, etc. pass it a few times through the bunsen flame without allowing them to become red hot. However, this method is unsatisfactory.

(ii) *Hot air oven:* Hot air oven is the most widely used method of sterilisation by dry heat. It is used to sterilize glassware, forceps, scissors, scalpels, all glass syringes, swabs, paraffin, sulphonamides, dusting powder, fats, greases, etc. Glassware should be perfectly dry before being placed in the oven. Test tubes, flasks, etc. should be plugged with cottonwool and other glassware like petridishes and pipettes should be wrapped in craft paper before being placed in the hot air oven.

Hot air oven consists of an insulated double walled cabinet heated by electricity and it is constructed to withstand a high temperature. There is a fan fitted either at the bottom or back to circulate hot air inside the chamber. A thermostat regulates the temperature at the desired level and a thermometer is fitted for recording the temperature.

Once the glassware and other articles to be sterilised are loaded onto the hot air oven, it should be switched on and the temperature should be slowly raised and at a particular temperature it should be maintained for a given time. The sterilisation time is always inversely proportional to the temperature as shown in the Table 23.1.

Table 23.1 Relationship between temperature and time for sterilisation in a hot air oven.

Temperature (°C)	Time (minutes)
120	480
140	180
150	150
160	120
170	60
180	20

Sterilisation by dry heat is not as efficient as by moist heat.

(b) *Moist heat:* The mechanism by which organisms are killed by moist heat is different from that of dry heat. The killing effect of moist heat is due to the denaturation and coagulation of protein. When the steam condenses on the cooler surface, it liberates a latent heat, which raises the temperature of that surface. In the case of the spore, steam condenses on it, increases its water content with ultimate hydrolysis and breakdown of the bacterial protein.

There are many methods of moist heat sterilisation. They are:

(i) *Temperatures below 100°C:* Commonly, this method is known as pasteurisation, which is generally used for sterilisation of milk, serum, heat sensitive materials, etc. The temperature employed is either 63°C for 30 minutes called holder method, or 72°C for 15-20 seconds called flash method, followed by cooling quickly to 15°C or lower temperature. By this method all nonsporing pathogens such as mycobacteria, brucellae and salmonellae are destroyed. This method is not suitable for sterilisation of spore producing bacteria or pathogens.

(ii) *Temperature at 100°C (Boiling):* Vegetative bacteria are killed almost immediately at 90°C to 100°C, but spore producing bacteria requires considerable period of boiling. The material to be sterilised should be immersed in the water and boiled for a period of 10 to 30 minutes. The lid of the sterilizer should not be opened during the period.

(iii) *Steam at atmospheric pressure (100°C):* A medium which may decompose if subjected to higher temperatures, can be sterilised in the atmosphere of free steam. In this method container and medium are simultaneously sterilised, and the apparatus requires little or no attention. One exposure of ninety minutes usually ensures complete sterilisation but the media containing sugars or gelatin requires exposure of 100°C for 20 minutes on three successive days. This is known as intermittent sterilisation. In this method first exposure kills all vegetative bacteria and spores present being in a favourable medium will germinate and be killed on the subsequent occasions.

(iv) *Steam under pressure:* The principle of the steam steriliser is that water boils when its vapour pressure equals that of the surrounding atmosphere. When pressure inside a closed vessel increases, the temperature at which water boils also increases. Saturated steam has greater penetration power. When steam comes into contact with a cooler surface it condenses to water and gives up its latent heat to that surface, raising the temperature of that surface. The condensed water ensures moist conditions for killing the microbes present. By using the appropriate temperature and time (Table 23.2), a variety of materials like laboratory ware, media, pharmaceutical products, dressings instruments, etc. can be sterilised by steam under pressure. Several types of steam sterilizer are in use like laboratory autoclaves, hospital dressing sterilisers, bowl and instrument sterilizers and rapid cooling sterilizers.

Table 23.2 Relationship between pressure and temperature of steam sterilisation

Pressure above atmospheric (lb/sq. inch)	Temperature (°C)
0	100
5	109
10	115
15	121
20	126
25	130
30	134

The time required for sterilisation by steam under pressure will depend on the material to be sterilised and the initial microbial load. Solid materials require longer time for sterilisation than liquid media. Also acidic materials require shorter period of sterilisation.

(4) *Filtration:* Filtration is a physical method of removing microorganisms from solutions oı gases. The ability of porcelain filters to remove bacteria from solutions was known in the 19th Century. The effectiveness of a filter will depend on the pore size. In recent years porcelain filters have been replaced by membrane filters composed of cellulose acetate, polycarbonates and other polymers. These filters are very thin and are available with known pore sizes. They are relatively inert and absorb very little fluid. These filters are used to sterilize liquids which cannot be heat sterilised such as beer,

wine, other alcoholic beverages, antibiotic solutions, serum and carbohydrate solutions used in the preparation of culture media.

(5) *Radiation:* The effectiveness of the radiation depends on the degree of susceptibility of microorganisms, presence of oxygen, water and other chemicals. The presence of enzymes within the organisms which can repair the damage also determines its effectiveness.

For sterilizing purpose, two types of radiation are used – non-ionising and ionising.

Nonionising radiation: e.g. Infrared and ultraviolet rays. In nonionising radiation, electromagnetic rays with wavelengths longer than those of visible light are used. Infrared radiation is used for rapid mass sterilisation of syringes. Ultraviolet radiation is used for disinfecting enclosed areas such as entryways, hospital wards, operation rooms, inoculation rooms and laboratories.

Ionising radiation: e.g. X-rays, gamma rays, cosmic rays, etc. These are highly lethal to DNA and other vital cell constituents and cause biological damage by producing hyper reactive ions. They have very high penetration power. Since there is no appreciable increase in temperature, this method is referred as cold sterilisation. Gamma radiation is used for sterilisation of plastics, syringes, swabs, culture plates, catheters, animal feeds, various types of rubber, cardboard, oils, greases, fabrics and metal foils.

(6) *Ultrasonic and Sonic Vibration:* Even though the ultrasonic and sonic vibrations have bactericidal activity, their results are variable. Microorganisms vary in their sensitivity to them and hence this method is of no practical value in sterilisation technique.

B. CHEMICAL METHOD OF STERILISATION

A variety of chemicals are used for sterilisation purpose. A variety of factors determine the potency of chemical agents such as concentration of the substance, time of action, pH of the medium, temperature, nature of the organisms and presence of extraneous materials. These chemical agents act in various ways such as protein coagulation, disruption of cell membrane, removal of free sulphydryl groups which are essential for functioning of the enzymes and substrate competition.

These chemical agents include halogens, heavy metals, phenolics, alcohols, etc.

(1) *Halogens:* Among the halogens, iodine and chlorine are the most important and are used in chemical sterilisation because both are strong oxidizing agents. Generally, chlorine is used in water purification in water supplies, swimming baths, food and dairy industries, etc. Chlorine have a wide spectrum of activity against viruses. Iodine is used as a local antiseptic. It is an active bactericidal agent with a moderate activity against spores.

(2) *Heavy metals:* Heavy metals cause cell inactivation. One of the first metals used was arsenic, but due to its toxic properties, it is not generally used now-a-days. Another effective disinfectant is mercuric chloride but due

to its toxicity, it is used only as a surface sterilizing agent. Other heavy metals used as surface disinfectants are silver nitrate, copper sulphate and potassium permanganate.

(3) *Dyes:* The two groups of dyes – aniline dyes and acridine dyes – are used extensively as skin and wound antiseptics. In higher dilution, both are bacteriostatic. The aniline dyes in use are brilliant green, malachite green and crystal violet. They are more active against Gram positive organisms. The acridine dyes are also more active against Gram positive organisms, but not as selective as aniline dyes. They are very little affected by the presence of organic matter. The important acridine dyes are proflavine, acriflavine, euflavine, aminacrine, etc. They impair the DNA complexes of the organisms and thus destroy the reproductive capacity of the cell.

(4) *Phenols:* These are obtained by distillation of coal tar. Lister first used phenol as disinfectant in surgery in 1865. Phenol inactivates cells irreversibly by disrupting the cell membrane and causes cell lysis. However, the susceptibility to the phenol varies with the concentration and the micro-organism. For example, vegetative cells are sensitive and spores are resistant. Phenol and its derivatives are commonly used in surgical and hospital sanitation environments. However, care must be taken while using phenolics since these can damage the skin and the tissues. Lysol and cresols are markedly toxic to man.

(5) *Alcohols:* Most commonly used alcohols for disinfection are ethyl alcohol and isopropyl alcohol. These are mild disinfectants and are non-toxic when used for external application as antiseptics. They act by denaturing bacterial proteins. Ethyl alcohol is both a cleansing agent as well as a mild disinfectant. Isopropyl alcohol is preferred to ethyl alcohol as it is a better fat solvent, more bactericidal and less volatile. It is used for the disinfection of clinical thermometers.

Methyl alcohol is effective against fungal spores and is used for treating cabinets and incubators affected by them. But methyl alcohol vapour is toxic and inflammable.

(6) *Aldehydes:* Formaldehyde is one of the oldest disinfectants known. It inactivates microbial proteins by oxidation of the amino groups. In aqueous solutions, it is markedly bactericidal and sporicidal and also has a lethal effect on viruses. It is used to preserve anatomical specimens and for destroying anthrax spores in hair and wool. However, because of its irritating vapour, formaldehyde is less commonly used.

(7) *Gases:* Since a long time, the burning of sulphur for fumigation has been known. Similarly, formaldehyde gas has been recommended for decontamination of operation rooms and other rooms. After sealing windows and other outlets, formaldehyde gas is generated by adding 150 $KMnO_4$ to 280 ml formalin for every 1000 cu. ft. of room volume. Heat resistant vessels should be used because the reaction produces considerable heat. After starting

the generation of formaldehyde vapour, the doors should be sealed for 48 hours.

Beside formaldehyde gas and sulphur gas, ethylene oxide is also used for sterilisation. This is the most effective gaseous disinfecting agent. Because of its penetrating power, it interacts with amino, carboxyl, or sulphydryl groups of organic substances. Its effectiveness depends on the temperature and relative humidity. Because of its explosive nature, the use of ethylene oxide is restricted to special chambers where temperature, pressure and humidity are controlled. Commonly, in practice, a sterilisation time with ethylene oxide for about four hours is followed by an additional period of several hours to allow the release of the absorbed gas. Beside these gases, Betapropiolactone which is a condensation product of ketone and formaldehyde is also used. Even though it has a low penetration power, it is said to be more efficient for fumigating purpose than formaldehyde. It has a rapid biocidal action but it has a *carcinogenic activity*. It is capable of killing all microorganisms and is very active against viruses.

SUMMARY

Since few microorganisms cause diseases, infection and contamination, it becomes necessary to remove or destroy them from materials or from areas. This is the objective of sterilisation. There are mainly two methods of sterilisation – physical method and chemical method. Many physical agents are used for sterilisation like moist heat, dry heat, filtration, radiation, ultrasonic and sonic vibrations. Many chemicals are also used for sterilisation like halogens, heavy metals, alcohols, phenolic compounds, etc.

EXERCISE

1. What is sterilisation? Discuss chemical sterilisation.
2. Discuss the sterilisation methods used in laboratories.
3. What is sterilisation? Discuss moist heat sterilisation.
4. Write short notes on:
   (i) Filtration
   (ii) Phenols
   (iii) Sterilisation at temperature below 100°C
   (iv) Flaming

24

Microbial Taxonomy

The term taxonomy may be defined as the classification of living organisms according to their natural relationship. As the biological world has great diversity, scientists find it useful to place organisms into groups based on their phylogeny and to name these organisms. The word phylogeny has originated from Greek 'phylon' meaning tribe and 'genesis' meaning origin. One of the aspects of taxonomy is *classification,* i.e. ordering of organisms into groups or taxa.

Taxonomy is also concerned with *nomenclature,* i.e. assigning names to the units described in a classification system and *identification,* i.e. applying the system of classification and nomenclature to assign the proper name to an unknown organism and to place it in its proper position within the classification system.

An organism can be classified only after its morphological, cultural, physiological and pathogenic characters have been exhaustively studied. After such a study, the organism must be given a name, otherwise reference to it cannot be made. Generally, two kinds of names are used, namely common or casual names and scientific or international names. Usually, the common names for the same organism will be different in each language and therefore the common names are responsible for considerable confusion in biology. On the other hand, scientific names are international in meaning and therefore it is better to refer to organism by their scientific names which are the same in all countries.

The method followed in naming plants was first introduced by the Swedish botanist Karl von Linne and is known as binomial system of nomenclature. Later on the same system was used by the zoologists and microbiologists. According to binomial system of nomenclature, each distinct kind of organism is called a *species.* Each species is given a name which consists of two words, the first word is the name of the genus and the second word is the name of the species. The genus name is taken from the Latin or Greek or a new word compounded from Latin or Greek roots or in rarer cases from some other language. The genus name is a noun, always written in capitals and should be used only in singular. e.g. Bacillus, Micrococcus, Streptococcus, etc. The species name generally may be an adjective but is never capitalized. e.g. subtilis (Bacillus Subtilis).

GENERAL PRINCIPLES OF TAXONOMY

Taxonomic hierarchy is a way of organizing the classification of organisms with different levels of similarity. (Fig. 24.1). The species is the fundamental unit of the taxonomic hierarchy. The next level is the genus which will contain one or more closely related species. Closely related genera are organized into families; families into classes; classes into divisions or phyla and phyla into kingdom.

Although species are the basic taxonomic units, the genetic variability of microorganisms permits a further division into subspecies or types that describe the specific clone of cells. The subspecies or types may differ physiologically (biovar), morphologically (morphovar) or antigenically (serovar). It is often important to differentiate the subspecies of a given microorganism. For example, one strain of a bacterial species may produce a toxin and be a virulent pathogen and other strain of a same species may be nonpathogenic. A *strain* is a population of cells that are descendents of a single cell. The ability to distinguish correctly between such strains and subspecies of a particular microbial species is of obvious importance in medical and industrial microbiology.

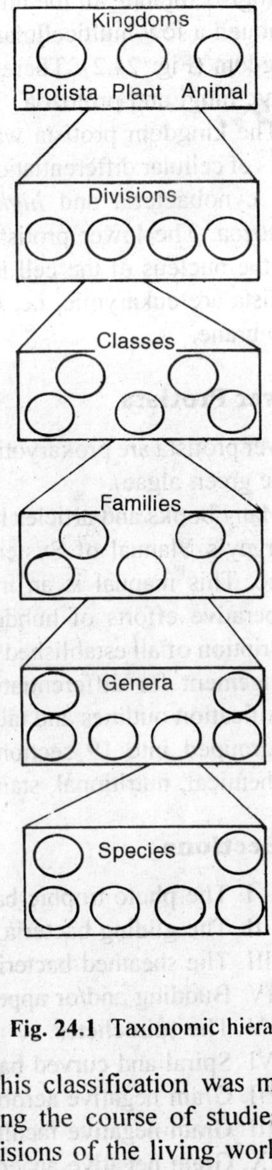

Fig. 24.1 Taxonomic hierarchy

Classification of microorganisms

In the past, biologists divided living things into plants and animals, which were later accepted as two broad kingdoms of living creatures. This classification was made before the discovery of microorganisms. During the course of studies, it becomes evident that the old and common divisions of the living world is insufficient because microorganisms show the similarity with plants (some are green and do not eat other organisms) on the one hand and on other hand with animals (highly motile). Bacteria were included in plants. Slime molds were regarded as middle path between plants and animals because they were considered as plants by botanists and animals by zoologists.

In 1866, Ernst Haeckel, a German biologist, created a new kingdom as *protista.* This kingdom included all those microorganisms which did not possess any extensive development of tissue. On the basis of this character, biologists include all the unicellular microscopic organisms in this kingdom, although a few multicellular organisms have also been included in the same kingdom (Fig. 24.2). Therefore, the kingdom protista includes the bacteria, algae, fungi and protozoa.

The kingdom protista was further divided into two large groups on the basis of cellular differentiation, namely *lower protista,* which includes bacteria and cynobacteria and *higher protista,* which includes algae, fungi and protozoa. The lower protista are prokaryotic in their cellular organisation, i.e. the nucleus of the cell is devoid of nuclear membrane while the higher protista are eukaryotic, i.e. the nucleus of the cell is surrounded by nuclear membrane.

Lower Protista

Lower protista are prokaryotic in nature and include bacteria and cynobacteria (blue green algae).

Many books and articles have been written that describes the bacteria, but "Bergey's Manual of Systematic Bacteriology" is unique due to its broad scope. This manual is an international reference work, resulting from the cooperative efforts of hundreds of microbiologists. It not only contains a description of all established genera and species, but also provides a practical arrangement for differentiating these organisms together with appropriate classification outlines and tables. According to this classification the bacteria are grouped into 19 sections on the basis of structure, genetic data and biochemical, nutritional, staining and ecological characteristics (Fig. 24.3).

19 Sections

I The photo trophic bacteria.
II The gliding bacteria.
III The sheathed bacteria.
IV Budding and/or appendaged bacteria.
V The spirochetes.
VI Spiral and curved bacteria.
VII Gram negative aerobic rods and cocci.
VIII Gram negative facultatively anaerobic rods.
IX Gram negative anaerobic bacteria.
X Gram negative cocci and cocco bacilli.
XI Gram negative anaerobic cocci.

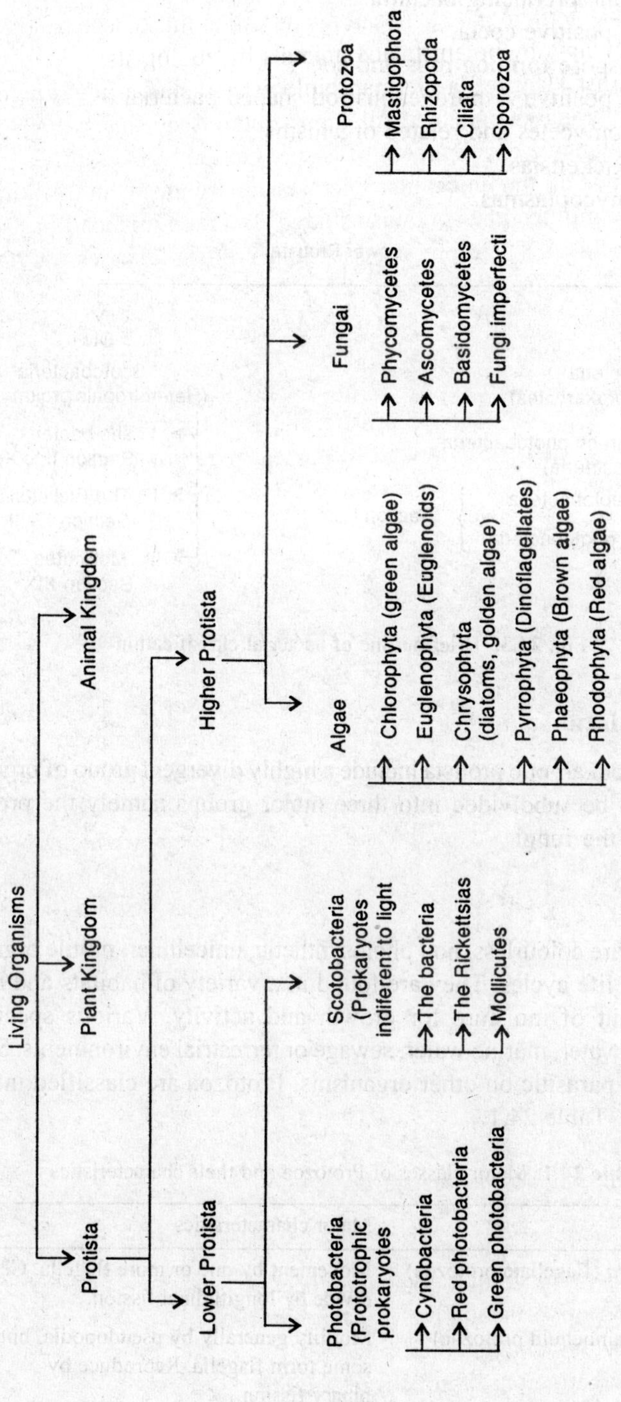

Fig. 24.2. Broad classification of microorganisms

XII Gram negative chemolithotrophic bacteria.
XIII Methane producing bacteria.
XIV Gram positive cocci.
XV Enclospore forming rods and cocci.
XVI Gram positive Asporogenous rod shaped bacteria.
XVII Actinomycetes and related organisms.
XVIII The Rickettsias.
XIX The mycoplasmas.

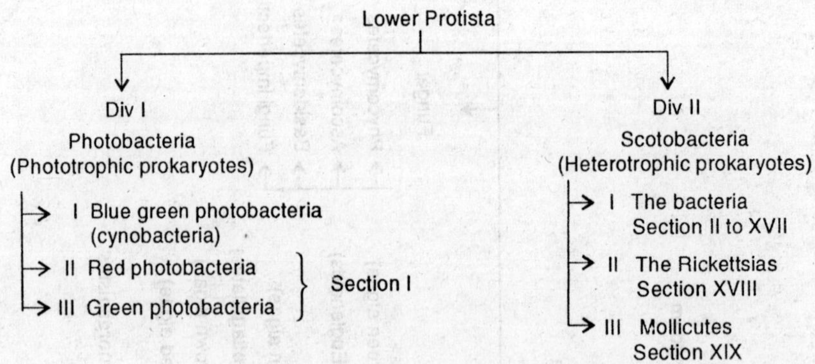

Fig. 24.3. Brief outline of bacterial classification

Higher Protista

The higher or eukaryotic protista include a highly divergent group of organisms and these can be subdivided into three major groups namely the protozoa, the algae and the fungi.

Protozoa

The protozoa are colourless, non photosynthetic, unicellular, motile organisms with complex life cycles. They are found in a variety of habitats and require a large amount of moisture for growth and activity. Various species are found in freshwater, marine water, sewage or terrestrial environments. Several protozoas are parasitic on other organisms. Protozoa are classified into four major classes. Table 24.1.

Table 24.1 Major classes of Protozoa and their characteristics

Class	Major characteristics
1. *Mastigophora* (flagellate protozoa)	Movement by one or more flagella. Cells divide by longitudinal fission.
2. *Rhizopoda* (amoeboid protozoa)	Motility generally by pseudopodia, but some form flagella. Reproduce by binary fission.

3. Ciliata (Ciliate protozoa)	Motility by numerous cilia. Divide by transverse fission. Cell contains two nuclei.
4. Sporozoa	Generally non motile, some show creeping or gliding motility. Multiple by multiple fission. Parasitic.

Algae

The algae are a ubiquitous group of aerobic photosynthetic organisms characterized by the presence of photosynthetic pigments and their ability to carry out a plant like photosynthesis. They are abundant in water, soil and other terrestrial environments. It occurs in a variety of forms such as rhizoidal, filamentous, mucilagenus colonies. Different algae species secrete various kinds of mucilages which are polysaccharides in composition. The mucilage may aid in motility or provide anchor to the organism. Based on the structure, pigments, cell wall composition, motility, reproduction and nature of reserve material, algae can be classified into six major classes. (Table 24.2).

Table 24.2 Major classes of algae and their characteristics

Class	Major Characteristics
1. Chlorophyta (Green algae)	Found in fresh and salt water, soil, etc. Chlorophyll a, b, carotene and xanthophyll present. Stores starch as reserve material. It divides vegetatively or asexually to form flagellated zoospores.
2. Euglenophyta (Euglenoids)	Contain both fresh and marine water forms. Unicellular flagellates. Chlorophyll a, b, carotene present. Stores fat and starch like carbohydrates. It divides by longitudinal binary fission and rarely by sexual method.
3. Chrysophyta (diatoms, golden algae)	Consists of a group of several diverse non-flagellar, unicellular algae. Some have calcium containing cell walls and others contain silica in their cell wall (diatoms). Stores oils and Lecucosin. Reproduce by asexual or sexual method.
4. Pyrrophyta (Dinoflagellates)	Primarily marine organisms. Two flagella with different structure and arrangement. Most are unicellular and contain carotenes and xanthophylls. Asexual reproduction and cell stores oils and starch.
5. Phaeophyta (Brown algae)	Multicellular and contain xanthophyll and carotene pigments. Found in sea water and structurally complex. Reproduce asexually by zoospores and sexually by gametes.
6. Rhodophyta (Red algae)	Contains phycobilins and chlorophyll a. Nonflagellate, unicellular or multicellular with complicated modes of sexual reproduction. Stores starch. Grows in slat water. Polysaccharides agar-agar, carrageenan and alginic acid are obtained from red algae.

Fungi

Fungi include the yeast and molds. They are characterised by the presence of a typical eukaryotic cell, a mycelial (thread like) thallus, and are nonphotosynthetic. Most are saprophytes but some are parasitic on plants and animals. Most are mesophilic but some are thermotolerant. Fungi are aerobic, while some yeasts are facultatively anaerobic. Reproduction occurs both sexually as well as asexually.

Their classification is based on morphology, physiology, the mode of reproduction and the nature of reproduction and the nature of the reproductive structures. They are divided into four major classes (Table 24.3).

Table 24.3 Major classes of fungi and their characteristics

Class	Major Characteristics
1. *Phycomycetes*	Mycellium nonseptate, aquatic or terrestrial, sporangiospores (asexual spores) motile or non motile. Sexual spores are oospores or zygospores.
2. *Ascomycetes*	Mycelium well developed and septate. Sexual spores are borne in asci. Parasitic and saprophytic.
3. *Basidomycetes*	Characteristics spores borne on basidia (club shaped structures) - basidiospores.
4. *Fungi imperfecti*	Reproduction by asexual process only. Mostly found on decaying wood.

Viruses

Viruses are the most important microbial enemies of human beings. They are very simple and tiny. Viruses are an intermediate stage between living and non-living. Viruses are called acellular as they do not have cellular organisation like other living organisms. The genetic material in viruses is either DNA or RNA but never both. It is an obligate parasite and it can not be cultured on media. But it can be cultured on cell media like bacteria, chick embryo cells, etc. Viruses are made up of only nucleic acid and proteins. But they lack the enzymes necessary for their synthesis. For this viruses depend on the host enzymes. They can be crystallized.They can be transmitted from one host to another. The capsid of viruses is mostly made up of proteins. Except in some animal viruses, polysaccharides are also present.

Classification of viruses is based on a variety of criteria. Based on their hosts, viruses were classified into following types:

1. Plant viruses - Infect plants
2. Animal viruses - Infect animals
3. Bacteriophages - Infect bacteria
4. Cyanophages - Infect cyanobacteria
5. Myco viruses - Infect fungi
6. Mycoplasma viruses - Infect mycoplasma
7. Phyco viruses - Infect algae

Beside phylogenitic classification, the microorganisms are classified on various criteria, as mentioned below.

Classification based on the forms of Microorganisms

Bacteria is the simplest prokaryotic microorganism. Depending on their forms, bacteria are classified into four main groups called spherical, rod like, spiral and pleuomorphic.

(a) *Spherical bacteria:* This is the simplest form of bacteria in which bacteria appears like a minute sphere. It is also called *cocci* (singular coccus). The term cocci has originated from the Greek word *'Kokkos'* meaning grain or Kernel. Cocci appears in different arrangements and on the basis of arrangements they are further divided as follows (Fig. 24.4).

1. *Coccus* – Single spherical bacteria
2. Diplococci – two spherical bacteria
3. Tetra cocci – cluster of four spherical bacteria
4. Staphylococci – bunch of spherical bacteria
5. Streptococci – chain of spherical bacteria
6. Sarcina – Cubical arrangement of eight or more spherical bacteria.

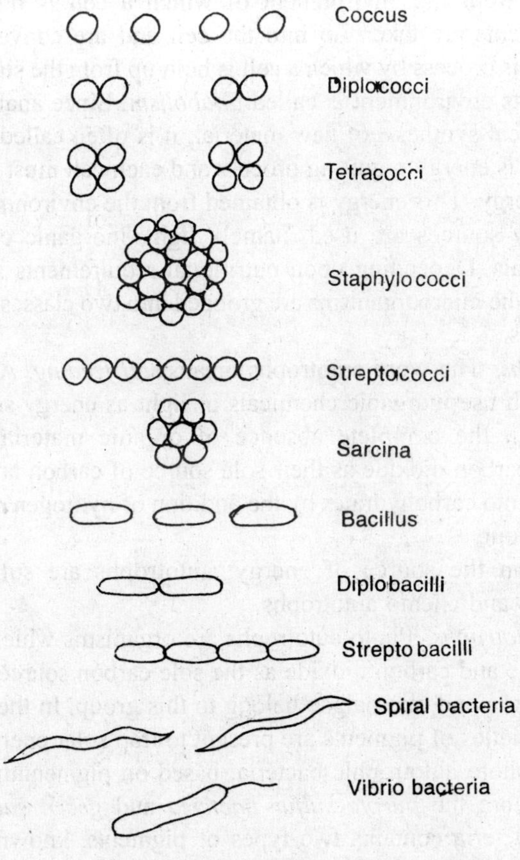

Fig. 24.4 Various forms of bacteria

(B) *Rod shaped bacteria:* They are also called bacilli (singular bacillus) and are the commonest in microbial world. The word bacilli is originated from Latin '*bacillum*' meaning *stick* or a rod. They are often straight and like rod but sometimes they are also slight curved. Generally they occur single and are called bacillus. But occasionally they occur in pairs called diplobacilli or they may occurs in chain called streptobacilli (Fig. 24.4).

(C) *Spiral bacteria:* These bacteria are generally spiral or helicoidal shaped. The word spiral is originated from Greek '*spira*' meaning coil. They appear like a *cork-screw*. A spirillum has more than one turn of helix while vibrio has less than one complete turn. Superficially, vibrio appears like '*comma*' shaped.

(D) *Pleomorphic bacteria:* Some bacteria exhibit several shapes or forms, called pleomorphic bacteria.

e.g. Azatobacter, Rhizobium, Acetobacter, etc.

Nutritional Classification of Microorganism

The chemicals from the environment of which a cell is built are called *nutrients*. Nutrients are taken up into the cell and are converted into cell constituents. This process by which a cell is built up from the simple nutrients obtained from its environment is called *anabolism*. Since anabolism results in the biochemical synthesis of new material, it is often called *biosynthesis*.

Biosynthesis is energy requiring process and each cell must have a means of obtaining energy. This energy is obtained from the environment and three kinds of energy sources are used, namely light, inorganic chemicals and organic chemicals. Depending upon nutritional requirements and source of energy utilized, the microorganisms are grouped into two classes—Autotrophs and heterotrophs.

(1) *Autotrophs:* The word autotrophs means *self feeding*. Autotrophs are organisms which use inorganic chemicals or light as energy source and are able to grow in the complete absence of organic materials. They use environmental carbon dioxide as their sole source of carbon and reduce this carbon dioxide into carbohydrates by the addition of hydrogen obtained from water or other sources.

Depending on the source of energy, autotrophs are subdivided into photoautotrophs and chemo autotrophs.

(a) *Photo autotrophs:* Photo autotrophs are organisms which use light as an energy source and carbon dioxide as the sole carbon source. Most of the pigmented bacteria and microalgae belong to this group. In these organisms one or more varieties of pigments are present to trap solar energy. There are two groups of photo autotrophic bacteria, based on pigmentation and other features. These are the *purple sulfur bacteria* and *green sulfur bacteria.* Purple sulfur bacteria contains two types of pigments, known as bacterio-chlorophyll and yellow carotenoid, while the green sulfur bacteria contains chlorobian chlorophyll or bacteriviridin.

In all photoautotrophic microorganisms the source of hydrogen is water, but in photoautotrophic bacteria the source of hydrogen is hydrogen sulphide. These photoautotrophic bacteria usually live in sulfur rich environments like sulfur springs.

(b) *Chemoautotrophs:* Chemoautotrophs obtain their energy from inorganic chemical sources whose oxidation provides the necessary energy. These inorganic chemicals are ammonia, nitrates, non-metalic materials, etc. which are easily available in nature. Bacteria absorb inorganic molecules as the substrate inside the body where the chemical bonds of these substrates are broken and energy is released. Depending on the inorganic compound used in oxidation, these bacteria are classified into sulfur bacteria, iron bacteria, nitrifying bacteria, hydrogen bacteria, etc. Chemoautotrophic group is entirely made up of bacteria and generally they are pigmentless.

(2) *Heterotrophs:* The word heterotrophs means feeding on sources other than oneself. Heterotrophs are organisms which obtain their readymade food directly from any external agency like organic compounds. This group includes many bacteria, fungi, protozoa, etc. These organisms cannot grow with carbon dioxide as the sole source of carbon. The carbon must be supplied in organic form, i.e. they grow on organic compounds formed by animals or plants. Depending upon their source of energy they are subdivided into photoheterotrophs and chemoheterotrophs.

(a) *Photoheterotrophs:* These utilise the light as energy source and organic materials as a carbon source. These organisms are regarded as the intermediate forms between phototrophs and heterotrophs. The most important example of this category are *non-sulfur purple bacteria.* Like purple sulfur bacteria, these bacteria also possess bacteriochlorophyll and yellow carotenoid.

Purple non sulfur bacteria absorb organic raw materials from the environment but they cannot be directly used as food. Instead these organic materials only serve as a source of hydrogen and this released hydrogen is then used for reducing the carbon dioxide. These bacteria can synthesise food only if light is present and oxygen is absent.

(b) *Chemo heterotrophs:* These microorganisms use organic compounds as an energy source as well as a principal carbon source. It is characteristic of this category that both energy and carbon requirement can be obtained by organic compound. The example of this group is many bacteria, protozoa, fungi, etc.

Heterotrophs are pigmentless organisms that obtain food either by parasitic mechanism or by saprophytic mechanism.

(i) *Parasitic organisms:* This group of microorganisms obtains its nutritional requirements by parasiting other organisms like plants and animals. Parasitic bacteria may be either pathogenic or non-pathogenic. The pathogenic bacteria secrete toxins which bring about the disease. Non-pathogenic bacteria which live in host are in fact helpful. For example, the intestinal bacteria synthesize vitamin B complex, which could be used by human beings.

(ii) *Saprophytic organisms:* Saprophytic microorganisms obtain food anaerobically by oxidizing organic matter. They grow on dead organic matter. Depending upon the nutritional source they may be fermenting microorganism or putrefying bacteria. Fermenting microorganisms use carbohydrates as the substrate. They anaerobically respire and incompletely oxidise the carbohydrates, resulting in the formation of alcohols. Putrefying bacteria make use of protein or nitrogenous matter as the nutritional source and liberate ammonia, marsh gas (methane), etc.

SUMMARY

Taxonomy is concerned with identification, nomenclature and classification of an organism. A Binomial System of nomenclature is most widely accepted for naming an organism.

In the taxonomic hierarchy, the species is the fundamental unit and the next level is genera, families, classes, Divisions and kingdoms. All the microorganisms come under the kingdom protista. The lower protista includes bacteria and cyanobacteria while higher protista includes protozoa, algae, fungi.

Nutritionally, microorganisms are divided into autotrophs and heterotrophs. Autotrophs grow on inorganic substances, while heterotrophs grow on organic substances. Autotrophs are of two types, namely photoautotrophs and chemoautotrophs. Similarly, heterotrophs are of two types, namely photoheterotrophs and chemoheterotrophs. Based on the form, bacteria are of different types namely bacillus (rod shaped), cocci (spherical), spiral and pleomorphic.

EXERCISE

1. What is taxonomy? Add a note on the importance of taxonomy.
2. What is Binomial System of nomenclature?
3. Explain the taxonomic hierarchy.
4. Describe nutritional classification of microorganisms.
5. Classify bacteria based on its forms.
6. What is Protista? Add a note on lower protista.
7. Write a note on higher protista.
8. Write short notes on the following.
   - (i) Cyanobacteria
   - (ii) Bacilli
   - (iii) Autotrophs
   - (iv) Spiral bacteria
   - (v) Chemoheterotrophs
   - (vi) Species
   - (vii) Viruses.

25

Prokaryotic and Eukaryotic Microbial Cell

Cells are the tiny, but potent biological compartments. They are the structural and functional units of living organisms. Except viruses, all living organisms are cellular. These organisms may be unicellular or multicellular.

Early in the history of biology, all the living organisms were divided into two groups—plants and animals. After the discovery of microorganisms, algae were placed in the plant kingdom and protozoa were placed in the animal kingdom. But many microorganisms were having the features of both plants and animals. Therefore, in 1866, Haeckel proposed a third kingdom called *Protista,* to include all the microorganisms. Based on the degree of complexity of the cellular organisation by microscopic observation and by biochemical analysis, protista is further divided into prokaryotic cells and eukaryotic cells.

Prokaryotic Cell

Bacteria and cyanobacteria are the examples of prokaryotic cell. Prokaryotic cell is very simple as compared to eukaryotic cell. It is generally smaller and varies in size in different members. In mycoplasma it is about 0.12 mm while in *Oscillatoria,* a filamentous cyanobacteria, it is 40 × 5 mm. However, a majority of them are about 1.0 mm in size.

Under electron microscope the ultrastructure of bacteria (typical Prokaryotic cell) can be studied (Fig. 25.1). It is observed that many bacteria can be surrounded by loose slimy layer, called capsule. Beneath the capsule is a cell wall which gives shape to the cell. Internal to cell wall is cell membrane. The folded invagination of the cell membrane is called mesosomes, which are present in many bacteria. Some bacteria have one or more flagella, which may arise from one or both ends of the cell or may arise from all over the cell. The nucleoid is in the form of a highly folded circular DNA without a nuclear envelope. The cytoplasm contains 70's ribosome, an internal membrane system, storage granules, vacuoles, inclusion bodies and in photosynthetic bacteria chromatophores.

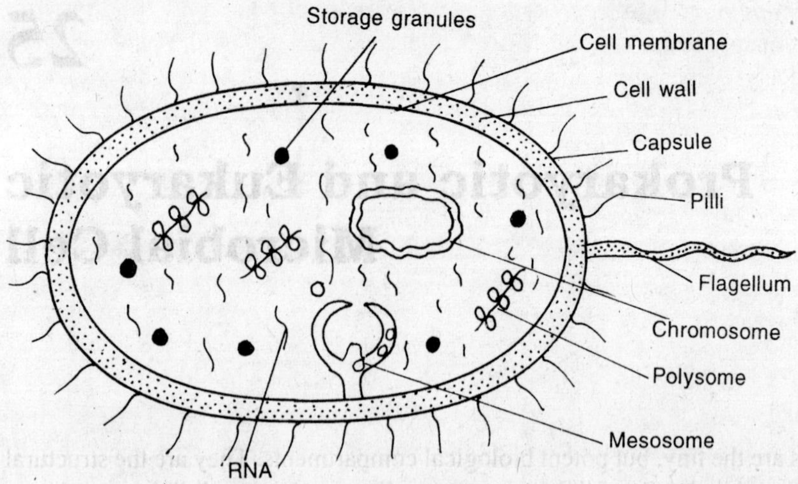

Fig. 25.1. Ultrastructure of Bacteria (A typical prokaryotic cell)

Capsule: Many bacterial species like Diplococcus, pneumococcus, etc. are surrounded by a capsule. It is a viscous or gelatinous loose aggregates of substance secreted surrounding the cell wall. It is not strictly a part of the organism. Chemically the capsules are made up of proteins, polysaccharides and lipids but more particularly polysaccharides. The capsule is not essential for the survival of the organisms under favourable conditions but it provides protection in unfavourable conditions. The presence of a capsule is important in the disease causing ability, i.e. virulence of some bacteria.

Capsules are divided into two kinds, namely macrocapsules and microcapsules. Macrocapsules are at least 0.2 μm thick and can be seen under the light microscope, while microcapsules cannot be seen under light microscope.

Besides protection to the cell, the capsule has many benefits for bacteria like it gives lightness to cell, which permits the bacteria to float near the water surface where aerobic conditions are greater. The capsule may also function in the storage of food substances and a site for disposal of waste material.

Bacterial capsules are species specific and therefore can be used for immunological distinction of closely related species. They are also used as laboratory gels for filtration.

Cell wall: Cell wall is the outermost boundary in a majority of bacteria, while in some bacteria and mycoplasma it is absent. It is non-living and does not have the property of selective permeability but it acts as a sieve as it is porous. The cell wall is thin about 10-25 nm thick and provides rigidity to the cell. It is visible only under electron microscope. It accounts for 20-30% total dry weight of the cell.

Chemically cell wall is made up of strong fibres of heteropolymers called *mucopeptides* or *peptidoglycans*. Mucopeptides are also known as glycopeptide, murein, etc. Mucopeptides are made up of alternative chains of N-acetyl glucosamine and N-acetyl muramic acid molecules with 1-4 linkage. These chains are cross linked by short peptide chains consisting of alamine, glutamic acid and either lysine or diaminopimelic acid and muramic acid. The narrow spaces left by these chains are filled by various other chemical compounds which vary with the species.

Depending upon cell wall composition, bacteria are divided into two types namely *Gram positive* bacteria and *Gram negative* bacteria (Fig. 25.2). The cell wall of gram positive bacteria is simple and contains around 90-95% peptidoglycan and 5-10% teichoic acid. In gram negative bacteria cell wall composition is more complex. It is made up of several layers. It contains at least four different proteins components, lipoprotein, lipopolysaccharides and very less amount of peptidoglycan. Liprotein, lipopolysaccharide and other proteins form the outer portion of membrane, while peptidoglycan occurs in between cell wall and plasma membrane. The outer membranous portion is joined to the peptidoglycan region by many links of lipoproteins.

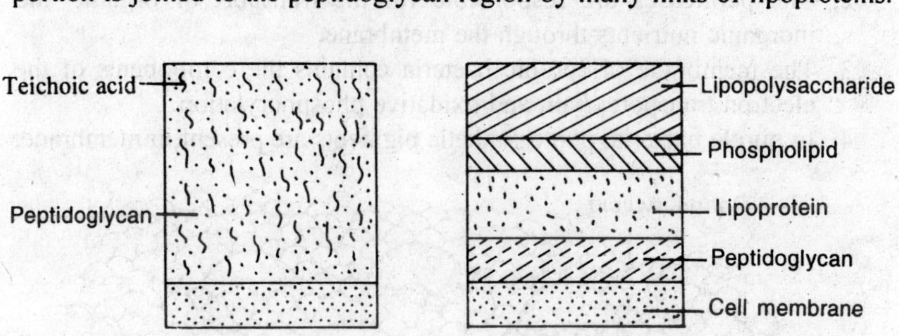

Fig. 25.2 A schematic representation of gram positive and gram negative cell wall

Table 25.1 Comparison between gram positive and gram negative cell wall

Gram positive cell wall	Gram negative cell wall
1. Cell wall is thick.	1. Cell wall is thin.
2. Few amino acids present.	2. Many amino acids present.
3. Sulfur containing amino acids absent.	3. Sulfur containing amino acids present.
4. Teichoic acid present.	4. Teichoic acid absent.
5. Peptidoglycan present.	5. Peptidoglycan present in less quantity.
6. Lipoproteins absent.	6. Lipoproteins present.
7. Lipopolysaccharide absent.	7. Lipopolysaccharide present.
8. Polysaccharide present.	8. Polysaccharide absent.

Cell membrane/plasma membrane

Internal to the cell wall, the plasma membrane or cell membrane is present. Cell membrane forms the outer boundary of the cytoplasm and is selectively permeable and thus regulates entry and exit of molecule into cytoplasm. It is chemically made up of lipoproteins (70% protein and 30% lipid) and no carbohydrates.

The major structural features of the cytoplasmic membrane were explained by *fluid mosaic model* of Singer and Nicolson (1972). Now it is a widely accepted model for explaining the properties of cell membrane. According to this model, there is a continuous bilayer of phospholipid molecule in which globular proteins are embedded (Fig. 25.3).

Prokaryotic cells contain relatively fewer membranes than eukaryotic cells. Therefore, the cell membrane carries out many functions in prokaryotic cell, which are distributed among the various cell organelles of eukaryotic cells. It is an important centre of metabolic activity. In bacteria cell wall performs following functions:

1. The cell membrane contains the enzymes which synthesise the components of bacterial cell membrane and cell wall such as phospholipids, peptidoglycans, teichoic acids, lipopolysaccharides and simple polysaccharides.
2. The permeases are responsible for the transport of organic and inorganic nutrients through the membrane.
3. The membrane of aerobic bacteria contains the components of the electron transport chain and oxidative phosphorylation.
4. In purple bacteria, photosynthetic pigments are present in membrane.

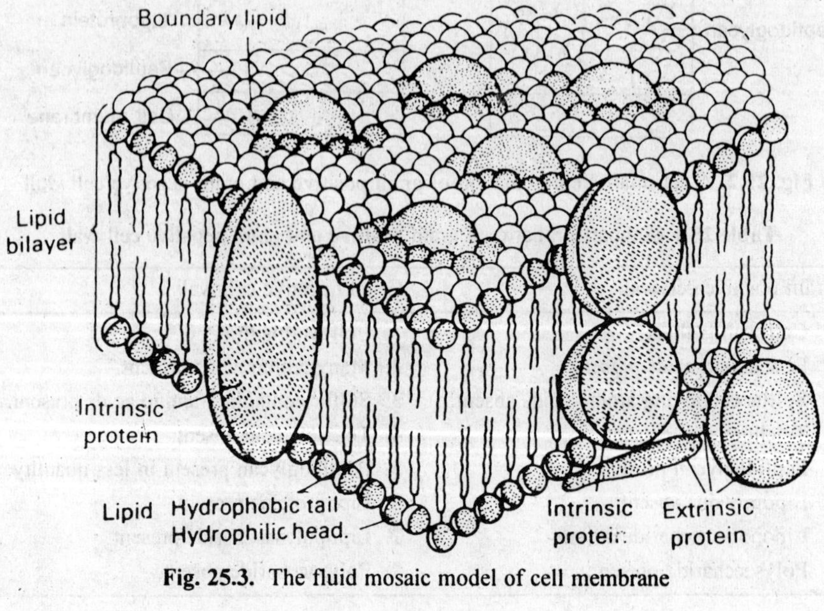

Fig. 25.3. The fluid mosaic model of cell membrane

5. It provides attachment sites for chromosomal and plasmid DNA.
6. It functions as an osmotic barrier.

Mesosomes: In some bacteria, the cell membrane forms vesicles or pocket like infoldings into cytoplasm at several regions, which are called *mesosomes*. Mesosomes are more frequently found in gram positive bacteria than gram negative bacteria.

The mesosome consists of many vesicles, tubules or lamellar whorls. The lipid component of the mesosome appears to be similar to the cell membrane, but its protein differs (Fig. 25.4).

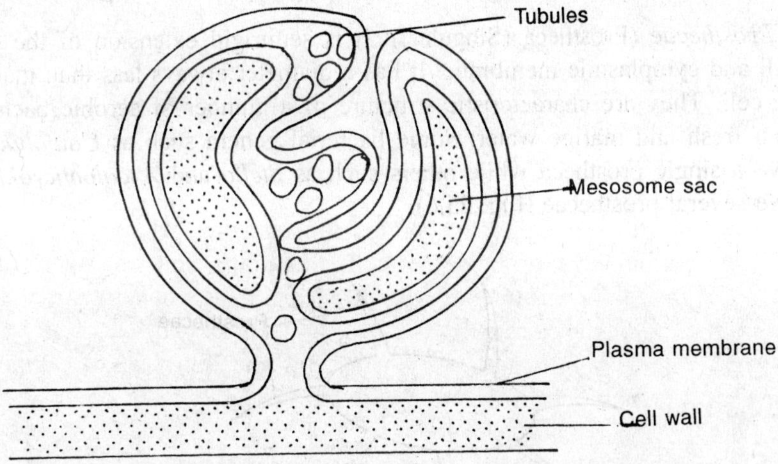

Fig. 25.4. The structure of mesosome

Mesosomes are found in large numbers in organisms that have a higher respiratory activity as the nitrogen fixing bacteria namely Azotobacter or during the logrithmic phase of growth. It is believed that mesosomes are equivalent to the mitochondria of higher cells. And they were thought to be the centre of respiratory activity. But, now it is observed that respiratory enzymes are present throughout the cell membrane.

In photosynthetic bacteria, the extent of membrane infolding has been related to pigment content and photosynthetic activity. In sporulating bacteria, the appearance of such infolding is a prerequisite for sporulation. According to Peberdy (1980) mesosomes are also involved in cell wall synthesis and are necessary for cross wall formation during cell division.

Sheaths: Some species of bacteria, particularly from freshwater and marine water, form chains or trichomes [a single row of distinct cells in which there is a large area of contact between the adjacent cells], that are enclosed by a *hollow tube* called *sheath* (Fig. 25.5). This structure is most readily visualised when some of the cells have migrated from it. Some examples of sheathed bacterial genera are Sphaerotilus, Leptothrix, Haliscomenobacter, Streptothrix, Lieskeella, Crenothrix, etc.

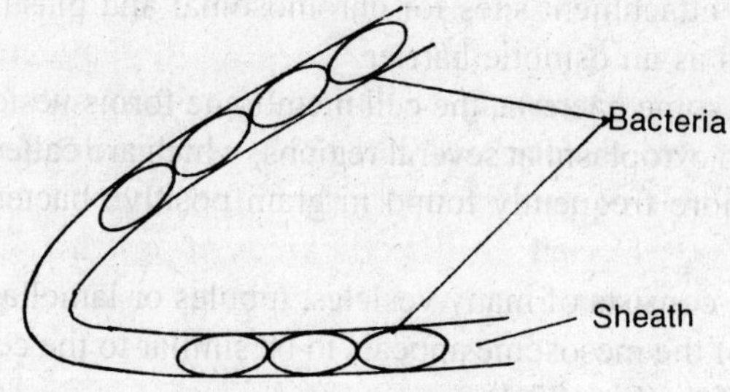

Fig. 25.5. Sheathed bacteria

Prosthecae (Prostheca (Singular)): It is semirigid extension of the cell wall and cytoplasmic membrane. It has a diameter always less than that of the cell. They are characteristic structure of a number of aerobic bacteria from fresh and marine water. Some bacterial genera such as *Caulobacter* have a single prostheca while others such as *Stella* and *Ancalomicrobium* have several prosthecae (Fig. 25.6).

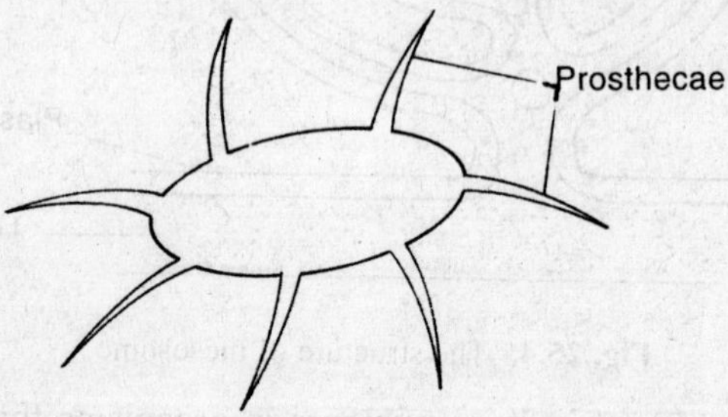

Fig. 25.6. Ancalomicrobium with several prosthecae

Prosthecae increase the surface area of the cells for nutrient absorption, which is advantageous in dilute environments. Some prosthecate bacteria may form a new cell (bud) at the end of a prostheca. Some bacteria have an adhesive substance at the end of a prostheca that helps in attachment to surface.

Stalks: Sometimes the term *stalk* is used interchangeably with the prostheca or hypha. But it is certain non-living ribbon like or tubular appendages that are excreted by the cell. These stalks help in attachment of the cell to surface. Some examples of genera having stalks are Gallionella, Planctomyces, etc.

Cytoplasm: Cytoplasm is a viscous substance like in eukaryotes but without streaming movements. It includes ribosomes, nuclear material, protein, amino acids, various enzymes, inorganic ions, sugars, vitamins and other water soluble and reserve food materials. Plasmids and episomes are also found in cytoplasm.

Ribosomes: Ribosomes are dispersed throughout the cytoplasm, which involved in protein synthesis. The number of ribosomes per cell varies and it is high during increased activity of protein synthesis. Prokaryotic cell contains 70S ribosome, which is made up of two subunits 50S and 30S. The 50S subunit consists of one molecule of 23S rRNA, one molecule of 5S-rRNA and 32 different kinds of protein while the 30 S subunit consists of one molecule of 16 S - rRNA and 21 different kinds of proteins. During protein synthesis many ribosomes are held together by mRNA chain, called *polyribosomes* or *polysomes*.

Nuclear Material: In prokaryotes, the nuclear material is present in the cytoplasm, without nuclear membrane and nucleolus. It is called nucleoid. Nucleoid is made up of single molecule of double stranded, circular DNA. In many bacteria, the nuclear material is in supercoiled state in the centre. The bacterial DNA molecule is often referred as the bacterial chromosome.

Plasmids and episomes: In many bacteria, the extra chromosomal double stranded, circular DNA is present, which are called plasmids or episomes. Plasmids are free of the nucleoid chromosome while episomes may be free or integrated to nucleoid chromosome. Generally plasmids contains gene for drug resistance, toxin producing ability, etc. beside genes for its replication.

Flagella: Except cocci, bacilli and spirillum have long, thread like appendages attached to the cells, called flagella. In moist medium, flagella bring about locomotion. It is thin, about 0.02 μm in diameter and 3 to 20 μm long.

Flagella is invisible under light microscope. Electron microscopic studies show that it is made up of three parts namely filament, hook and basal body.

(1) *Basal body:* It is spindle shaped structure that joins the hook to the cell membrane. In gram negative bacteria, basal body is complex and made up of two pairs of rings connected by a rod. These rings are 'M' (membrane) ring, 'S' (super membrane) ring, 'P' (peptidoglycan) ring and 'L' (lipopoly-saccharide) ring from inner to the outer side (Fig. 25.7).

The outer pair (L and P) rings are fixed to the cell wall, while the inner pair (S and M) rings are attached to the plasma membrane.

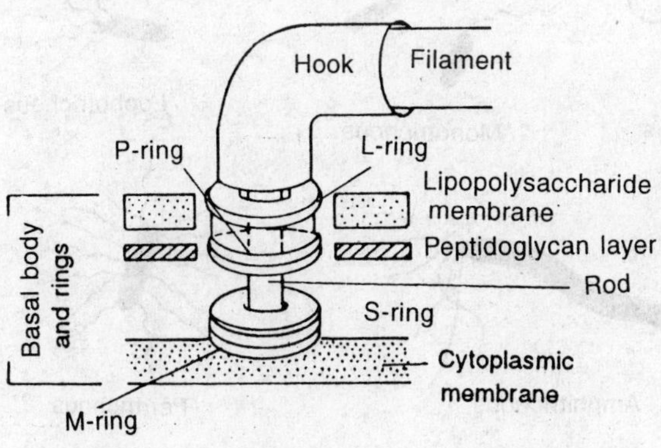

Fig. 25.7. Flagellum of gram negative bacterium

In gram positive bacteria, the basal body is much simpler and the outer set of rings are absent. The 'M' ring is embedded in the inner membrane as in gram negative bacteria while the 'S' ring may be attached to the inside of the thick peptidoglycan layer (Fig. 25.8).

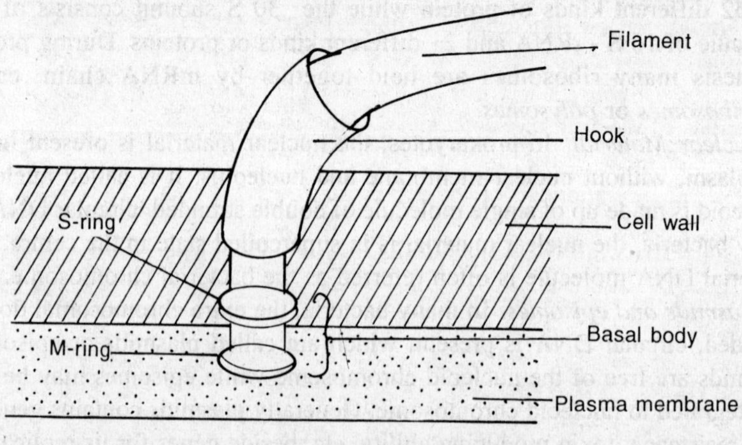

Fig. 25.8. Flagellum of gram positive bacteria

(2) *Hook:* The hook connects the basal body with the filament. It is thicker than the filament and it penetrates the cell wall. It is bent. The hook of the gram positive bacterial flagellum is longer than that of the gram negative bacterial flagellum.

(3) *Filament:* Filament is made up of protein fibers of about 13-14 nm in diameter. These proteins are called *flagellin*. The molecular weight of flagellin ranges from 30,000 to 60,000 and it is made up of 304 amino acid molecules.

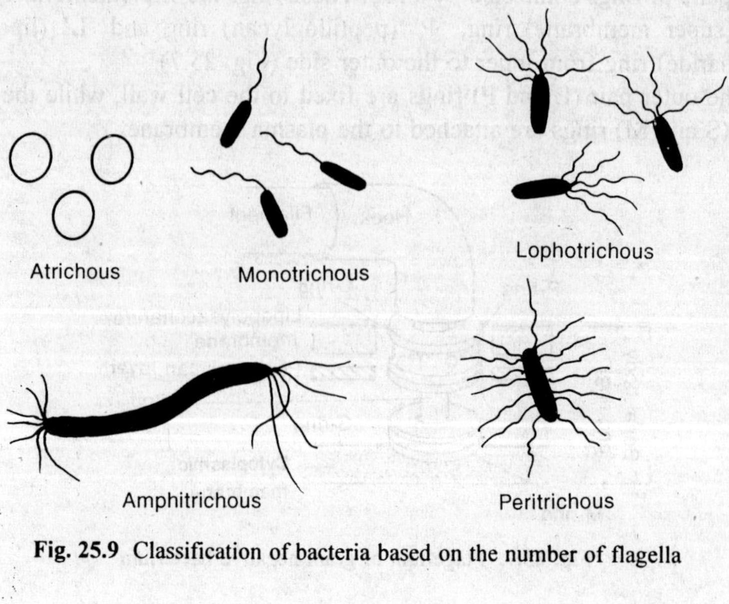

Fig. 25.9 Classification of bacteria based on the number of flagella

On the basis of the number and position of flagella, the bacteria are classified into following types (Fig. 25.9).

(a) *Atrichous:* A bacteria without flagella are called atrichous. e.g. cocci.

(b) *Monotrichous:* Bacteria with single flagellum present at one end. e.g. *vibrio.*

(c) *Lophotrichous:* Bacteria with many flagella present at one pole. e.g. *Spirillum undula.*

(d) *Amphitrichous:* Bacteria with two tufts of flagella attached at either poles of the cell. e.g. *spirilla.*

(e) *Peritrichous:* Bacteria with many flagella are present all round the cell. e.g. *Salmonella typhosa.*

Fimbriae (Pili): Many gram negative bacteria have short, very fine, hair like appendages called fimbriae or pili. These are originated from plasma membrane. Pili are present only in gram negative bacteria. For example Enterobacteriaceae, Caulobacter, Pseudomonadaceae, etc.

Fimbriae of bacteria differ from flagella in being thinner, straighter, and less rigid. Their number is also much more than flagella. Their length varies from 0.2-20 µm and width varies from 30°A-140°A; some have up to 250°A.

The fimbriae are made up of protein molecules which consist of at least 163 amino acids. In some cells, some pili are longer than remaining pili; such pili are called sex pili and cells are called donor or male cells. The sex pili have a greater diameter, around 60-135°A, and an axial hole around 25-30°A in diameter. The number of sex pili is very less as compare to pili.

Functions of pili: The fimbriae of bacteria serves multiple functions like

(1) It is mainly associated with adhesive properties, which enable the organism to attach themselves to natural substrates and to other individuals of their species.

(2) The sex pili may provide the means of chromosome transfer during conjugation, by acting as conjugation tube.

Eukaryotic cells: Eukaryotic cell is more complicated than prokaryotic cell. It is characterised by the presence of a multiplicity of unit membrane system which are structurally and topologically distinct from the cytoplasmic membrane. These membrane systems enable the separation of different eukaryotic cytoplasmic functions into specialized organelles.

The plant, animal, fungi, algae, etc. are the examples of eukaryotic cells. The eukaryotic cell contains cell membrane, which is outermost membrane in animal cells while in plant cells cell wall is outermost membrane. Inside the cell membrane it contains cytoplasm. In the cytoplasm various cell organelles like endoplasmic reticulum, gogi complex, mitochondria, ribosome, lysosomes and chloroplast (plant cells) is present. The DNA or genetic material in eukaryotic cell is surrounded by nuclear membrane. Such structure is called nucleus, which is the characteristic feature of eukaryotic cell.

Cell wall: The cell wall is the outermost boundary of plant cells. It is absent in animal cells. It is made up of non-living matter, which is believed to be secreted by the plasma-membrane. It is very rigid and maintains the shape of the cell. The structure and composition of cell wall varies in different plants and also in different types of cells of the same plant.

The cell wall is made up of three layers namely primary cell wall, middle lamella and the secondary cell wall (Fig. 25.10). The middle lamella separates the primary cell wall of adjacent cells. Secondary cell wall is deposited in few special cells, after a certain period of development.

The primary cell wall is about 1 to 3 μ in thickness, which is secreted by plasma membrane. Chemically the primary cell wall is composed of cellulose, hemicellulose and pectic compounds.

The middle lamella is cementing substance, which binds together the adjacent cells. It is chemically composed of pectin, cellulose, calcium and magnesium sulphate.

The secondary cell wall is very rigid and does not alter its shape, which is deposited on the primary cell wall after cell maturation. The thickness of cell wall varies from 5-10 μm. Generally, it has three layers namely outer layer, middle layer and inner layer.

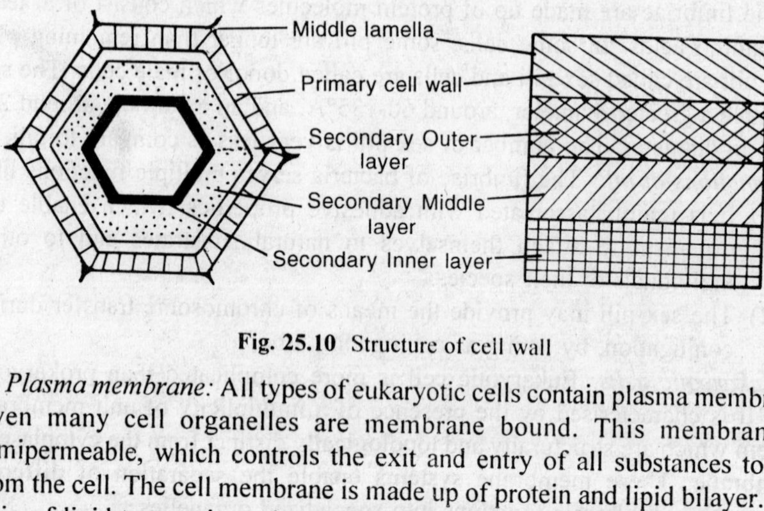

Fig. 25.10 Structure of cell wall

Plasma membrane: All types of eukaryotic cells contain plasma membrane. Even many cell organelles are membrane bound. This membrane is semipermeable, which controls the exit and entry of all substances to and from the cell. The cell membrane is made up of protein and lipid bilayer. The ratio of lipid to protein varies depending on the type of membrane, type of organism and the type of cell. Plasma membranes have three different types of proteins, namely, structural proteins, functional proteins (enzymes) and carrier proteins. The structural proteins maintain the structural framework of the membrane. The enzymes or catalytic proteins catalyze various metabolic activities, which are found in the membrane of mitochondria, endoplasmic reticulum, chloroplast, etc. Carrier proteins form a complex with the substances and help them to pass through the membrane.

Plasma membrane have three different types of lipids, namely sterols, phospholipids and glycolipids. But phospholipids are present abundantly in most membranes.

The fluid mosaic model by Singer and Nicolson (1972) was the most accepted model for membrane structure (Fig. 25.3), which satisfactorily explains all its known functions.

Functions of plasma membrane

1. It is the boundary for cytoplasm and cell contents.
2. It regulates the entry and exit of many substances to and from the cell.
3. Receptor molecules (surface substances) mediate and control the message from one cell to other, from environment to the cell and from hormones to the cell.
4. As it is a living boundary, it is responsible for effective coordination of actions between cells.
5. It is associated with interconversion of energy. For example, enzymes present in mitochondrial membrane convert chemical energy (glucose) into energy currency (ATP).
6. It provides structural framework for the location of various enzymes.
7. It is involved in cell movement.
8. It acts as electrical insulator.

Cytoplasm: Cytoplasm is a colloidal substance which presents internal to plasma membrane. The cytoplasm together with the plasma membrane is known as the protoplast. Generally cytoplasm is granular and has many substances either dissolved or suspended in it. Cytoplasm has a number of inclusions like nucleus, mitochondria, plastids, golgi complex, endoplasmic reticulum, ribosome, lysosomes, etc. Around 90% of cytoplasm contains water.

Nucleus: Nucleus is the most important cell organelle, which regulates all the cell activities. Usually it is present at the centre of the cell. Generally it is spherical or ellipsoidal, but its shape varies depending on the cell.

Structurally nucleus is made up of nuclear membrane, nucleoplasm, chromonemata, nucleolus (Fig. 25.11). In all eukaryotic organisms, genetic material is present in nucleus, which is bounded by the nuclear membrane. It is a double membrane called outer and inner membrane. The nuclear membrane have number of pores, which are octogonal in shape. These nuclear membrane pores help in effective communication between nucleus and cytoplasm.

Inside the nuclear membrane, there is a homogenous matrix of fluid substances containing a chromatin fibre called nucleoplasm. In the cytoplasm, the chromatin fibre is suspended. During cell division, the chromatin fibre condenses into distinct structure called chromosomes. There are two types of chromatin called euchromatin and heterochromatin. Euchromatin is loosely condensed while heterochromatin is highly condensed region of chromatin. It is rich in DNA.

Nucleolus is spherical body, without any membrane, which is situated in the central or peripheral part of nucleus. Generally two or more chromosomes are associated with nucleolus. Nucleolus is rich in RNA. It has three regions called granular region, fibrillar region and amorphous region. Nucleolus is the active site of RNA synthesis. It is the source of ribosomal RNA.

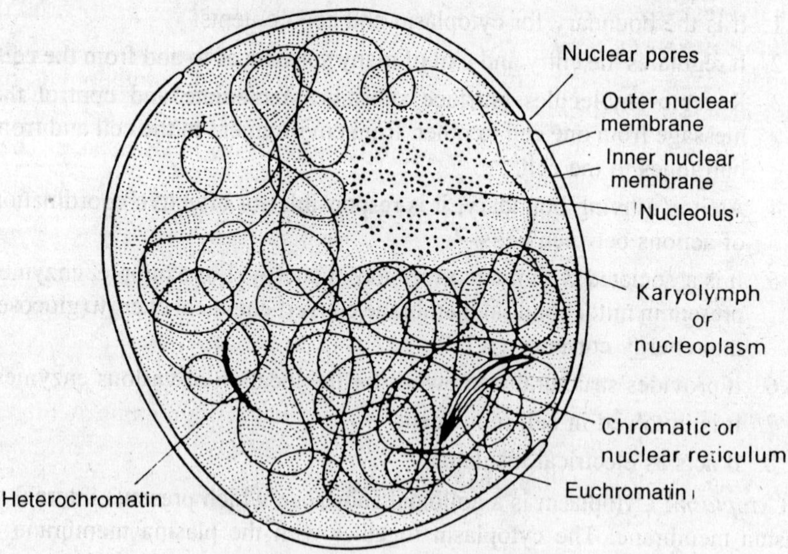

Fig. 25.11. The structure of nucleus

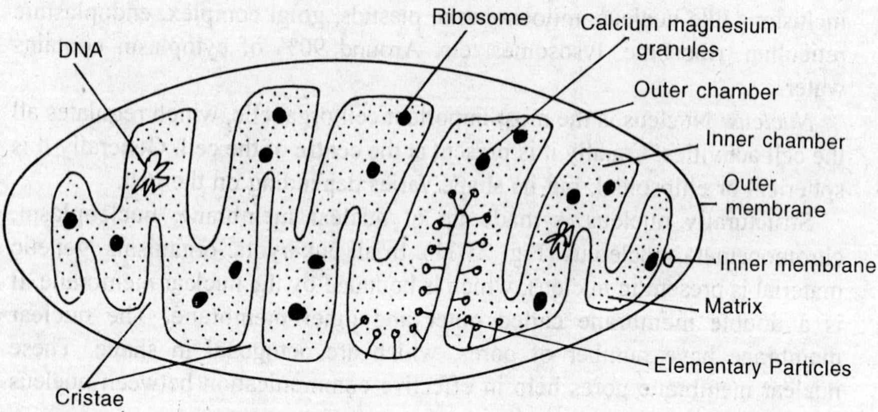

Fig. 25.12. The ultrastructure of mitochondria

Mitochondria: Generally mitochondria is called 'power house' of the cell. It is the site for cellular respiration, where ATP molecules are synthesized. Mitochondria have two membranes called outer membrane and inner membrane. The space between outer and inner membrane is filled with a fluid. Mitochondrial matrix is present inside the inner membrane. Inner membrane is folded into finger like projections called *cristae*. On the inner side of inner membrane many small F_1 particles are present, which contains enzymes for respiration (Fig. 25.12). Mitochondrial matrix is proteinaceous material, in which 70 S ribosome and circular DNA molecules are suspended.

Chloroplast: In autotrophic eukaryotes, the photosynthetic pigments are present in a special cell organelle called chloroplast. In photosynthetic eukaryotes, the solar energy is converted into chemical energy in the chloroplast. The number of chloroplast per cell varies. The chloroplast contain carbohydrates, lipids, proteins, chlorophyll, carotenoid, xanthophylls and nucleic acid.

Chloroplast is double membranous with an inner membrane space. These membranes are smooth without any perforations or particles. Inner chamber or space is filled with a granular matrix and an intricate system of membranes. These membranes are called *lamellae* and the surrounding matrix is called the *stroma.* These lamellae are packed like stacks and are called thylakoids. Thylakoids form highly compact bundles called granna. Some thylakoids of granum extend into the stroma and maintain contact with other grana, which is called stroma lamellae or stroma thylakoids (Fig. 25.13).

The chloroplast of algae lack the granum arrangement, while in cyanobacteria, the thylakoids lie naked in cytoplasm without any envelope.

In addition to lamellae, the stroma has granules, lipid droplets, starch grains and vesicles.

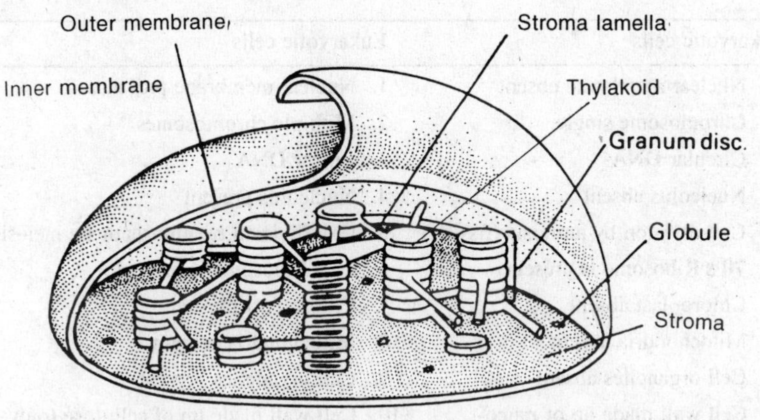

Fig. 25.13. The ultrastructure of chloroplast

Endoplasmic reticulum: The cytoplasm of eukaryotic cell is transversed by a network of fibrillar structures called endoplasmic reticulum. The endoplasmic reticulum is connected to the outer nuclear membrane. It is also connected with the plasma membrane.

Endoplasmic reticulum consists of three types of structures namely cisternae, vesicles and tubules. The cisternae are elongated cylindrical unbranched structures, which are arranged parallel to one another. It is 40-50 µm in diameter. The vesicles are oval bodies, attached to the membrane. It is

25-500 μm in diameter. The tubules are irregularly branched structures with diverse shapes. It is 50-100 μm in diameter. All these three structures are bounded by a unit membrane of 50-60°A thickness. Depending on the nature of membrane, endoplasmic reticulum is of two types called smooth and rough endoplasmic reticulum. The rough endoplasmic reticulum has a granular surface due to presence of ribosomes attached to it, while the membranes of smooth endoplasmic reticulum are not granular due to the absence of ribosomes.

The endoplasmic reticulum has many functions like storage and transport of metabolities, formation of nuclear envelop, intracellular signal conduction, etc.

Golgi complex: Golgi complex has a very complex structure and varied functions. In some lower eukaryotes like fungi, golgi complex is absent. Structurally, it is made up of cisternae, tubules, vesicles and vacuoles.

The cisternae are flattened sacs filled with a fluid. Each cisternae is enclosed by trilaminar membrane 6 μm in thickness and is separated from the other by a 100-150°A. The tubules arise from the peripheral region of cisternae. The vesicles are globular structures attached to the tubules. Vesicles are of two types called smooth and coated vesicles. Vacuoles are large spaces occurring at the distal end of the cisternae.

Table 25.2 Difference between prokaryotic and eukaryotic cells

Prokaryotic cells	Eukaryotic cells
1. Nuclear membrane absent	1. Nuclear membrane present
2. Chromosome single	2. Multiple chromosomes
3. Circular DNA	3. Linear DNA
4. Nucleolus absent	4. Nucleolus present
5. Cell division by amitosis (fission)	5. Cell division by mitosis or by meiosis
6. 70 s Ribosome is present	6. 80 s Ribosome is present
7. Chloroplast absent	7. Chloroplast present
8. Mitochondria absent	8. Mitochondria present
9. Cell organelles absent	9. Cell organelles present
10. Cell wall made up of muco-peptides	10. Cell wall made up of cellulose (only plants)
11. In flagella, no definite arrangement of fibrils	11. In flagella, 9+2 fibrillar arrangement present
12. Amoeboid movement of cytoplasm absent	12. Amoeboid movement of cytoplasm present
13. Generally unicellular	13. Generally multicellular

Golgi complex performs different functions like absorption of compounds, site of enzyme and hormone production, site of protein, storage, formation of acrosome during sperm formation, formation of plant cell wall, etc.

Ribosomes: The eukaryotic cells contain 80s ribosomes, having a molecular weight of 4×10^6 daltons. The 80s ribosomes have two sub units, namely 60s and 40s units. The number of ribosomes per cell varies. Ribosomes may occur singly (monosomes) or may be in clusters (polyribosomes) in association of mRNA.

Lysosomes: It is also called as 'suicidal bags' of cells, due to the presence of hydrolytic enzymes in it. The main function of lysosome is of intracellular substances. Lysosomes are sac like structure with single membrane filled with digestive enzymes. Basically there are two types of lysosomes called primary and secondary lysosomes. Primary lysosomes are sac like structure produced directly from endoplasmic reticulum or golgi complex. The secondary lysosomes are of three types namely phagosomes, residual bodies and autophagic vacuoles.

Glyoxysomes: It is found in the cells of yeast and some fungi. The glyoxysomes contain enzymes for fatty acid metabolism, glyoxylate cycle, etc.

SUMMARY

All the living organisms are broadly classified into prokaryotic cells and eukaryotic cells. In prokaryotic cells, nucleus is absent, while in eukaryotic cells nucleus is present.

The prokaryotic cells contain cell wall, plasma membrane, cytoplasm, circular, single, large, double stranded chromosome, plasmids, 70 s ribosome, flagella, pili, capsule, etc. The eukaryotic cells contain plasma membrane, cell wall (only in plants), cytoplasm, mitochondria, nucleus, chloroplast, endoplasmic reticulum, golgi complex, lysosomes, 80 s ribosome, etc.

EXERCISE

1. Describe the broad classification of living organisms.
2. Explain prokaryotic cells.
3. Describe the typical eukaryotic cell.
4. Differentiate between prokaryotic and eukaryotic cells.
5. Write short notes on:
   (i) Bacterial flagella
   (ii) Sheathed bacteria
   (iii) Plasma membrane
   (iv) Mitochondria
   (v) Capsule
   (vi) Endoplasmic Reticulum
   (vii) Nucleoid

26

Microbial Growth

One of the unique properties of living organisms is their ability to grow and reproduce. This unique property has been responsible for the maintenance of life on this planet, ever since it originated a few billion years ago. The growth of an organism can be defined as an orderly increase in all cellular as well as structural components of cell or cells. However, an increase in cell mass alone need not reflect growth since cells can simply take up water or deposit storage products such as glycogen or lipids and thus increase in size and weight of the cells. Therefore growth is followed by cell division, resulting in an increase in cell number. Under favourable conditions of growth all cells (plants, animals and microorganisms) multiply by the asexual processes of cell fission. If food and energy sources are available adequately and environmental conditions are favourable, then asexual reproductive process can continue indefinitely.

In multicellular plants and animals, growth of body cells results only in increase in size of the individual organism and not in increase in the number of organisms. But the term *growth* applied to unicellular microorganisms refers to not the increase in size of individual organism but increase in number of microorganism. In other word growth means same thing in multicellular organism. For example, a child grows into an adult human being not because its cells have become larger, but because the adult has more cells.

Most bacteria multiply by transverse, binary fission, in which cell divides in two cells after developing a transverse cell wall. Each of these cells are assumed to be nearly identical in all properties. Replication of a cell is the more or less accurate doubling of the amount of all various components of cell and then their partition into two daughter cells, each of which is a replica of the mother cell. For example, if a cell contains 15,000 ribosomes, as the cell goes through its life cycle the number of ribosomes will increase to about 30,000. And when the cell divides, it will produce two daughter cells, each containing about 15,000 ribosomes.

Microorganism can create a complete duplicate of itself within a short period. Bacterium usually replicate in as short as 20 minutes. The life cycle of the microorganisms consists of the time between one cell division and the next cell division. Therefore, the degree of morphological changes within this life cycle is slight as compared with higher eukaryotic organisms, which develop from a fertilized egg to a complex organism. But, in the microbial life cycle, complex biochemical events occur, which require the synthesis of millions of parts.

Factors affecting growth of microorganisms

A variety of factors may affect microbial growth, which are nutrients, pH, temperature, osmotic pressure, concentration of oxygen, etc. The effect of these factors are briefly summarized below:

(1) *Nutrients:* Microorganisms, to grow, must obtain all the substances that they require for the synthesis of all the cell components and for the generation of energy, from the environment. These substances are called nutrients. In nature, microorganisms obtain all the necessary nutrients from their habitats. But for culturing microorganisms, these nutrients must be made available in the medium. Water plays a key role in the culture of microorganisms. It is essential for all the biological reactions which is required for the survival and growth of microbial cells.

Microbial nutritional requirements are qualitatively as well as quantitatively extremely diverse. Therefore, a culture medium must have all the required material for the growth of a particular microorganism. Thousands of media have been proposed for the cultivation of microorganisms, which contain specific nutrients according to the requirement. Therefore a culture media should be designed to fulfil the minimal nutritional requirements which are essential for the synthesis of cell components.

(2) *Temperature:* Growth is the net result of a large number of enzymatic reactions. Thus, the rate of growth would have a direct relationship with the rate of reactions. Again, it is well known that temperature affects the rate of enzyme catalysed reaction. Hence, growth can also be related with temperature. With an increase in temperature, the growth rate will increase but falls sharply at both the upper and lower limits. The sharp fall in the growth rate at high temperature is due to thermal denaturation of enzymes. Every organism, therefore has a optimum temperature for growth. At the optimum temperature for growth, the organism shows maximum growth. On the basis of optimum temperature for growth, the microorganisms are broadly divided into three groups namely the psychrophiles, the mesophiles and the thermophiles.

(3) *pH or H^+ ion concentration:* Depending upon the group or species, the microorganisms have a wide range of optimum pH. Many yeasts and molds and certain bacteria grow well in acidic media (pH 5.0 or lower),

while most bacteria require a pH near neutrality (pH 6.5 to 8.0) for maximum growth. The pH of the media profoundly affects the growth of microorganisms, since enzymes are sensitive to change in pH. Each organism, therefore, has an optimum pH for normal growth. The growth rate is maximum at the optimum pH and becomes slower at lower or higher pH values.

(4) *Oxygen requirement:* Microorganisms vary remarkably in the amount of oxygen required for optimum growth. For example, the aerobes are dependent on the presence of oxygen while the anaerobes are sensitive to oxygen. In between these two types, there are facultative anaerobes which can grow either in the presence or in the absence of oxygen. The growth rate of each organism will depend on the optimum aeration conditions. However, increase in the degree of aeration will increase the rate of growth up to only a certain point. Thus, every organism has a need for optimal aeration for obtaining a maximum growth rate.

(5) *Osmotic pressure:* The maintenance of a relatively constant ionic strength within the microbial cell is of great physiological significance since the stability and behaviour of enzymes are greatly influenced by this factor. Microbes have the ability to grow in media with varying osmotic pressures, but to most microorganisms a high sugar or salt concentration may be lethal. However, some yeasts and molds have the capacity to grow in medium containing a high concentration of sugar (40-70% sucrose) called *osmophiles*. Also, some bacteria of marine origin require high concentrations of sodium chloride (more than 20%) for their optimal growth and are termed as *halophiles*.

Bacterial Growth Curve

The growth of microorganisms may be studied by inoculating a culture medium with a small number of microorganisms. When all the nutritional and other factors required for growth are ideal, the cell continues to multiply. But when one or the other necessary factors are exhausted, the growth gets retarded. If during this period, the culture medium is not replaced with fresh culture medium or waste products of microbial metabolism are not removed, the growth in such a system is called *batch culture*. If there is continuous addition of fresh growth medium and subsequently withdrawal of equal volume of the bacterial culture from system, it is called continuous culture of microorganism.

A batch culture is very helpful in studying the growth character of microorganisms. When bacteria are transferred from a siant culture to a known volume of liquid medium, the population undergoes a characteristic sequence in its rate of increase in cell number. The increase in cell number of microorganism may be ascertained by counting them in a specific amount at periodic intervals. When a logarithm of the counted cells is plotted against time a growth curve is obtained (Fig. 26.1). Basically this growth curve is sigmoid which shows four recognizable phases—lag phase, log phase, stationary phase and the death phase. While the above phases are clearly recognizable phases, there is no sudden change from one phase to another phase. Since all bacterial cells will not be in the same stage of growth, the change from one phase to another phase is gradual and there will be several phases between the four standard phases of growth. These different phases of microbial growth curve are as follows:

1. Lag phase or initial slow phase
2. Phase of accelerated growth
3. Log phase or exponential phase
4. Phase of decrease in growth rate
5. Stationary phase
6. Phase of increase in death rate
7. The declining phase
8. The survival phase

(1) *The lag phase or initial slow phase:* In the lag phase, there is no increase in the number of viable cells. However, cell growth occurs as indicated by the increase in cell mass. During this period cells increase in size as a result of extensive macromolecular synthesis. This stage represents a period of active growth without cell division and the cells prepare for division. The length of the lag phase depends on a variety of factors such as the age of the inoculum, the composition of the growth medium and the environmental factors such as pH, temperature, aeration, etc.

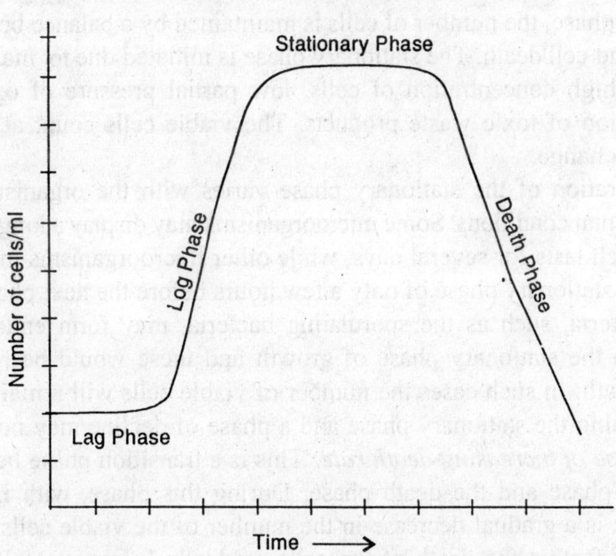

Fig. 26.1. Growth curve of bacteria in a batch culture

(2) *Phase of accelerated growth:* This is the intermediate phase between lag phase and log phase. Once the cells are sufficiently physiologically

mature at the end of lag phase, the cells start dividing. However, all the cells are not dividing simultaneously because all of them are not completing the lag phase at the same time. Therefore, in the phase of accelerated growth, the number of cells dividing will increase gradually.

(3) *The log phase or the exponential phase:* In the log phase, the cell number increases logarithmically and the cells divide at the maximum rate. The cell number during this period increases as a function of the exponent (2^1, 2^2, 2^3, 2^4, ...). Therefore, at this phase, the log of the number of cells plotted against time will show a straight line.

During this phase, cells are actually smaller in size as they are constantly dividing. The log phase is the most suitable period for studying biochemical and physiological properties of cells, because the cells are metabolically more active during this phase. The log phase is also extremely sensitive to the change in environmental factors like pH, temperature, aeration, concentration of nutrients, etc.

(4) *Phase of decrease in growth rate:* As the microbial cells increase continuously during log phase, it results in gradual decrease in nutrients quantity and also there is accumulation of toxic waste produced by microbial metabolism. These factors will reduce the cell division and the cells do not multiply with the same rate as during the log phase. This phase is the transition phase between log phase and stationary phase.

(5) *The stationary phase:* In the stationary phase, the cells can no longer divide as the amount of available nutrients is almost negligible. In the stationary phase, the number of cells is maintained by a balance between cell division and cell death. The stationary phase is initiated due to many factors, like very high concentration of cells, low partial pressure of oxygen and accumulation of toxic waste products. The viable cells count at this stage shows no change.

The duration of the stationary phase varies with the organism and the environmental conditions. Some microorganisms may display a long stationary phase which lasts for several days, while other microorganisms may show a very short stationary phase of only a few hours before the next phase begins. Some bacteria, such as the sporulating bacteria, may form endospores as they reach the stationary phase of growth and these would be resistant to lysis or death. In such cases the number of viable cells will remain constant after attaining the stationary phase and a phase of decline may not be seen.

(6) *Phase of increasing death rate:* This is a transition phase between the stationary phase and the death phase. During this phase, with increase in time, there is a gradual decrease in the number of the viable cells. Due to a number of factors the death of the cells gradually increases and reaches a maximum at the end of this phase.

(7) *The death phase or the declining phase:* The death phase or the phase of decline is characterized by an exponential decrease in the number of viable cells. A cell during this phase is considered to have died or to have

become inviable when it is no longer capable of multiplying. The death rate of bacteria is not uniform like that of log phase where the increase in number is uniformly exponential, because few species of bacteria (e.g. Neisseria gonorrhea) die rapidly as the cells are very susceptible to autolysis.

(8) *The survival phase:* During this phase the rate of the cell death is balanced by the rate of cell division, and very less population of cells is maintained. The survival of the cell depends on the type of organism and the environmental conditions. Some microbes die within a few days while some microbes might remain viable for months and even for years. This is also a stage where the mutant forms induced recently find the favourable condition for rapid growth.

Methods of determining microbial growth

Microbial growth can be determined by a variety of techniques. The choice of technique for determination of microbial growth depends on many factors like nature of microorganism, characteristics of the microorganism, purpose of study, etc. These techniques falls into following categories—

1. Determination of cell numbers
   (a) Total count
   (b) Viable count
2. Determination of cell mass
   (a) Direct method
   (b) Indirect method.

1. Determination of cell numbers

As mentioned in the introduction, the growth is followed by cell division and results in increase in cell number. Therefore, counting the number of microbial cells in the culture medium is essential for the determination of microbial growth. There are two methods for counting the microbial cells.

(a) *Total count:* The total count of microorganisms include the total number of living as well as dead cells. Hence it is called a non viable count. The total count of microorganisms can be determined either microscopically or electronically. In the microscopic method, the number of cells in an appropriately diluted sample can be determined using a haemocytometer. Then the total count is calculated. In the electronic method, a special instrument called the coulter counter is used for direct counting of cells from a suspension. The electronic method can count thousands of cells in a few seconds. But the microscopic method has the advantage in that cells are viewed and counted while in electronic method, the counter counts even dust particles. The main disadvantage of total count is that it cannot distinguish between living and dead cells.

(b) *Viable count:* In the viable count only the living microbial cells are taken into consideration. The principle behind viable count is that all viable cells or spores, under suitable conditions, multiply and form a colony. The number of colonies is equal to the number of viable cells present in the sample. In viable count method, a known quantity of diluted sample is plated on an appropriate medium and after an incubation period, the number of colonies are determined.

The difference between the total cell number and the viable cell number will give the number of nonviable cells.

2. Determination of cell mass

In this method, the weight or mass of the cells is estimated as an indicator of increased mass. There are two methods of determining the mass of cell—direct method and indirect method.

(a) *Direct method:* In the direct method, the cell mass is actually weighed. Direct method is of following types:

(i) *Measurement of dry weight:* In this method, the culture suspension is centrifuged and the pellet obtained is dried. Then this dried pellet which consists of cells is weighed.

(ii) *Measurement of cell nitrogen:* In this method, the microbial growth can be determined in terms of the quantity of nitrogen content, because a major cell component is protein of which nitrogen is an important ingredient. This method is useful with dense microbial suspension where the amount of growth is large.

(b) *Indirect method:* The most popular indirect method of determining growth in bacteria and yeast is by the use of colorimeter or turbidometer. In this method, the density of cell suspension can be determined by measuring the varying intensity of light as it passes through a cell suspension. The light absorbed or reflected by a bacterial suspension is directly proportional to the concentration of cells in culture.

Another indirect method for determination of microbial growth is measurement of biochemical activity. The logic behind this method is more number of cells produce more amount of metabolic product. But this method is not reliable.

SUMMARY

Growth is the unique property of living organisms. In multicellular organisms, growth will result only in increase in size while in unicellular microorganisms growth results in increase in number of organisms. Many factors influence the microbial growth. They are nutrients, pH, temperature, osmotic pressure, oxygen concentration, etc. Bacterial growth curve, which is obtained by plotting number of cells against time, is always sigmoid. Bacterial growth curve shows four distinct phases—lag phase, log phase, stationary phase and death phase. There will be intermediate phases between these four standard phases. There are many methods available for determining microbial growth.

They are determination of cell number by total count and viable count and determination of cell mass by direct and indirect methods.

EXERCISE

1. What is growth? Describe the factors affecting microbial growth.
2. Mention the difference between growth of unicellular microorganism and growth of multicellular organism.
3. Describe the bacterial growth curve.
4. Describe the methods for analysis of microbial growth.
5. Write short notes on
   - (i) Stationary phase
   - (ii) Log phase
   - (iii) Viable count
   - (iv) Direct method of microbial growth counting
   - (v) Lag phase
   - (vi) Microbial growth
   - (vii) Optimum temperature

27

Pathogenic Microorganisms

Generally, depending on their relationship to the host, microorganisms can be classified as saprophytes and parasites. Saprophytes are free-living microbes, which live on dead organic matter. Generally, they are not capable of multiplying on living tissues and are thus not associated with infectious diseases. While parasites are established and multiply in living hosts. But a majority of parasitic microorganisms are harmless to the host. For example, the large number of microorganisms living on the skin, on the mucous membrane and in the digestive tract of man and animals are not causing any visible harm to the host. These parasitic microorganisms are called *normal flora* of the body or commensals, which live in complete harmony with the host without producing disease. While some parasitic microorganisms that are able to produce disease in the host are called *pathogens*. sometimes, when the resistance of the host is lowered, the normal flora of the body behave as facultative pathogens.

Definition of some terms used in this chapter is as follows:

Infection: Infection is the entering and multiplication of a parasite in the host without necessarily producing the disease.

Infectious disease: It results from the interaction of parasites.

Pathogenicity: It is the ability of microorganisms to cause disease.

Virulence: It is the degree of pathogenicity exhibited by virulent organisms when introduced into the host in very small quantity or, in other words, it is the capacity of a given strain of microbial species to produce disease.

Primary pathogens: Primary pathogens are parasites, which are capable of penetrating or evading the normal host defence mechanism of a healthy susceptible individual.

Secondary pathogens or opportunistic parasites: These parasitic microorganisms are not capable of penetrating the normal defence mechanisms of a healthy susceptible host. But it can infect or cause disease when host defence mechanism has been broken as by wounds, prolonged disease, old age, poisons, etc.

Generally, there are two types of pathogens. They are extracellular and intracellular parasites. *Extracellular parasites* usually resist phagocytic engulfment and multiply in body fluids and tissue spaces, while *intracellular parasites* multiply inside the host cells. Extracellular and intracellular parasites produce characteristically different symptoms. For example, *Streptococcus pneumoniae,* an encapsulated extracellular bacterium, produces a rapidly progressing acute pneumonia which quickly kills a person. The patient who survives this, rapidly recovers because of the production of antibodies that neutralize the capsule's antiphagocytic property. While intracellular parasites, usually produce chronic diseases with less rapid onset, much longer recovery time is required even with chemotherapy and a greater likelihood for clinical relapse. e.g. *Mycobacterium tuberculosis* (intracellular parasite). These characteristics of intracellular infections are largely due to the protection provided by the host cell to the parasite. Humoral antibodies have little access to intracellular pathogens and fail to eradicate infection and thus aid patient recovery. Intracellular pathogens are attacked by CMI (Cell Medical Immunity) which eliminates the microbes by killing the infected host cells. Some of these pathogens may escape destruction by CMI and cause recurrences months or years after recovery. Intracellular infection also is poorly controlled by chemotherapeutic agents that either fail to penetrate the infected host cells or are ineffective in cytoplasm.

Unlike typical intracellular parasites, most viral pathogens produce rapidly progressing acute diseases that usually subside spontaneously several days following onset of symptoms. In the course of infection, most viruses antigenically alter host cells, which are recognized as foreign and thus destroyed by CMI. Viruses also stimulate interferon production which aids in recovery.

Many microorganisms are pathogenic to human beings. They may be bacteria, fungi, viruses, protozoa, mycoplasma, Rickettsiae, etc. Different types of microorganisms and diseases caused by them are given in Table 27.1.

Table 27.1 Some microorganisms and diseases caused by them

Microorganism	Diseases caused
A. Bacteria	
1. Vibrio comma	Asiatic cholera
2. Spirillum minus	Rat-bite fever
3. Pseudomonas aeruginosa	Secondary infections of urinary tract and wounds
4. Escherichia coli	Infantile diarrhoea, meningitis, urinary tract infections

5. Klebsiella aerogenes	Pneumonia
6. Salmonella typhosa	Typhoid
7. Shigella dysenteriae	Bacillary dysentery
8. Pasteurella pestis	Plague
9. Bordetella Pertussis	Whooping cough
10. Corynebacterium diptheriae	Diphtheria
11. Bacillus anthracis	Anthrax
12. Clostridium botulinum	Botulism
13. Clostridium tetani	Tetanus
14. Treponema pallidum	Syphilis
B. Viruses	
1. HIV	AIDS
2. Yellow fever virus	Yellow fever
3. Encephalitis virus	Encephalitis, Dengue
4. Rhabdo viruses	Rabies
5. Polio viruses	Polio
6. Hepatitis A virus	Hepatitis
7. Hepatitis B virus	Serum hepatitis
8. Enteric viruses	Gastroenteritis
C. Fungi	
1. Epidermophyton floccosum	Ringworm of the groin or dermatomylosis.
2. Microsporum audouini	Ringworm of scalp
3. Microsporum gypseum	Ringworm
4. Trichophyton tonsurans	Hair infections, lesions of the nails
5. Trichophyton rubrum	Lesions of skin and nails
6. Coccidioides immitis	Coccidiomycosis
7. Candida albicans	Moniliasis or thrush
8. Aspergillus fumigatus	Aspergillosis
D. Rickettsia	
1. Chlamydia psittaci	Psittacosis
2. Coxiella burnetii	Q fever
3. Rickettsia rickettsii	Rockey mountain spotted fever
4. Rickettsia akari	Rickettsial pox
5. Rickettsia prowazlkii	Epidemic typhus fever
E. Protozoa	
1. Entamoeba histolytica 2. Entamoeba coli 3. Endolimax mana 4. Iadoamoeba biltschlii	Amoebic dysentery
5. Plasmodium vivax 6. Plasmodium falciparum	Malaria

7.	Leishmania donovani	Kala-azar
8.	Tryponosoma gambiense	Sleeping sickness
9.	Leishmania tropica	Oriental sore
10.	Tryponosoma rhodesiense	Trypanosomiasis

Source of Infection

The following are the various routes through which pathogenic organisms generally pass from one host to another.

1. *Ingestion:* This is the most common route for transfer of infection from one host to another. The members of the enterobacteriaceae group such as Salmonella are adopted to live in intestine. These intestinal organisms are excreted in the faeces and contaminate the environment like water, soil, food, etc. Since these organisms remain viable for many weeks or many months, they are likely to be ingested by other susceptible hosts and to colonise in the intestinal tract after ingestion.

There are many species of bacteria which may be transmitted by ingestion and excreted in the faeces, e.g. *Bacillus anthracis* (which cause anthrax); Clostridium species; Leptosprial; Pseudomonos; Vibrio cholrae (cholera); Shigella; etc.

2. *Inhalation:* This is another most common route of infection especially in crowded condition, which helps the inhalation of air borne microorganisms for respiratory diseases caused by Klebsella, Pseudomonas. These pathogens remain viable after inhalation in various locations in the respiratory tract or in lymph nodes and cause infection. For example certain viral diseases such as influenza, small pox and chicken pox are transmitted by the respiratory route. A few example of pathogens transmitted by inhalation are Bordetella pertusis, Corynbacterium diphtheriae; Mycobacterium tuberculosis; Streptococcus pneumoniae, etc.

3. *Sexually transmitted diseases:* The important diseases spread by this method are AIDS, Gonococcal urethritis, etc. The transmission of these diseases may also indirectly be affected by freshly contaminated bedding coming in contact with susceptible animals. Different pathogens infecting sexually are Herpes virus, HIV virus, Neisseria gonorrhoea, Trichomonas vaginalis etc.

4. *Diseases transmitted by skin wounds:* This is common route of entry for pyogenic bacteria like Staphylococcus, Streptococcus, Corynebacterium, Spherophorus, etc. Many toxigenic species of clostridium like C. tetani (tetanus), C. oedematiens C. welchii type A (gas gangrene) infect the host through the skin wound. Again for few pathogens like Actinobacillus lignieresii, Actinomyces bovis, Nocordia farcinia, Bacillus anthracis, etc., skin wounds produce good source of entry into the host tissues.

5. *Insect bites:* Many flies and insects transmit a number of diseases from one host to another. The most important diseases transmitted by insect bites

are malaria, yellow fever, Dengue, etc. And the important diseases transmitted by non-biting insects and flies are typhoid, cholera, sleeping sickness, kala azar, spotted fever, dysentery, etc.

The insects which transmit infection may be infected themselves or may act only as a passive or mechanical carrier of the pathogens. These are also called vector. These insects transmit infection by inoculation into or through the skin or mucous membranes by biting or by deposit of infective organisms on the skin, food or other objects.

The development of disease in an animal depends upon the interaction between three factors, namely the etiological agents, the host and the environment. The virulence capacity of these etiological agents (parasites) is generally attributed to three factors, namely infectivity, invasiveness and toxigenicity.

1. *Infectivity:* The *infectiousness* of a microorganism depends upon its ability to get established upon or within the host by overcoming the defensive barriers of the host such as antibodies or phagocytes. Infection does not necessarily involve pathogenicity or virulence and thus it takes place without any visible sign of disease.

2. *Invasiveness:* Once microorganisms infect the host, its existence depends on the production of chemical components, metabolic products and enzymes that can interact normal body defences. When parasite invades, complex changes occur and produces an infectious disease in the host. Invasiveness is frequently enhanced by the production of extracellular enzymes by the organisms that destroy the defence mechanisms of the host. These enzymes are called *aggressins.* The enzyme aggressin usually possesses proteolytic or lipolytic activity and is therefore, able to break down cell or connective tissue components. For example the aggressins enzyme '*hyaluronidase*' is produced by Clostridium, Pneumococci, haemolytic streptococci, etc., which destroys hyaluronic acid. Hyaluronic acid is an important intercellular cementing substance, which binds tissues into their normal structure and thus opposes the invasion of microorganisms through the tissues. Therefore, the hyaluronidase is also called as spreading factor.

3. *Toxigenicity:* The normal cellular components or metabolic products of pathogenic organisms, which interfere with the activity of host cells are called toxins. Some highly pathogenic species are not very invasive but their main property is toxigenicity. Microbial toxins are secreted into the surrounding fluids as *exotoxins* or remain attached to the producing cell, called *endotoxins*.

Exotoxins are soluble proteins secreted by a few types of pathogenic microorganisms. Generally, the toxin producing parasites may remain localized at the primary site of infection but the toxin produced by them is disseminated through the body and causes generalized symptoms and lesions of various types. Few examples of exotoxins producing microorganisms are Clostridium botulinum, Clostridium tetani, Corynbacterium diphtheriae, Staphylococcus aureus, vibrio cholerae, etc.

Endotoxins are toxins which remain attached to the producing cell. Endotoxins are the part of cell wall structures of gram negative microorganism. It has complexes of lipid, polysaccharide and proteins. Endotoxins are released by autolysis of cells. Endotoxins are resistant to heat, alcohol and dilute acids. Generally, endotoxins are not very potent poisons but they are emetic and pyrogenic. Endotoxins also produce some other symptoms like diarrhoea, weakness, paralysis of limbs, lowered arterial blood pressure, intestinal haemorrhage, etc. The example of endotoxin producing bacteria are Enterobacteriaceae,

Development of Symptoms

Pathogenic microorganisms injure the host body by various ways. Pathogenic microorganisms produce some well recognised symptoms of diseases in infected individuals. Various symptoms of different infections are:

1. *Toxemia:* It is the poisoning of the host body by toxic substances produced by the pathogens. In many diseases, it is the common method of injury. Mild toxemia leads to nervous irritability while heavier toxemia, particularly by exotoxins, results in the death of tissue, paralysis, pain, fever and even death of the infected individual.

2. *Fever:* Generally all infectious diseases are accompanied by a rise in the general body temperature called fever.

3. *Loss of appetite:* It is also called anorexia, which is a well known symptom of disease, especially during a fever.

4. *Pain:* Pain is another common symptom of infectious diseases. It results in physical discomfort, headache.

5. *Inflammation:* Inflammation is often accompanied with wound infection. In the inflamed region, blood capillaries are enlarged with blood and distended. There is redness, swelling, tenderness and pus formation.

6. *Necrosis:* Necrosis is the death of tissues. The tissues are generally killed by the absorption of an exotoxin.

7. *Hypertrophy:* Hypertrophy is an abnormal growth of cells, which results due to stimulation of cells to divide that had ceased to divide by pathogenic microorganisms. Hypertrophy results in the formation of nodules.

Defence mechanism against microorganisms

There are wide range of infectious agents that can use the human body as a sanctuary to raise their offspring. The immune system provides a defence mechanism. The defence mechanism against microorganisms involves special type of cells called lymphocytes. Lymphocytes are of two types namely T-lymphocytes (cell-mediated immunity) and B-lymphocytes (Humoral immunity). In the adults, both types of lymphocytes are derived from common stem cells in the bone-marrow.

Immune system has three main functions:

(a) Defence against microorganisms
(b) Homeostasis, removal of damaged cells
(c) Surveillance, recognition and destruction of mutant cells.

Humoral immunity

When B-lymphocytes come into contact with foreign materials, they differentiate into *plasma cells* or *immunocytes.* The plasma cells secrete antibodies. A particular plasma cell produces only one type of antibody.

Conversion of B-lymphocytes into plasma cells requires many successive cell generations and about five days. With each cell generation there is a progressive development of the endoplasmic reticulum and ribosomes. Around the fifth day, the endoplasmic reticulum cisternae are filled with antibody molecules and plasma cell starts secreting antibodies.

Antibodies are also known as immunoglobulins. Basically, it is made up of four polypeptide chains (Fig 27.1), where two identical chains are heavy ('H' chains) and remaining two identical chains are light chains ('L' chains). Each polypeptide chain contains an amino acid end and carboxyl acid end. The amino acid end of polypeptide chain is part of the variable or 'V' region and the carboxyl end is the part of the constant or 'C' region.

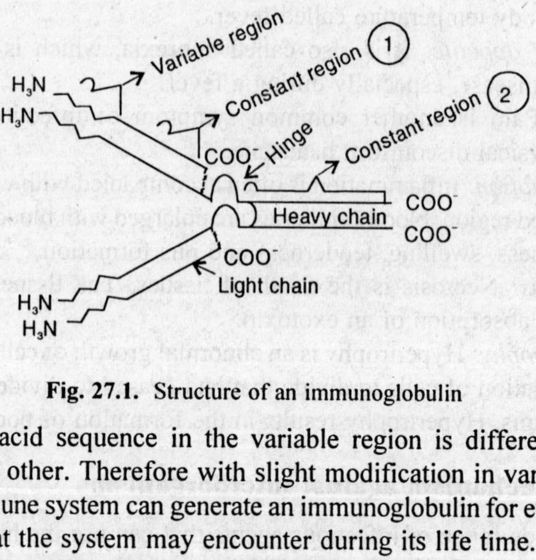

Fig. 27.1. Structure of an immunoglobulin

The amino acid sequence in the variable region is different from one antibody to the other. Therefore with slight modification in variable region, the normal immune system can generate an immunoglobulin for every possible immunogen that the system may encounter during its life time.

There are five major classes of immunoglobulins, namely IgG, IgD, IgE, IgM and IgA. Among five different immunoglobulin classes, IgG constitute around 75% of the total serum immunoglobulins. It plays a vital role in the defence against infection. It is the only immunoglobulin that crosses the placenta in human; therefore, it is responsible for the protection of the neonate during the first few months of life. Once IgG recognises the antigen, it forms complex with antigen, which attract macrophages via their Fc (fraction crystallizable) receptors, thus enhancing phagocytosis.

IgM is the largest immunoglobulin in size. It is a pentamer of the basic four chain immunoglobulin, which is held together by disulphide bonds and a small molecule called 'J' chain. IgM is an important defence mechanism against bacteria.

IgA is actively secreted by mucosal associated lymphoid tissue (MALT). Therefore IgA appears selectively in sero-mucus secretions such as saliva, tears, nasal fluids, colostrum and in the secretions of the lung, genito-urinary and gastro-intestinal tracts. It is a dimer, where two immunoglobulins are held together by 'J' chain. IgA provides defence mechanism against attack by microorganisms to the exposed external surfaces of the body.

IgE is present in very low concentrations in serum. IgE have a high affinity for mast cells and binding occurs via its FC portion. Its main physiological role is protection of external mucosal surfaces.

IgD is present in trace amounts in serum. The main function of IgD has not yet been determined. It is believed that they may operate as antigen receptors and controls the lymphocytic activation and suppression.

Cell mediated immunity

When T-lymphocytes are stimulated by antigen, they are transformed into lymphoplasts. The T-lymphocytes migrate to the thymus, where under the influence of the hormone thymosin they become immunologically competent. In cell-mediated immunity there is a direct reaction of lymphocytes with foreign material. Cell mediated immunity has an important role in the defence mechanism against viral, bacterial and fungal infections; in transplant rejection; destruction of tumour cells, delayed type of hypersensitivity; and in some autoimmune processes.

Different types of cells are involved in cell mediated immunity. They are Helper T cells, cytotoxic T Cells, Activated macrophages or Natural Killer (NK) cells and K cells.

T-helper cells: When T helper cells are stimulated by surface bound foreign antigen, they secrete various soluble mediators called *lymphokines.*

Cytotoxic T cells: Cytotoxic T cells kill virally infected target cells by direct cell lysis. The cytotoxic T cells recognize antigen in conjunction with MHC class I molecules, which are stimulated by T helper cells and then turn cytotoxic.

Activated macrophages: Activated macrophages are the effector cells in the cell mediated immune response to various microorganisms. It ingests the microorganism, processes the antigen and then presents the antigen on its surface, which activate the helper T cells.

Natural killer cells: Natural killer cells are large granular lymphocytes which have both anti tumour and anti microbial activity. Natural killer cells do not require prior exposure to antigen to become cytotoxic. Natural killer cells attach to their target cells by a receptor and cause lysis of the target cell by releasing cytotoxic factors capable of forming pores in the target cell membrane. These cytotoxic factors are called perforins.

K cells: A number of different cell types including lymphocytes termed K cells can mediate antibody dependent cell mediated cytotoxicity (ADCC). Generally IgG binds to an Fc receptor on the K cell and to the target cell (antigen bearing cell). Thus antibody acts as a bridge between the target cell and a killer cell. The mechanism of killing involves release of basic proteins and other molecules from its granules, which results in the formation of pores in the target cell membrane. The target cell dies because of the leaky membrane.

SUMMARY

Parasites are establish and grow in the living host. Generally many parasites are harmless to the host, but some species of parasitic microorganisms cause diseases in the host. Different species of bacteria, fungi, protozoa and viruses cause various diseases in the human being. The source of these disease causing organisms are ingestion, inhalation, sexual transmission through skin wounds, flies and insect bites, etc. The virulence capacity of parasites is generally attributed to three factors, namely infectivity, invasiveness and toxigenicity.

The host human has different defence mechanisms against these parasites. The main defence system is immune system, where lymphocytes play an important role. Two types of lymphocytes, namely B lymphocytes and T lymphocytes, are responsible for humoral and cell mediated immunity respectively.

EXERCISE

1. What is immune system? Add a note on different types of immune systems.
2. Write a note on pathogenic microorganisms.
3. Describe the source of infection.
4. Describe the various symptoms produced by pathogenic microorganism.
5. Write short notes on
   (i) Parasites
   (ii) Toxemia
   (iii) Invasiveness
   (iv) Normal flora
   (v) Natural killer cells
   (vi) Immunoglobulins

28

Microbial Metabolism

Energy is the most basic requirement of all living organisms. Without energy life activity is not possible. Autotrophic organisms trap solar energy, convert it into chemical energy (carbohydrates). These are oxidised by living organisms and the energy is released, which is utilized to carry on all vital processes of living beings. All the reactions which take place in living organisms are called *metabolism*. Metabolism is of two types, namely anabolism and catabolism. Anabolism is the synthesis of complex molecules from simple molecules. It is also known as biosynthesis. Catabolism is the oxidation of complex molecules into simpler molecules. The example of anabolism is photosynthesis and the example of catabolism is respiration.

Microbial Respiration: Respiration means any chemical reaction, which occurs in a living cell that results in a release of energy. The term respiration is derived from Latin word 'respirare' meaning breathe. Respiration occurs universally in all forms of life, because each cell obtains its own energy by the oxidation of substrate material present within. Respiration is a complex process, which involves degradation of substrates, release of energy, formation of many intermediate products which play different roles in metabolism. Energy released in respiration is used for various metabolic processes. During respiration insoluble food in converted into soluble form and potential energy is converted into kinetic energy.

Depending upon the involvement of atmospheric oxygen during respiration, it is of two categories, called aerobic and anaerobic respiration. In aerobic respiration, atmospheric oxygen participates and substrate is completely broken down into carbon dioxide and hydrogen. In aerobic respiration final hydrogen acceptor is free oxygen. In anaerobic respiration, the final hydrogen acceptor is some compound other than oxygen. In anaerobic respiration, substrate is incompletely oxidized. Besides carbon dioxide, end product includes some intermediate products like ethanol, acids etc. When the starch is the reserve food, it should be hydrolysed into simple sugars before it can enter into the respiratory cycle. In the hydrolysis process, starch is broken down into simple sugars with the addition of water.

Glycolysis

The pathway in which glucose is degraded into pyruvic acid is called glycolysis. This process occurs in the cytoplasm. It is also called as Embder-Meyerhof (EMP) pathway or common respiratory pathway or cytoplasmic pathway. The reaction of glycolysis occurs during aerobic as well as anaerobic respiration.

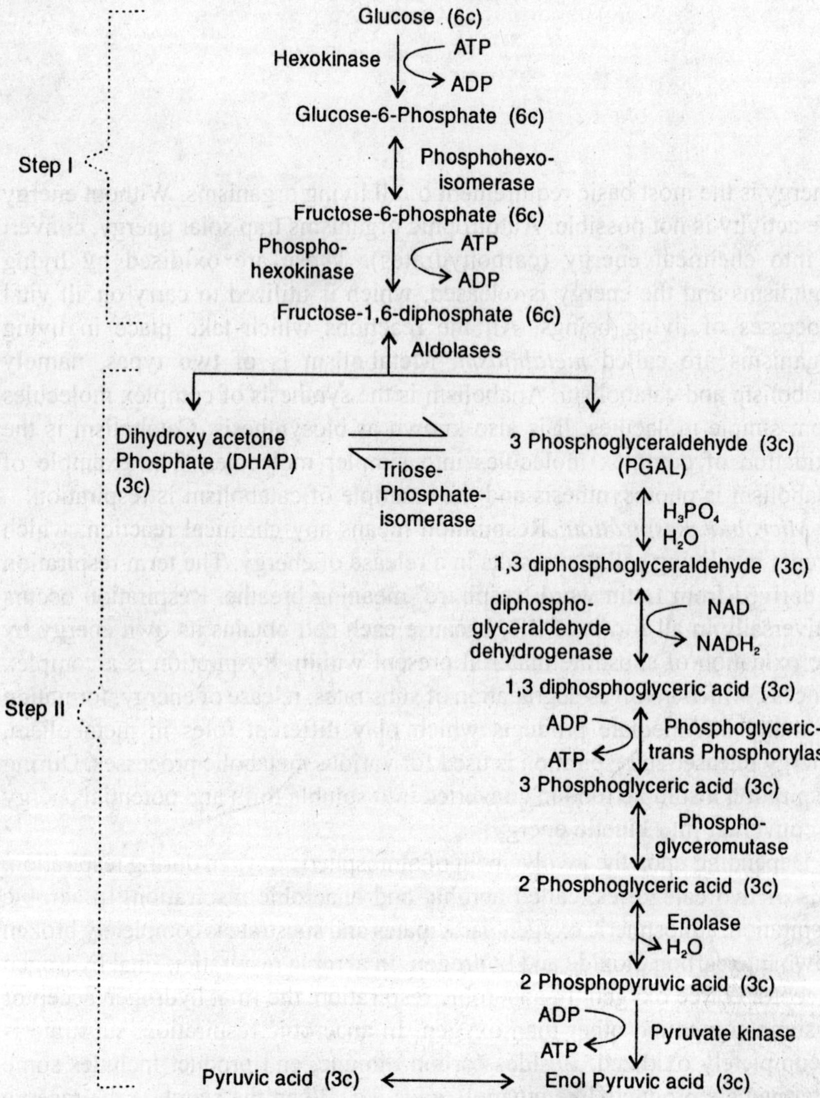

Fig. 28.1. The glycolysis pathway

Glycolysis may be subdivided into two steps—(1) Conversion of Glucose to fructose 1, 6 diphosphate. It requires 2 ATP molecules. (2) In the second step, fructose, 1, 6 diphosphate splits into two molecules of pyruvic acid in which four ATP and two $NADH_2$ molecules are produced.

Step I Preparatory phase

(1) *Phosphorylation of glucose:* Glucose is a fairly stable compound. The activation of glucose takes place by a reaction called oxidative phosphorylation. A phosphate group is attached to glucose by a low energy phosphate bond and thus glucose-6-phosphate is formed in presence of enzyme *hexokinase.* This is an energy requiring step and energy is derived from ATP molecule.

(2) *Isomerization of glucose-6-phosphate to fructose-6-phosphate:* In this reaction, glucose-6-phosphate isomerizes to fructose-6-phosphate. This reaction is catalysed by *phosphoglucoisomerase.*

(3) *Second phosphorylation:* In this reaction, fructose-6-phosphate is phosphorylated to fructose 1, 6 diphosphate. The enzyme *phosphofructokinas* catalyses this reaction. The energy required for this reaction is obtained by hydrolysis of ATP molecule.

(4) *Cleavage of fructos 1, 6 diphosphate:* Fructose 1, 6 diphosphate splits into two 3 carbon molecules by the action of enzyme *fructose diphosphate aldolase.* The two molecules are—dihydroxy acetone phosphate (DHAP) and glyceraldehy-3-phosphate (PGAL). DHAP is readily converted to PGAL in presence of enzyme *triose phosphate isomerase.*

Step II Oxidative phase

All the subsequent reactions must be considered as operating twice.

(5) *Phophorylation and oxidative dehydrogenation:* This is first reaction in glycolysis which yields energy. PGAL undergoes combined phosphorylation and oxidative dehydrogenation, which is catalysed by the enzyme *glyceraldehydephosphate dehydrogenase.*

(6) *Formation of ATP from 1, 3 DPGA:* In this reaction, high energy phosphate group in 1, 3 diphosphoglycerate is transferred to ADP, resulting in the formation of 3-phosphoglycerate and ATP molecule. *Phophoglycerate Kinase* catalyses reaction.

(7) *Isomerization:* The enzyme *phosphoglyceromutase* isomerizes 3-phosphoglycerate into 2-phosphoglycerate.

(8) *Dehydration of 2 phosphoglycerate:* Enzyme enolase dehydrates phosphoglycerate and yields phosphoenol pyruvic acid, which has a high energy enolic phosphate group.

(9) *Formation of ATP and pyruvic acid:* It is final reaction of glycolysis. Here pyruvic acid and molecule of ATP is synthesized by phosphoenolpyruvic acid and ADP. The enzyme *pyruvate kinase* catalyses this reaction.

All the reactions in glycolysis are reversible, except the reaction catalyzed by the enzyme (1) hexokinase, (3) phosphohexo-kinase and (10) pyruvate kinase, the former two requiring ATP and the last producing ATP.

Table 28.1 Energy production in glycolysis

	Reaction	ATP used	ATP gained
1.	Hexokinase	1	–
2.	Phosphohexokinase	1	–
3.	By reoxidation of the two $NADH_2$ in the respiratory chain	–	$2 \times 3 = 6$
4.	Phosphoglyceric transphosphorylase		2
5.	Pyruvate Kinase		2
			Total gain = 8 ATP molecules

The pyruvic acid formed by glycolysis goes into three alternative catabolic pathways.

(1) In aerobic organisms or tissues, under aerobic conditions, pyruvate is oxidised with loss of its carboxyl groups as CO_2 to acetyl CoA. This Acetyl CoA is then completely oxidized to Co_2 by the Kreb's cycle.

(2) In the second route, pyruvate is reduced to lactic acid via *lactic acid fermentation.* Vigorously contracting skeletal muscle, which functions aerobically and some anaerobic microbes reduces pyruvate to lactate. Certain tissues like retina, brain, erythrocytes, etc. convert glucose to lactate even under aerobic conditions.

(3) Some plant tissues, a few invertebrates and some microorganisms convert pyruvic acid anaerobically into ethanol, which is called ethanol fermentation.

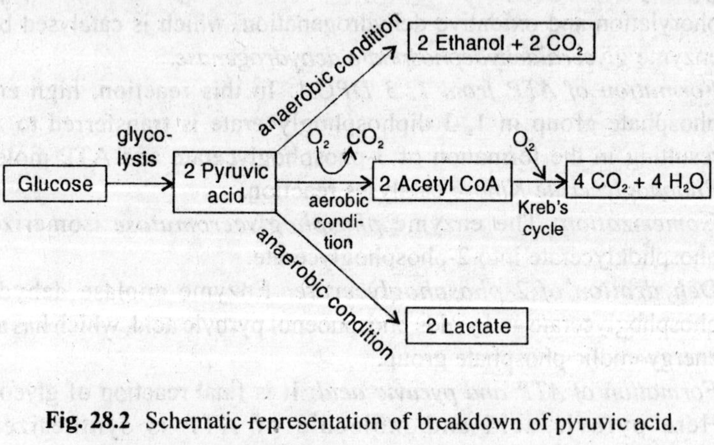

Fig. 28.2 Schematic representation of breakdown of pyruvic acid.

1. Aerobic respiration

Citric Acid Cycle (Kreb's Cycle)

This Metabolic pathway is known as Kreb's cycle because it was first suggested by Hans Kreb in 1937. It is also called as citric acid cycle because the first product of this cycle is citric acid. It is also known as tricarboxylic acid cycle (TCA).

The largest proportion of energy available from glucose (about 60%) comes from this pathway. This is the aerobic pathway and all the enzymes involved in this pathway are located mainly in the mitochondria of enkaryotic microorganisms while in bacteria it is present in cell membrane.

Before entering the citric acid Cycle, pyruvate oxidatively decarboxylates to acetyl-CoA (Fig. 28.3).

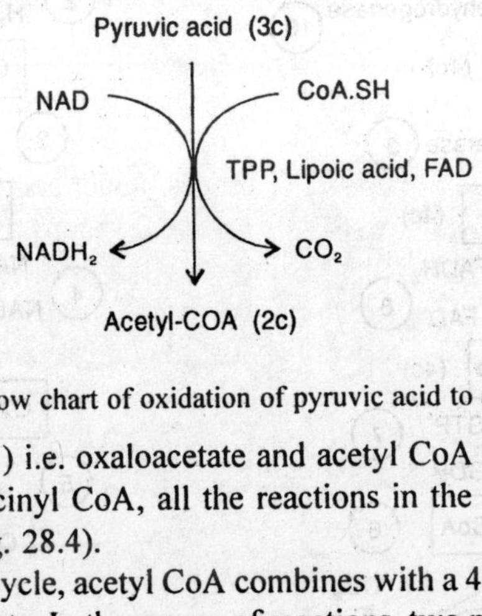

Fig. 28.3 Flow chart of oxidation of pyruvic acid to acetyl CoA.

Except reaction (1) i.e. oxaloacetate and acetyl CoA to citrate and (6) α-Ketoglutarate to succinyl CoA, all the reactions in the citric acid cycle are freely reversible (Fig. 28.4).

In the citric acid cycle, acetyl CoA combines with a 4 carbon oxaloacetate to form 6 carbon citrate. In the course of reactions, two molecules of CO_2 are lost and oxaloacetate is regenerated.

Table 28.2 Energy Production in Kreb's Cycle

Reaction	Coenzyme oxidised	No. of ATP formed
(1) Isocitrate dehydrogenase	$NADH_2$	3
(2) α-Ketoglutarate dehydrogenase	$NADH_2$	3
(3) Succinate thiokinase	GTP	1
(4) Succinic dehydrogenase	$FADH_2$	2
(5) Malic dehydrogenase	$NADH_2$	3
		Total no. of ATP = 12

Total energy output from one molecule of glucose by aerobic respiration is 38 ATP

During glycolysis	8 ATP
Conversion of 2 pyruvate into 2 Acetyl CoA	6 ATP
Oxidation of 2 acetyl CoA in TCA cycle	24 ATP
Total	38 ATP molecule

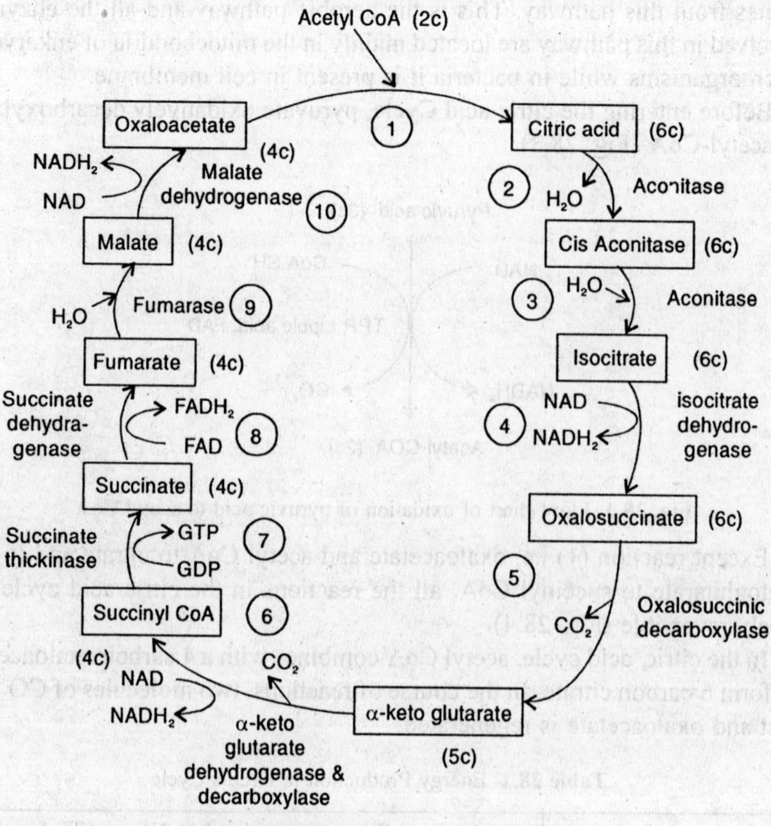

Fig. 28.4. Flow chart of citric acid cycle.

2. Anaerobic Respiration

In the absence of atmospheric oxygen, anaerobic microorganisms break down pyruvic acid into alcohol (alcoholic fermentation) and Lactic acid (Lactic acid fermentation). Thus Fermentation is anaerobic respiration found in microorganisms like fungi and bacteria. During fermentation, less energy is released as compared to aerobic respiration. Generally, the end product of fermentation may be alcohol but some time it may be butyric acid, lactic acid, etc.

Alcoholic fermentation: When respiratory substrates are sugars, many microorganisms during anaerobic respiration ferment them and produce ethanol. Generally yeast and fungi are ethanol producers. Again, a number of anaerobic and facultative aerobic bacteria produce ethanol during fermentation of sugars.

$$C_6H_{12}O_6 \longrightarrow 2C_2H_5OH + 2CO_2$$

In yeast, conversion of pyruvic acid to ethanol takes place in a few steps. In the first step pyruvate is decarboxylated by pyruvate decarboxylase to acetaldehyde. Acetaldehyde is further reduced to ethanol in the presence of enzyme alcohol dehydrogenase. Here $NADH_2$ donates hydrogen molecule.

Lactic acid fermentation: The lactic acid bacteria belongs to family lactobacteriaceae and Enterobacteriaceae. The members of lactobacteriaceae generally requires many accessory factors in addition to the basic nutrients. Depending on the fermentation products, these organisms are grouped into two types namely heterofermentative and homofermentative. Homofermentative bacteria produce only lactic acid as end product, while the heterofermentative bacteria produce many other compounds in addition to lactic acid.

Microbial Photosynthesis

Many microorganisms are heterotrophic. Only a few bacteria are autotrophic or photosynthetic bacteria.

In photosynthetic bacteria, the pigments are organized into some plasma membrane systems. Phototrophic purple bacteria are rich in intracytoplasmic membrane system, which appear as tubules, vesicles and plates. Such system is called *'Chromatophores*, which is capable of photophosphorylation and contains chlorophylls, carotenoids, cytochromes, quinones, proteins, lipids, etc. Depending upon the organism, chromatophores appear in various shapes like disc, cupshaped, hemispherical or spherical.

Cyanobacteria contains photosynthetic apparatus in the peripheral region of cytoplasm. Here photosynthetic apparatus consists of few membrane layers which extend in concentric sheets below the cell membrane and surround the central zone. Cyanobacteria contains the pigments chlorophyll a having absorption peak at 670, 680 and 700 nm. Beside chlorophyll a, it contains some carotenoids like βcarotene, Xanthophylls, etc.

The photosynthetic bacteria contains different pigments like Bacterio chlorophyll a, b, c, d, e and g, Chlorobium Chlorophyll (bacteriviridin) and carotenoids.

The prokaryotic phototrophs include a number of pigmented organisms like purple non sulfur bacteria (Rodospirillaceae); purple sulfur bacteria (chromatiaceae); green sulfur bacteria (chlorobiaceae); Cyanobacteria; etc. Photosynthetic bacteria generally utilize carbon dioxide from environment as a source of carbon. But the reducing agent may be inorganic or organic sources. In photolithotrophic organisms, inorganic substance like H_2S or thiosulphate is the source of reducing agent. While in photoorganotrophic organism, organic substances like malate or succinate is a reducing agent.

$$2H_2A + CO_2 \longrightarrow CH_2O + 2A$$

In algae and plants H_2A is H_2O while in sulphur bacteria H_2A is H_2S and in non-sulphur bacteria H_2A may be ethanol, succinate, etc.

Photosynthesis mechanism: Like plants, in bacteria also the photosynthetic process is divided into two types namely light reaction and dark reaction. In light reaction ATP molecule is generated. During light reaction oxygen is released or not released depending on the source of reducing power (H_2A). There are two types of organisms, anoxygenic bacteria which do not release oxygen while oxygenic bacteria release oxygen during light reaction.

SUMMARY

All the chemical reactions which take place in living organism are called metabolism. Mainly two types of metabolic reactions namely respiration and photosynthesis take place in an organism. Respiration takes place in all organisms while photosynthesis takes place only in photo-autotrophic organisms.

In all microorganisms, glycolysis, where glucose is oxidised into two molecules of pyruvic acid, is common pathway of respiration. In various microorganisms, this pyruvic acid is oxidised by three different pathways, namely lactose fermentation, alcohol fermentation and Kreb's cycle.

Like plants, photosynthetic microorganism traps solar energy and converts it into chemical energy. In microorganisms also photosynthetic reaction takes place in two types, namely light reaction and dark reaction.

EXERCISE

1. What is metabolism? Describe different types of metabolic reactions which take place in microorganisms.
2. Write a note on microbial respiration.
3. Describe the glycolysis in microorganisms.
4. What is anaerobic respiration? Explain.
5. Write short notes on the following:
   (i) Alcoholic fermentation
   (ii) Photosynthetic microorganisms
   (iii) Lactic acid fermentation
   (iv) Respiration

29

Interaction Among Microbial Population

Many microorganisms live much of their lives in a special ecological relationship. An important part of their environment is as organisms of other species. Any organism that spends a portion or its complete life in association with another organism of different species is called a *symbiont* or symbiote and the relationship is designated as *symbiosis* (Greek "Sym" means together and "bios" means life).

In other words symbiosis is the living together in close association of two or more dissimilar organisms. There are two types of symbiotic association, namely ectosymbiosis and endosymbiosis. In ectosymbiosis one symbiont remains outside the other organism, while in endosymbiosis one symbiont is present within the other. There are two types of symbiotic relationship, namely commensalism and mutualism.

Commensalism (Latin "com" = together and "mensa" = table) is a relationship in which one organism, the *commensal,* benefits while the other, the host is neither harmed nor benefited. The spatial proximity of the two parterns permits the commensal to feed on substances captured or ingested by the host. Generally the commensal also obtains shelter by living either on or in the host. The commensal is not directly dependent on the host metabolically and causes it no particular harm. When the commensal is separated from its host experimentally, it can survive without being provided any factors of host origin. For example, the bacterium *Escherichia coli* lives in the human colon and benefits from the host nutrients, warmth and shelter found there, but usually causes no disease or discomfort.

Mutualism (Latin 'mutuus' means borrowed or reciprocal) defines the relationship in which some reciprocal benefit occurs to both partners. In this relationship, the mutualist and the host are metabolically dependent on each other.

Rhizobium

One of the best examples of endosymbiosis is symbiotic nitrogen fixers like Rhizobium, Bradyrhizobium, Azotobacter, Azospirillum, Acetobacter, etc.

The genus Rhizobium is estimated to fix around 200 million tons of nitrogen globally per year. The term Rhizobium originates from Greek 'rhizo' meaning root. Rhizobium is an endosymbiont, which lives in the root nodules of legumes. Rhizobium is gram negative, motile, aerobic, non-spore forming rod shaped bacteria. But when Rhizobium is isolated from nodules a variety of morphological shapes are observed.

Rhizobium species invade the root hairs of leguminous plants by penetration. Once the Rhizobium invades the root cell, it stimulates the multiplication of infected cells, resulting in the formation of nodules. In this endosymbiotic association, both Rhizobium and plant get benefited. The bacteria obtain nutrients and shelter from the plant and, in turn, fix atmospheric nitrogen and make it available to the plant.

Rhizobium contains nitrogen fixing gene, also called *nif gene,* which produces enzyme *nitrogenase.* An enzyme nitrogenase is sensitive to oxygen.

Rhizobium strains nodulating one type of legume are different from those nodulate other type of legumes. Therefore certain degree of specificity exists between the Rhizobium strains and legumes. As the number of proper strain of Rhizobium associated with a particular legume may be present in very less amount or absent in the soil, this will limit the productivity of these crops. To avoid this limitation, inoculate legume seeds before planting with strains of specific Rhizobium.

Mycorrhizae

Mycorrhizae is another good example of organism which fixes nitrogen symbiotically. The term Mycorrhizae is derived from the Greek word *'mykos'* meaning fungs and *rhiza* meaning root.

Mycorrhizae occurs in different plant roots like Acacia nilotica, casuarina equisetifolia, Gmelina arborea, Michaelia champaca, Prosopis cineraria, Tectona grandis, Indigofera, maize, wheat, pigeonpea, chickpea, ginger, black pepper, tomato and many species of medicinal plants. The fungus gets nutrients from the plant, while fungus in turn helps plant in several ways. For example it fixes atmospheric nitrogen and makes it available to the plant. By extending its mycelium into the rhizosphere, fungus helps anchor the plant and increases the surface area for capturing water from dry soils and minerals from poor soils. Plants with mycorrhizae can inhabit severe habitats more successfully than plants without them. Plants with mycorrhizae also have reduced fungal and nematode diseases.

The formation of mycorrhizae begins with the invasion of the plant root by a soil fungus. The fungal mycelium penetrates the root cells by means of projections called haustoria and then develops intracellularly. Depending on the degree of integration of fungus with root, there are four types of mycorrhizal associations as mentioned below.

(a) *Ectomycorrhizae:* In ectomycorrhizae, the fungal cells form an extensive sheath around the outside of the root with very little penetration into the root tissue. It is mainly found on forest tree roots.

(b) *Endomycorrhizae:* In the endomycorrhiza, the fungal hyphae enter the root cells of the host plant and thus penetrate the host tissue. It is found in the roots of orchids.

(c) *Vesicular arbuscular mycorrhizae (VAM):* In this type of mycorrhizal association, root cortex contains specialized structures such as vesicles and intracellular branching, called arbuscules, formed by the fungal hyphae. It is found in most major agricultural crop plants like wheat, maize, potatoes, strawberries, apples, oranges, grapes, cotton, tea, coffee, tobacco, etc. VA mycorrhizae have a loose network of hyphae in soil and extensive growth within the cortex of the pant.

(d) *Peritrophic Mycorrhizae:* In this form of mycorrhizal interaction, fungi form clusters of hypae around the roots, but the hyphae do not penetrate the epidermis of the root. Peritrophic mycorrhizae produces favourable chemicals which are used by the plants.

Antagonism

Among microbial interactions, when one species adversely affects the other species it is called *antagonism* or negative interactions. It is a mechanism which controls population densities. In some cases, antagonism totally eliminates a population of a species which is not well adapted for continued existence within the community in a given habitat. As antagonistic organisms often produce antibiotics or some other inhibitory substances which affect the normal growth processes or survival of other organisms, these are of great practical importance. It provides self regulatory mechanism which prevents overpopulation and destruction of the habitat's resources.

Broadly there are four different types of negative interaction, namely competition, antibiosis, parasitism and predation.

1. *Competition:* Competition among microorganisms occurs when the populations of two or more species are mutually limited because of their joint dependence on a common nutritional factor. In other words, it can be defined as a rivalry between two species for a limiting factor in the environment. In competition, the less active species is suppressed and the more active species is not affected.

For example (a) The phytophthora parasitica, a fungus causing damping off of tomato seedlings, may be virtually rid by inoculation with penicillium patulum. Here penicillium patulum is antagonistic to the parasitic fungus.

(b) In a mix culture of two protozoans such as Paramecium aurelia, and Paramecium caudatum, the fast growing P. aurelia replaces P. caudatum.

2. *Antibiosis (Amensalism):* In antibiosis, population of one species produces a antagonistic substances that inhibit or kill susceptible species sharing same habitat. In antibiosis, first species gains a competitive advantage over the second species. For example, when competing for the same nutrients, antibiotic producing species have an advantage over the antibiotic sensitive species. Experiment of Alexander Fleming is the classical example of antibiosis in which the strain of *Penicillium notatum* inhibits the growth of *staphylococcus aureus.*

Some examples of antibiosis are: (a) The production of lactic acid by Lactobacillus and Lactococci eliminates pathogens. The Lactococci produces *nisin,* an antibiotic which inhibits the growth of coliforms in milk.

(b) An antibiotic penicillin, which is produced by Penicillum species, suppresses the growth of most Gram positive bacteria.

Beside antibiotics, various other chemicals like inorganic compounds such as ammonia, mineral acids, hydrogen sulphide and organic compounds such as alcohols, fatty acids, etc. are produced by different species, which inhibit the growth of other species.

3. *Parasitism:* When one species harms or lives at the expense of another organism (the host), it is a parasitic organism and the relationship is called parasitism. The parasitic organism is generally smaller and metabolically dependent on the host.

Some of the examples of parasitism among microorganisms are:

(a) Rhizoctonia parasitizes other fungi.

(b) Bdellovibrio bacteriovorus parasitizes E. coli.

(c) Bacteriophoges.

4. *Predation:* In predation one organism totally engulfs and digests another organism. The organism which engulfs is called *predator* and the organism which is engulfed is called *prey.* Generally predator is larger than the prey. The predator species derives nutrition from the prey species.

In water, protozoa are the major microbial predators which feed on bacteria and unicellular algae. Sometimes a predator is smaller than the prey. For example, the protozoan *Didinium* can engulf and consume the larger protozoan Paramecium. Some fungi are also able to trap and consume much larger nematodes.

SUMMARY

In the complete life cycle, microorganisms interact with many other organisms like other species of microorganisms, plants and animals. These interactions may be positive or negative. In positive interaction or symbiosis, the symbiont is benefited. Symbiosis is of two types, namely commensalism and mutualism. In commensalism, one organism is benefited and the other organism is unaffected. In mutualism, both the organisms are benefited.

In negative interaction or antagonism, one species adversely affects the other species. It is of four types, namely competition, antibiosis, parasitism and predation.

Rhizobium and mycorrhizae are the best examples of endosymbiosis. Rhizobium is a gram negative bacteria which lives in legume root nodules. The mycorrhizae is a fungus which lives in association with various plant roots.

EXERCISE

1. What is microbial interaction? Add a note on symbiosis.
2. What is antagonism? Explain different types of antagonism.
3. Write short notes on:
   (i) Predation
   (ii) Commensalism
   (iii) Antibiosis
   (iv) Competition
   (v) Mutualism
   (vi) Parasitism
4. Write a note on importance of Rhizobium in agriculture.
5. Describe the different benefits available to the plant by mycorrhizal interaction.

30

Fermentation

Since prehistoric times, microbial processes have been used in the preparation of food and drink. But it was only in the middle of the nineteenth century that the role of microorganisms in the transformation of organic matter was recognized. The word fermentation was derived from the Latin word 'fervere' meaning "to be boiling". In the beginning, the word fermentation was applied only for production of wine. But in modern days, the word fermentation is defined as a process in which chemical changes are brought about in organic substances like carbohydrates, fats, proteins, etc. through the action of some enzymes by specific microorganisms (Table 30.1). In nature microorganisms like bacteria, fungi perform an important role in the decomposition of organic materials. These microorganisms decompose complex organic substances into simple inorganic substances. During decomposition of organic substances, some organisms release CO_2 and other gases. These organisms are called *fermentors* or *ferments.* Although many genera and species of yeast exist in nature and many of them are used industrially, the strains of Saccharomyces cerevisiae are very important. The microorganisms used in fermentation technology may be aerobic or anerobic.

Table 30.1 Some microorganisms used in Fermentation

Microorganisms	Raw Material	Product
1. Saccharomyces cerevisiae	Glucose, Sugar molasses	Ethanol
2. Acetobactor	Dilute Alcohol	Vinegar
3. Saccharomyces cerevisiae	Barley + malt + starch + warm water	Beer
4. S. elliopsiodens	Grape juice	Wine
5. S. crevisiodae	Grain mash + malt	Whisky

(Contd.)

Microorganisms	Raw Material	Product
6. Penicillium chrysogenum	Corn steap liquor + salt	Antibiotic-Penicillin
7. Aspergillus niger	Sugar molasses	Citric acid
8. Streptomysis olivaceus	Sugar rich medium + cobalt chloride	Vitamin B_{12}
9. Clostridium acetobutylicum	Glucose	Riboflavin
10. Acetobactor suboxydans	Glucose	Vitamin C

(A) Alcohol fermentation

Ethanol can be prepared by fermentation of corn, wheat, potatoes, molasses, etc. For ethanol fermentation many microorganisms are used, but Saccharomyces cerevisiae is the most important one which can tolerate high concentration of alcohol.

Preparation of medium (mash): Molasses contains about 90% of sugar. It should be diluted to 14-18% by adding warm water. Then adjust its pH to 5.5 by using H_2SO_4 or HCl and then add nutrients like ammonium sulphate and ammonium phosphate. This nutrient rich acidic molasses is called *'mash'*.

Preparation of yeast inoculum: High yielding stains of yeast Saccharomyces cerevisiae is selected, which must be aciduric and alcohol resistant. Selected strain of yeast is grown by inoculating it in the sterile mash under controlled conditions. Generally for industries, inoculum is required in large quantities. From starting with a small pure culture, the inoculum is increased to hundreds of litres in a number of stages.

Sterilization: Fermentation apparatus and mash is sterilized by heat and then allowed to cool.

Fermentation: The sterile mash is pumped into a large steel or wooden fermentation tank. For the total quantity of mash, add 5% to 10% inoculum. Fermentation is allowed for about 48 hours at 25°C. In the initial period, to promote rapid growth of yeast, aeration is required. But after certain period anerobic condition should be provided to promote accumulation of alcohol and to stop oxidation of alcohol into CO_2 and water.

$$\underset{\text{Sucrose}}{C_{12}H_{22}O_{11}} + H_2O \xrightarrow[\text{Invertase}]{\text{Yeast}} \underset{\text{Glucose}}{C_6H_{12}O_6} + \underset{\text{Fructose}}{C_6H_{12}O_6}$$

$$\downarrow \text{Yeast Zymase}$$

$$\underset{\text{Ethanol}}{2\,C_2H_5OH} + 2\,CO_2 + 26 \text{ Kcal heat}$$

Yeast secretes an enzyme *invertase* which hydolyses sucrose, present in molasses, into glucose and fructose. Another enzyme *zymase* converts the glucose into alcohol and CO_2.

During fermentation, about 26 Kcal of heat is produced and thus increases the temperature of the medium, which will affect the production. Therefore cooling the medium to 25° C is essential.

Removal of yeast: After fermentation is completed, yeast which is present in the medium is separated by filtration or settling. The separated yeast can be used in the preparation of inoculum for next batch.

Distillation of alcohol: The fermentation liquor contains 8 to 10% alcohol. This is subjected for distillation by pumping into distillation columns. Alcohol vapours are condensed in condenser, which contains about 50 - 60% alcohol. It is again subjected to fractional distillation to remove other impurities like aldehydes and volatile oils. Pure alcohol (95%) can be obtained by rectification in rectifying columns. Absolute or anhydrous alcohol can be obtained by distillation of rectified alcohol with benzene.

Uses of ethyl alcohol

1. Used as solvent, extractant, antifreezer.
2. In the preparation of alcoholic beverages.
3. In the preparations of medicines, dyes, detergents, lubricants, cosmetics, explosives, etc.
4. Used as antiseptic solution.
5. It is used in synthesis of many organic chemicals.

Brewing (Beer production)

Beer is made by the yeast fermentation of malt grains to ethanol and carbon dioxide. Brewing is a complex process because multiple products do not results from microbial activity and exact nature of some products is not known. Brewing involves five major steps, namely malting, mashing, fermentation, maturing and finishing.

In malting and mashing, malt (beer medium) is synthesized by soaking the barley in water for two days. Then drain the excess water and incubate four to five days for germination. The germination allows synthesis of α and β amylases and proteolytic enzyme, which converts starch into fermentable form such as maltose or glucose. Then add adjuncts (malt supplement) to dilute protein and also add hops. Hops are dried female flower of Hop plant. Hops are added for flavour, colour, aroma and for mild antibacterial activity to prevent the growth of spoilage bacteria. Then ground this mixture (mashing) and remove the husk and other grain residues. Otherwise, it interferes fermentation procedure. This is called wort. Before fermentation, filter the wort and boil it with continuous agitation for two to three hours and add hops.

After cooling wort, a large inoculum of selected strains of yeast is added to bring a vigorous fermentation. Generally the yeast, Saccharomyces cerevisiae is used for brewing. For fermentation closed tank can be preferred. Beer fermentation takes place at 6-12°C. Beer fermentation can complete in 7-9 days.

After fermentation, fermented wort is refrigerated at 0°C for few weeks to many months to remove the harsh flavour and other undesirable characteristics. This procedure is called maturing or ageing. Finishing process consists of carbonation, pasteurizing, cooling, filtering and dispensing aseptically into sterile cans and bottles.

Wine fermentation

Wine is produced by normal alcoholic fermentation of grape juice. Other than grape, wine is also produced by other fruits like peach (peach wine); orange (orange wine); cherry (cherry wine), etc.

Wine fermentation is a very simple process. As many fruits have the wine yeast Saccharomyces cerevisiae var. ellipsoideus on them, fermentation starts spontaneously after crushing the fruits. But to avoid undesirable or average product, many wine makers destroy natural yeasts by adding sulphur dioxide (in the form of sodium metabisulphite) to the raw juice. This fruit juice is called *must.* Then must is inoculated with a starter culture of a selected strain of S. Cerevisiae var. ellipsoideus. Initially aerate the must to promote vigorous yeast growth. Later on, maintain anaerobic condition for rapid production of ethanol. Generally, fermentation requires 25-30° C temperature and around 10-15 days. The ethanol produced varies from 7 to 15%. After fermentation allow to settle yeast and other debris, filter it, pasteurize it. Later on keep it aside in container for years for ageing. Ageing develops a characteristic flavour and aroma.

Vinegar Fermentation

Vinegar is produced in two steps; in first step sugar or starch is converted to alcohol by alcoholic fermentation and in second step vinegar is produced from alcohol by acetic acid fermentation. In the first step, alcohol is produced by yeast S. Cerevisiae and in second step alcohol is oxidized to acetic acid by *Acetobacter*. Fig vinegar contains at least 4% acetic acid and small amounts of alcohol, glycerol, esters, reducing sugars, pentosans, salt, etc.

$$\underset{\text{glucose}}{C_6H_{12}O_6} \xrightarrow[\text{S.Cerevisiae}]{\text{Yeast}} \underset{\text{ethanol}}{2C_2H_5OH} + 2CO_2$$

$$\downarrow + O_2 \text{ Acetobacter}$$

$$\underset{\text{Acetic acid}}{CH_3COOH + H_2O}$$

Vinegar fermentation is of two types—slow process and quick process. In slow process, wooden vats are filled one-third with a good grade vinegar as starter. Then at weekly intervals add 10-15 litres wine. After five weeks, 10-15 litres of vinegar are drawn off each week with addition of same amount of wine.

In quick method, generators of various sizes and shapes are used, which is aerated and filled with beech wood shavings. Then add alcohol. The acetic acid bacteria grow as a thin film over the wooden shaving. Thus beechwood shaving provide large surface area for cells for the oxidation of alcohol to acetic acid. The Oxidation of alcohol to acetic acid evolves heat. Therefore, cooling coils are necessary to keep the temperature within the range of 25-30°C. Several passages through a generator are required to produce vinegar of desired strength.

Raw vinegar is opaque due to presence of Acetobacter. Therefore, after filtration and pasteurization, store it in air tight bottles.

Production of Vitamin (B_{12})

Vitamins are required by all living organisms. Deficiency of these vitamins causes various diseases. Deficiency of vitamin B_{12} causes anaemia and the only source of this vitamin is microbial productions. In nature vitamin B_{12} is synthesized by wide range of bacteria.

Industrially, vitamin B_{12} is produced at commercial scale by two methods—

1. Recovery of vitamin B_{12} as by-product of the streptomycin fermentation.
2. Fermentation processes using Bacillus megatherium, Streptomyces olivaceus, etc.

In the first method, vitamin B_{12} can be obtained by adding cobalt compounds to the streptomycin fermentation medium, but high concentration of cobalt decreases the streptomycin production. In the second method, *Streptomyces olivaceus* and *Bacillus megatherium* can be used only to produce high amount of vitamin B_{12}. These organisms are incubated into glucose rich medium at 27°C for 3-4 days. To this medium cobalt chloride is added as a precursor.

After fermentation vitamin B_{12} can be recovered by following two methods.

1. In one method, before autolysis of microorganisms, they are separated from the medium by filtration and dried. The dried microorganisms are served as vitamin B_{12} enriched animal feed.
2. In the second method, the microorganisms are autolysed by acid, heat or alcohol treatment. By the autolysis vitamin B_{12} is released into the medium. From the medium, cell debris are separated by filtration and vitamin B_{12} is purified from liquid part by various processes and crystallized.

Production of food

The largest use of microorganisms for food purpose is baker's yeast or *Saccharomyces cerevisiae.* Beside this, many microorganisms are used as source of SCP and SCO.

Single cell protein (SCP): Single cell proteins are the microorganisms produced by fermentation which satisfy proteinaceous ingredients for animal feed or human food. In 1970 the SCP was produced from hydrocarbon wastes of the petroleum industry. Yeast, fungi, bacteria and algae are grown on hydrocarbon wastes and cells are harvested as source of protein. Recently SCP is obtained by growing microorganisms on waste materials such as wood shavings, sawdust, straw, sulphite liquor, corn cobs and other agricultural wastes, and industrial wastes. Basically SCP comprises proteins, fats, carbohydrates, ash ingredients, water, etc. The composition of SCP depends on the organism and the substrate on which it grows.

The choice of organism for SCP depends upon several factors including economics of management. Yeast has been used as SCP since the beginning of the twentieth century. Some common example of microorganisms used as SCP is given in table 30.2.

Table 30.2 Microorganisms used in production of SCP

Microorganisms	Raw Material used
(A) Yeast	
1. S. Cerevisiae	Molasses
2. Saccharomycopsis rapolytica	n-Paraffin hydrocarbons
3. Kluyveromyces fragilis	Cheese whey
4. Candida utilis	Starchy waste
5. Candida tropicalis	Cassava
(B) Fungi	
1. Agaricus campestris	Cane and beet molasses
2. Morchella esculenta	Maize syrups, dexture
3. Trichoderma species	Coffee processing wastes
4. Gliocladium species	Maize wet-milling wastes
5. Paecilomyces	Sulfite effluents
(C) Algae	
1. Spirulina Species	Open ponds and CO_2
(D) Bacteria	
1. Cellulomonas	Cellulose, wood shavings
2. Pseudomonas fluorescence var. cellulose	Cellulose, straw

Single cell oil (SCO): Many microorganisms, particularly yeast and molds, are used as a source of lipids. Few examples of microorganisms used as SCO are yeast like Candida pulcherrima, Torulopsis lipofera, Saccharomyces cerevisiae, Rhodotorula glutinis, etc. The example of molds used as SCO are Geotrichum candidum.

Generally, media for SCO production should have high carbon to nitrogen ratio, a good supply of phosphates and for most organisms a pH on the acid side. the source of carbohydrates are molasses, cellulose waste, hydrolyzed wood and spent sulphite liquor, while the source of nitrogen are ammonium salts, urea, urine, yeast waste, molasses slop, extracts of grains, etc. Salts to be added include potassium chloride, monopotassium phosphate and magnesium sulfate. Addition of small amount of alcohol or sodium acetate increases yields of lipids.

Generally, these microorganisms grow best under aerobic conditions. Their growth for fat production has been in thin layers on trays or other flat areas or in well aerated tanks when yeasts were grown submerged. The optimum temperature for SCO production varies from 15-20°C depending on organisms employed. A fairly long period is required for maximum fat production by most of these organisms. After fermentation, recovery of lipid is by extraction with solvents, with or without autolysis.

ANTIBIOTIC FERMENTATION

Penicilium Production

Introduction: Penicillin, because of the impetus of World War-II, was the first antibiotic to be produced on a large scale and it is still one of the best antibiotics available. *It is active against many Gram positive bacteria like Nocardia and Actionmyces,* but *not against most Gram negative forms except at higher dosage levels. It interferes with cell wall synthesis of sensitive organisms and it is active only against growing cells.* In addition, it is almost *non-toxic to mammals,* except for certain *allergic reactions.*

There are *several penicillins* which are all *closely related in structure* and also in *activity* against sensitive microorganisms. These penicillins have *common chemical nucleus and differ in the chemical structure of a side chain attached to the nucleus.* Various compounds resembling the side chains can be added as precursors to the fermentation medium; and these compounds through microbial action, are directly incorporated into the penicillin molecule. The side chain can be removed enzymatically, liberating the penicillin molecule and new chain can be added to get new penicillins. Several differential fungi produce one or more of the penicillins.

Penicillin production: Penicillin was first produced commercially in stationary mat culture. However from this technique penicillin yields suffered the problems commonly associated with the culture. Oxygen penetrates only poorly into the mycelial mat; portion of the mycelium submerged in liquid is deficient in Oxygen; and the aerial hyphae have only poor access to nutrients. The medium was sterilized in shallow layers in bottles, flasks, trays or pans made of glass or metals. Then large numbers of wet or dry spores were added as inoculum to this medium and the cultures were incubated at 24 to 28°C for 6-7 days.

Deep tank aerated fermentation for penicillin production developed during World War-II and it soon become apparent that the deep tank process was more promising from a commercial point of view. But still there was some problem with the design of fermenters like—sterilization of air; the shift to submerged aerated fermentations required for the strain of penicillium—these problems nevertheless were successfully solved for penicillin production.

Stock culture: Maintenance for procedures for high yielding strains of *P. chrysogenum* required serious study because that high yielding strains proved to be genetically unstable and the higher the yield capacity, the more unstable is the strain. As a result of this the frequent transfers of the mold on the agar medium tend to select those natural population which sporulate profusely—although heavy sporulation is not necessarily associated with high penicillin yields. Therefore agar media are chosen to allow minimum vegetative growth with rapid sporulation of the entire culture and these media are inoculated with high levels of spores. The primary stocks for these organisms are maintained by one of three methods: a well sporulated agar slant is frozen; soil-stocks are prepared and stored at 2-4°C; or; the spores are lyophilized.

For *inoculum production,* spores are suspended in water or in a dilute solution of non-toxic wetting agent such as 1 : 10,000 SDS. These spores are then added to flasks or bottles of wheat bran+nutrient solution and these are incubated for 5-7 days at 24°C to provide high sporulation. The resulting spores are used directly to inoculate inoculum tanks or in some instances they are pregerminated in separate medium to provide a mycelial of inoculum. The inoculum tanks are incubated for 24-48 hours with agitation and aeration to provide heavy mycelial growth—and the resulting inoculum is then used for production tank [or it is added to a 2nd or even a 3rd stage inoculum tank to produce inoculum for large scale fermentations]. Contamination in the inoculum tanks is tested both by microscopic observations and by subcultures to both medium.

The *production tanks* are inoculated by employing air pressure to force 10% by volume of inoculum into the tank. For actual penicillin fermentations, media is incubated at approximately 25-26°C for approximately 3-5 days. During production, periodic samples are removed for determination of penicillin yields and for contamination checks.

A *typical medium* described by Jackson (1958) contains

Cornsteep liquor solids	– 3.5%
Lactose	– 3.5%
Glucose	– 1%
Calcium carbonate	– 1%

Potassium dihydrogen phosphate
(NaH_2PO_4) – 0.4%
Edible oil and Penicillin precursor – 0.25%

The pH of this medium after sterilization is 5.5 to 6.0.

A typical production medium described by Sylvester and Coghill ((1954) is composed per liter of

Cornstreep liquor solids	– 30 gm
Lactose	– 30 gm
Glucose	– 5 gm
Zinc sulphate	– 0.044 gm
Sodium nitrate	– 3 gm
Magnesium sulfate	– 0.25 gm
Calcium carbonate	– 3 gm
and Phenyl antamide precursor	– 0.05 gm

Inoculum media are similar to production media except that lactose and precursor are not added. Glucose or sucrose serve as the carbon source.

The pH remains constant at the state of the fermentation while glucose, cornsteep liquor, carbon compounds and ammonia are being utilized. Ammonia is liberated by deamination of the amino acid of cornsteep liquor and the pH raises to 7-7.5 (the optimum range for penicillin production). Beyond this point the pH remains constant and the mold uses lactose to produce penicillins and little further the growth occurs. At the end of the fermentation the pH rises to 8 or higher.

During the initial stage of carbohydrate and cornsteep liquor utilization i.e. during the 1st 20 to 30 hrs, the fungal growth becomes very thick and heavy. This growth occurs as disperse strands or dumps of mycelium or as definite pellets of mycelium –0.5 to 2mm in diameter.

The final *penicillin yield* is in the approximately range of 3-5% based on carbohydrate consumed and is probably in excess of 1500 units per milliliter.

Penicillin is sensitive to pH values above 7.5 especially in the presence of ammonium ion. Because of these considerations, during the stage of actual penicillin production, the pH is maintained near neutrality by calcium and magnesium carbonate in the medium and by the phosphate buffer. In addition the pH may be controlled by adding either sodium hydroxide or H_2SO_4.

Penicillin Harvest and Recovery: Penicillin in the acid (anion) form is solvent extractable and the antibiotic, as dissolved in an organic solvent, can be back-extracted as a salt into aqueous solution. However the exact procedure of harvest and recovery of penicillin to be used depends somewhat on the particular production employed and on the final penicillin yields in this medium. High yields, in conjunction with a medium that does not interfere with recovery and purification, greatly simplify the procedures required to obtain a pure product.

At harvest, the completed penicillin fermentation culture is filtered on a rotary vacuum filter.

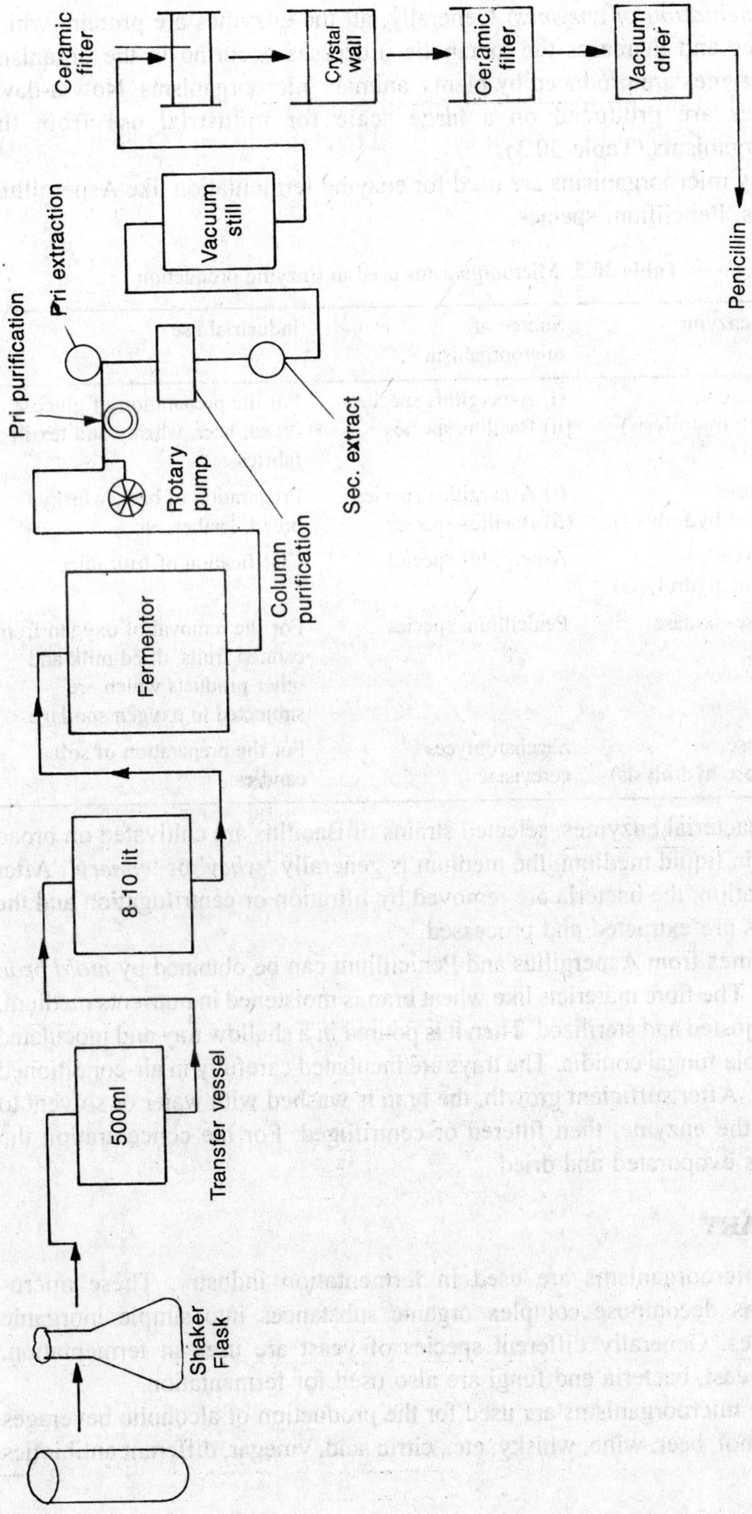

Fig. 30.1. Penicillin fermentation

Fermentation of enzymes: Generally, all the enzymes are protein, which regulates and increases the metabolic processes occuring in the organism. The enzymes are produced by plants, animals microorganisms. Now-a-days, enzymes are produced on a large scale for industrial use from the microorganisms (Table 30.3).

Many microorganisms are used for enzyme fermentation like Aspergillus, Bacillus, Pencillium species.

Table 30.3 Microorganisms used in Enzyme production

Type of enzyme	Source of microorganism	Industrial use
1. Amylase (Starch hydrolysis)	(i) Aspergillus species (ii) Bacillus species	For the preparation of glucose, bread, beer, whisky and textile fabrics.
2. Proteases (Protein hydrolysis)	(i) Aspergillus species (ii) Bacillus species	Preparation of beer, whisky, bread, leather, etc.
3. Pectinase (Pectine hydrolysis)	Aspergillus species	Clarification of fruit juice.
4. Glucose-oxidase	Penicillium species	For the removal of oxygen from canned fruits, dried milk and other products which are subjected to oxygen spoilage.
5. Invertase (Sucrose hydrolysis)	Saccharomyces cerevisiae	For the preparation of soft candies.

For bacterial enzymes, selected strains of Bacillus are cultivated on broad surface in liquid medium. the medium is generally '*whey*' or '*canery*'. After fermentation, the bacteria are removed by filtration or centrifugation and the enzymes are extracted and processed.

Enzymes from Aspergillus and Penicillium can be obtained by *mold bran process*. The fibre materials like wheat bran is moistened in nutrients medium, pH is adjusted and sterilized. Then it is poured in a shallow tray and inoculated by suitable fungal conidia. The trays are incubated carefully in air-conditioned cabinets. After sufficient growth, the bran is washed with water or solvent to remove the enzyme, then filtered or centrifuged. For the concentration the filtrate is evoporated and dried.

SUMMARY

Many microorganisms are used in fermentation industry. These micro-organisms decompose complex organic substances into simple inorganic substances. Generally different species of yeast are used in fermentation. Beside yeast, bacteria and fungi are also used for fermentation.

These microorganisms are used for the production of alcoholic beverages like ethanol, beer, wine, whisky, etc., citric acid, vinegar, different antibiotics like penicillin, streptomycine, etc., vitamin B_{12}, vitamin ascorbic acid, enzymes, etc.

EXERCISE

1. What is fermentation? Add a note on microorganisms used in fermentation.
2. Explain beer production.
3. What is penicillin? Describe penicillin fermentation.
4. What is single cell protein? Add a note on microorganisms used for single cell proteins.
5. Write a short note on
   (i) SCP
   (ii) Distillation of alcohol
   (iii) Penicillin
   (iv) SCO
   (v) Production of vitamin B_{12}

31

Extremophilic Microorganisms

The microorganisms are present in every nitch on the earth, except volcano and vacuum. These microorganisms have high adaptability. The majority of microorganisms survive and grow in normal conditions like room temperature, around neutral pH, etc. But some microorganisms survive and grow in extreme conditions, like very high or very low temperature, pH, salt concentration, etc. These are called extremophilic microorganisms. There are many extremophilic microorganisms like thermophiles, psychrophiles, acidophiles, alkalophiles, halophiles, etc.

All the processes of growth are dependent on chemical reactions and since the rate of these reactions are influenced by temperature, the pattern of bacterial growth can be drastically influenced by this condition. The temperature that allows for most rapid growth during a short period of time is known as *optimum growth temperature* (T_{opt}). On the basis of their optimum temperature, the bacteria are divided into three main groups namely psychrophiles, mesophiles and thermophiles.

Thermophiles: Organisms with T_{max} greater than 50°C were termed *thermophilic.* Generally thermophiles are prokaryotic organisms because up to now no eukaryotic organisms are known to grow above 50°C. Generally thermophiles can grow in the range of 55°C to 85°C. For example—Actinomycetes (62°C); Photosynthetic cyanobacteria (73°C); Clostridia (80°C); Acidothermus cellulolyticus (65°C), etc.

Thermophiles are generally isolated from hot springs and other geothermal environments. Beside this it is also present in some man made thermal systems like coal refuse piles, self heated hay stacks, piled organic waste materials such as compost heaps and hot effluent waters form power stations or industrial plant.

The growth range of many thermophiles extend into the mesophilic region. These species are called *facultative thermophiles,* while other thermophiles cannot grow in the mesophilic range, i.e. they grow strictly above 50°C.

These organisms are called obligate thermophiles.

Some thermophiles grow fastest at the upper temperature border of life, around 80°C to 100°C. Hyperthermophiles cannot grow below 60°C and some of them do not even grow at less than 80°C. Hyperthermophiles have been isolated from environments with apparent in situ temperature between 80-100°C. The low solubility of oxygen at high temperatures and the presence of reducing gases means that most hyperthermophiles are anaerobic. Some examples of thermophiles are Thermotogo maritima (90°C); Aquifex pyrophilus (95°C); Methanothermus sociabilis (97°C); Methanococcus Jannaschii (86°C); Methanophyrus Kandleri (110°C); etc.

Hyperthermophiles have good potential for use in novel biotechnological processes like heavy metal leaching, bioconversion of crude oil, desulphurization, etc. Again thermostable enzymes such as DNA polymerases, amylases, xylanases, proteases and lipases from these organisms are used in basic research and biotechnology processes.

Due to their great resistance to heat, they are the source of considerable trouble in the canning industry.

Psychrophiles: The word psychrophiles originated from Greek *'psychros'* meaning cold. Psychrotrophs (cold-seeking) are able to divide at 0°C and have optimum temperature around 20-25°C, whereas psychrophiles have optimum temperature below 16°C and they are able to grow up to 20°C. Facultative psychrophiles or psychrotrophs are able to grow best at temperature in the range of about 20-30°C. But obligate psychrophiles cannot grow at temperature above 20°C.

Straka and Stokes (1959) isolated psychrophilic bacteria from soil, sea waste and other materials obtained from Antarctica. These psychrophiles were isolated by cultivation at 0°C. They consist of rods, cocci and cocco-bacillary forms. And all are gram negative and non motile organisms.

During isolation of obligate psychrophiles it is usually necessary to maintain the source samples at cold temperature from the time they are collected and also to chill all media before isolation because strick psychrophiles usually die if they are even temporarily exposed to room temperature.

In cold ecosystems, these psychrophiles play an important role in the biodegradation of organic matter. These organisms produce extracellular proteases. Psychrophilic proteolytic microorganisms play an important ecological role in cold environment because they take part in the nitrogen cycle. Proteases from psychrophilic and psychrotrophic microorganisms have been shown to be quite active at 0°C. Extracellular proteases of psychrophiles have generally neutral or alkaline optimum pH for activity.

Proteases have great industrial importance. It is a very convenient tool whenever removal or degradation of protein is needed. Serine alkaline proteases of bacterial origin are used in laundering, leather tanning and food industry.

Mesophiles: (Greek 'meso' = middle). Mesophiles are not extremophiles. They grow best within a temperature range of approximately 25 to 40°C i.e. room temperature. The example of mesophilic organisms are all the bacteria that are pathogenic for human and warm blooded animals, which grow best at about body temperature (37°C). Another example is all saprophytes and plant parasites which grow best at 20-35°C.

For the survival and growth of living organisms, hydrogen ion concentration is one of the factors. The hydrogen ion concentration affects the ionic state and therefore the availability of many metabolites and inorganic ions to an organism. Very high concentrations of hydrogen ions (acid) or extremely low concentrations of hydrogen ions (alkaline) are normally toxic to most organisms. Most organisms live within the limits of pH 4 to 9 and have optimum pH around neutrality. But many organisms can grow or survive exposure outside these pH limits. Highly acidic and alkaline habitats harbour different microbial populations which can tolerate these environmental extremes. Depending upon the extreme pH, optimum for many micro-organisms, they are divided into *acidophiles, alkalophiles* and *neutrophiles.*

Acidophiles: Microorganisms which require pH value less than 5 for its maximum growth are called acidophiles. Acidophiles are generally found in natural environments of pH 3 to 4 like some pine soils, some lakes, acidic bogs, etc. Some examples of acidophiles are Bactoderma, caulobacter, Microcyclus, Planctomyces, Thiovobium, etc. which grow optimum at pH 3 to 5. *Acetobacter acidophilum* has optimum growth at pH 3.

Obligate extreme acidophiles occur in highly acidic environment. For example *Thiobacillus thiooxidans, Thiobacillus ferrooxidans,* etc. *Thiobacillus thiooxidans* is autotrophic and can produce metabolically useful energy from the oxidation of sulphur or sulphide minerals. It grows in the pH range 0.9 to 4.5. T. ferroxidans oxidizes reduced sulphur compounds as well as ferrous ion to ferric ion.

Extremely acidophilic hyperthermophiles have been found only in some terrestrial and marine solfataric fields and smouldering coal refuse piles. They are irregularly lobed cocci which may be aerobes, facultative aerobes or anerobes.

Currently indigenous populations of acidophiles are used in mineral leaching. T. ferrooxidans has been used to reduce sulfur content of coal in order to reduce sulfur emissions.

Alkalophiles: The bacteria which grow best in alkaline pH, i.e. pH 8 to 11.4 are called alkalophiles. These organisms are of two types, *alkaline tolerant* organisms which grow maximum in the pH range of 7.0-9.0 but cannot grow above pH 9.5 and obligate alkalophilic organisms, which grow maximum between pH 10.0 and 12.0 but cannot grow below pH 8.5.

All the extreme alkalophiles contain very high levels of respiratory chain components in the membrane, generally more acidic amino acid composition of proteins and a Na^+ cycle that facilitate solute uptake and pH homeostasis.

Most of the alkalophilic organisms are aerobic or facultatively anaerobic. Some examples of alkalophiles are spirulina sp. (8.0-11.0). Bacillus Strain (9.0-11.0); clostridium sp. (8.0-11.3); Methanobacterine (6.5-10.0); etc.

The alkalophiles have aroused much interest in biotechnology as they produce many enzymes of potential industrial use. Thermotolerant alkalophiles have been used for production of enzymes which might be alkalostable and thermostable. There are many applications for these enzymes which are stable at high pH.

The degradation of cellulose is an important problem in waste treatment. Extracellular cellulase from alkalophilic, Bacillus strain N-4 has been characterized and the gene encoding the enzyme has been cloned in E-coli and sequenced. Another cellulase from alkalophilic Bacillus strain 1139 has a pH optimum of 9.0.

Alkaline proteases are important as additive to detergents, in food processing and in leather tanning.

Halophiles: (Salt loving organisms) The halophilic bacteria are a distinctive group of prokaryotes clearly able to adopt to the high salt concentration and high light intensities of their natural habitat. Halophiles show optimum growth between 0.2 to 2.0 M NaCl. The ribosomes and enzymes of these bacteria requires high salt concentration in order to function effectively. A characteristic property of extreme halophiles is the presence of red pigment called bacteriorhodopsin and halorhodopsin in their cell membrane. These bacteria are known to produce a typical brick-red colour in sea water, pond water, etc, where it grows. This red pigment protects the halophiles against photochemical damage from the high light intensities.

Examples of this group are the *halobacteria* (Archaea); *Ectothiorhodospira* (phototrophic eubacterium) and Dunaliella (green algae). Dunaliella strains are potential producers of a range of valuable products like glycerols. It produces up to 30% w/w of its biomass as glycerol. Their extreme salt tolerance minimizes the risk of contamination, so they can be grown readily on an industrial scale in low cost open ponds.

Some gram positive and gram negative marine halophiles have been isolated which produce antibiotics, antiviral compounds and antitumour drugs. Extreme halophiles yield bioemulsifying agents and surfactants, which have wide applications in the food and pharmaceutical industry. Extreme halophiles are also present in leather curing vats and their enzymes may help soften the leather.

The salt resistance properties of halophiles are useful in the treatment of hypersaline waste streams such as slaughter house, waste, chemical brines, refinery wastes and some agricultural wastes.

SUMMARY

Many microorganisms grow in extreme conditions, called extremophilic microorganisms. Depending on the extreme condition tolerated by these microorganisms, they are of different types. Thermophiles grow and survive more than 55°C temperature. The psychrophiles cannot grow at room temperature, while they require temperature below 20°C for growth. The acidophiles require acidic conditions for growth while alkalophiles requires high alkaline pH for growth. The halophiles require very high concentration of salt for its growth.

EXERCISE

1. What are extremophilic microorganisms? Explain.
2. What is optimum temperature? Explain different types of microorganisms based on optimum temperature.
3. What is pH? Add a note on alkalophiles.
4. Describe halophilic microorganisms.
5. Write short notes on the following:
   (i) Acidophiles
   (ii) Psychrophiles
   (iii) Optimum pH
   (iv) Mesophiles
   (v) Optimum temperature

Part D
Microbiology Practicals

Microbiology Practicals

STUDY OF COMPOUND MICROSCOPE

The compound microscope is the example of bright-field microscope, which use direct light as source for illumination. Compound microscope is most commonly used for general laboratory observations.

A microscope is an optical instruments, which is used to magnify objects can not be seen by naked eyes.

The important characteristic of microscope is magnification power and resolving power.

Basically microscope has two parts namely mechanical parts and optical parts. The mechanical parts provides the structural framework to the microscope and also provides support to the optical parts. The optical parts are involved in the magnification of the objects.

A typical compound microscope consists of following parts.

(1) *Base and Pillar:* It provides support to the entire microscope. The base is generally 'U' shaped. It is generally made up of heavy metal to minimise vibration.

(2) *Inclination joint:* It joins arm to the pillar of the microscope. It permits the tilting of the arm of the microscope to any degree as desired by observer.

(3) *Arm:* It provides support to the body tube. At one end it is attached to the pillar by inclination joint and at the other end to the body tube.

(4) *Body tube:* It is cylindrical structure. It holds at one end occular lenses. It also maintains proper working distance between both lenses.

(5) *Draw tube:* It is fitted at the upper end of the body tube. The ocular lens fits into the draw tube.

(6) *Nose piece:* It is a circular rotating disc attached to body tube. The objective lenses are attached to the nose piece. It helps to brings desired objective at proper position.

(7) *Adjustment knob:* The adjustment knobs are used to bring the objects into proper focus. There are two sets of adjustment knobs namely coarse adjustment knob and fine adjustment knob. The coarse adjustment knob is used for greater adjustment while fine adjustment knob is used for fine adjustment.

(8) *Stage:* It is a small platform attached to the arm of microscope. It holds the slide in positions. It has hole in the centre exactly in the same line with body tube and condensor, to allow the light to pass through. It is also fitted with a mechanical device that helps in vertical and horizontal movement of slide.

(9) *Ocular lens:* The upper end of the body tube holds the ocular lens through which image is viewed.

(10) *Objective lens:* The objective lenses are attached to the revolving nose piece. Generally three objective lenses namely 10X (low power); 40X (high power) and 100X (oil immersion) are attached to microscope.

(11) *Mirror:* The mirror reflects light from the source of illumination via the condenser. The mirror have two sides i.e. concave and flat side. The concave side of mirror is used for natural day light and the flat side for the artificial light.

(12) *Iris diaphragm:* It controls the angle and amount of light used.

(13) *Condenser:* The condenser consists of several lenses that concentrate light on the slide.

PREPARATION OF GLASSWARES AND STERILISATION FOR MICROBIOLOGY EXPERIMENTS

Cleaning of Glasswares: Clean all glasswares properly with a brush using detergent and water. Then rinse the glasswares finally with distilled water and then invert them to drain water. Finally dry them in an oven.

For cleaning pipettes use chromic acid solution. Prepare chromic acid solution by dissolving 75g dichromate crystals in 500 ml of warm water. Allow it to cool and add 500 ml sulphuric acid slowly to the above preparation. Fill the tall cylinder with this solution and leave the used pipettes in this solution for overnight. Finally wash pipettes with distilled water.

Preparation of glassware for sterilisation: The glasswares such as test tubes, conical flask are plugged with cotton wool. Then cover it with paper before sterilisation. Plug pipettes at the mouth end with a little bit of cotton and place them in a stainless steel pipette-can for sterilisation. Similarly petri dishes can be placed in petricans.

The cotton used for making plug must be extremely clean. The plug should neither too tight nor too loose. A properly plugged tube can hold its own weight when suspended from the plug.

Sterilisation of glasswares: The glasswares are sterilised either by dry heat or by moist heat. By dry heat, sterilise all glasswares in a hot air oven at

160°C for 2-3 hours. After sterilisation do not open the oven. Allow it to cool slowly. Remove the materials and store them.

In alternative method, glasswares are sterilised by moist heat in the autoclave. In the autoclave 15 lbs pressure and 120°C temperature is maintained for 15 minutes. Then allow it to cool for a few minutes, release the steam pressure inside the autoclave and open the lid.

PREPARATION OF CULTURE MEDIA

Aim: To learn the principles and techniques of preparing media for culturing microorganisms.

Medium is defined as any substrate or material that will enable microorganisms to grow and multiply. A common nutrient medium used for culturing microorganisms are exist in three forms i.e. liquid, semisolid and solid medium. In liquid medium all nutrients are added to water. It is also called broth eg. nutrient broth. The liquid medium can be made into solid or semisolid medium by adding different quantity of solidifying agents like agar-agar, gelatine, silica gel, etc. Example: Nutrient agar, potato dextrose agar, martine rose bengal agar, etc.

The medium should contain all the essential substances required by an microorganism to be cultured. It should provide carbon, nitrogen, water sources in addition to minerals and other growth factors for the growth of microorganisms. The basal nutrient medium can be supplemented with different substances such as sugars, proteins, inorganic salts etc. to satisfy the requirement of a particular organism.

Beside providing different substances required by microorganisms, the pH of medium should be adjusted at the optimum pH of microorganism to be culture. If pH is not adjusted, the growth of the microorganisms may be inhibited. In solid medium, agar should be added only after adjusting the pH.

Types of Media: On the basis of chemical components used for preparation of media, it is classified into three categories. They are as follows:

(1) *Synthetic media*: Media in which all the constituents are chemically defined are called synthetic media. They are generally used to study the specific nutritional requirements of different microbes. e.g. Minimal medium

(2) *Complex media*: Media in which different components and their composition are incompletely defined are called complex media. e.g. Beef extract used in nutrient medium is chemically complex.

(3) *Natural medium*: Substances of natural origin that favours microbial growth are used in natural media e.g. Milk.

Media can also be classified based on their functional properties. They are as follows:

(1) *Simple Media*: It is a common media used for cultivation of most of the microorganisms. e.g. Nutrient agar.

(2) *Differential media*: These types of media are used to distinguish microorganisms that differ in their specific property e.g. EMB agar, Starch agar, Blood agar.

(3) *Selective media*: These types of media selectively allow the growth of particular type of organisms and prevent the growth of the most of other microbes. e.g. Martine Rose Bengal agar (it allows growth of fungi and inhibit the bacterial growth).

(4) *Selective-Differential Media*: These types of media allow only a particular types of organisms to grow that can also be further distinguish based on their property e.g. Mac Conkey agar.

Materials required: Beef extract, peptone, sodium chloride, agar-agar, conical flask, cotton wool, measuring cylinder, beaker, test tubes, glass rod, etc.

Procedure: Weigh various quantity of different components as given in following table and dissolve in water. Adjust pH to 7.2 (±0.2) with 1N NaOH using a pH meter.

Nutrient Agar

Chemicals Required	Quantity
(1) Peptone	5 g
(2) Beef extract	3 g
(3) Sodium Chloride	5 g
(4) Agar-agar	15 g
(5) Water	1000 ml

Plug the flask with cotton wool. Sterilize the medium at 15 lbs pressure for 15 min using a autoclave.

ISOLATION OF MICROORGANISMS

Microorganisms are abundant and ubiquitous in our environment. Microorganisms occur in air, soil, water, food, sewage, decomposing matter, living plants, and on body surfaces of animals including human being.

(A) *Isolation of microorganisms from soil by the serial dilution-agar plating method:* The serial dilution-agar plating method is one of the commonly used procedures for the isolation of microorganisms. This method is based upon the principle that when material containing microorganisms is cultured, each viable microorganism will develop into a colony. Thus the number of colonies appeared on the plates represent the number of living organisms present in the sample.

In serial dilution agar plate method, a known amount (1 gm or 1 ml) of sample is suspended in a known volume of sterile saline (9 ml) to make a microbial suspension (10^{-1}). Serial dilutions 10^{-2}, 10^{-3}, ..., 10^{-7} are made by adding 1 ml from previous dilution into 9 ml sterile saline. Finally 1 ml of various dilutions were added to sterile petri dishes, to which added sterile, cool (45°C), molten media. The number of colonies appearing on dilution plates are counted, averaged and multiplied by the dilution factor to find the number of cells or spores per gram of the sample.

$$\text{No. of cells/ml or g} = \frac{\text{Colonies} \times \text{Dilution factor}}{\text{Dry weight of soil}}$$

Materials Required: Soil sample, sterile nutrient agar, sterile petridishes, 9 ml sterile saline blanks (7), sterile 10 ml pipettes (7), etc.

Procedure: (1) Collect soil samples at random, minimum five places from a field and mix thoroughly.

(2) Label sterile saline blank tubes and petridishes as 10^{-1}, 10^{-2}, ..., 10^{-7}.

(3) Add 1 g of dried soil sample to first blank test tube to make 10^{-1} dilution.

(4) Vigorously shake the dilution to obtain uniform suspension of microorganisms.

(5) Transfer 1 ml of suspension from table one into saline blank tube two with a sterile pipette, to make 10^{-2} dilution.

(6) Repeat the same procedure to make 10^{-3}, 10^{-4}, ..., 10^{-7} dilutions.

(7) Transfer 1 ml sample each from 10^{-2} to 10^{-7} tubes into separate petridishes.

(8) Add around 15 ml of the cooled medium (45°C) to each petridish and mix the inoculum by gentle rotation of the petridish.

(9) Incubate the plates in an inverted position at 25°C for 24-48 hours.

(10) Observe the plates for number and distribution of colonies of bacteria and fungi from each dilution.

(11) Calculate the number of organisms per gram of soil.

(B) *Isolation of Microorganisms from air:* The atmosphere contains various groups of microorganisms, from algae to the viruses. Outdoor air, with the changing season contains a variety of pollen grains, spores of fungi, bacteria, viruses, etc. The presence of *air-spora* may pose great hazards to living beings, by causing diseases and loss in productivity. However, air is not a medium in which microorganisms can grow but it is a carrier of micro-organisms.

The sampling of air to determine its microbial content requires special apparatuses called solid impingement devices or liquid impingement devices.

But the *gravity slide* or *settle plate* technique has been most widely used method for studying the microbial contents of the atmosphere. In settle-plate technique, a petridish containing a suitable agar medium is horizontally exposed for 5-10 minutes. The plates are incubated at suitable temperature and the number of colonies are counted. Each colony represents a particular microorganism which has fallen on the agar surface.

Materials required: Nutrient agar media, sterile petridishes, etc.

Procedure: (1) Pour sterile, molten cool (45°C) nutrient agar medium in petridishes.

(2) Allow the plates to solidify.

(3) Remove the tops of petridishes and expose each plate to the air for 5-10 minutes in different places.

(4) Incubate the exposed petridishes at 37°C for 24-48 hours.

(C) *Isolation of microorganisms from water:* Water obtained from various source contains different types of microorganisms, may be pathogenic or non-pathogenic. Therefore, water for domestic use must be subjected to regulate microbial examination.

Materials Required: Sterile nutrient media, sterile petridishes, pipettes, sterile saline blanks, various water samples, etc.

Procedure: (1) Collect water samples from various sources in clean sterilised test tubes.

(2) Dilute the sample by serial dilution-agar plating method, as explained in previous experiment.

(3) Incubate the petridishes in an inverted position at 37°C for 24-48 hours.

(4) Observe and count the number of microorganisms and calculate the number of microorganisms present per ml water sample.

(D) *Isolation of microorganisms from body surfaces like skin, scalp and tooth scum:* Many microorganisms establish themselves inside the body and on the body surfaces of normal adults. The microorganisms which are permanently established and do not produce diseases under normal conditions are called normal flora. Microorganisms are not found throughout the entire body but are localised in certain regions like eyes, nose, mouth, skin, digestive system, etc. Once established the normal flora can benefit the host by preventing over growth of harmful microorganisms, but when this balance is upset, disease condition can result.

Material required: Nutrient agar plates, sterile distilled water, sterile cotton swab, etc.

Procedure: (1) Moisten a sterile cotton swabs in saline aseptically.

(2) Rub the different swab over the skin surfaces like palm, scalp, tooth scum, etc.

(3) Streak the agar plate with the above swab.

(4) Incubate the inoculated agar plate for 24-48 hours at 37°C in an inverted position.

Pure Culture of Bacteria

There is large population of microorganisms in our environment i.e. air, soil and water, which include many species of bacteria, fungi and algae. A study of the microorganisms in these habitats requires knowledge of the specific microorganisms present. This requires techniques for preparing species as pure cultures from the mixed population.

Pure culture is a population of organisms of a single species in the absence of living cells of any other species. The various techniques are used to isolate different species in a natural population and grown as pure culture. They are:

(1) Streak plate technique
(2) Pour plate technique and
(3) Spread plate technique.

Materials required: Nutrient agar, Nutrient broth, Petriplates, Sterile pipettes, Dilution blank, L-rod, test tubes, inoculation loop.

(A) Streak Plate Technique:

(1) Pour the nutrient agar into sterile petri plates allow it to solidify.
(2) Sterilize the inoculation loop by flaming it red hot and allow it to cool for 30 seconds.
(3) Holding the culture tube in your left hand, take the tube near the flame and remove cotton plug with your right hand and flame the month of the tube for a few seconds.
(4) With the help of inoculation loop take the culture.
(5) Place it on the agar plate and streak it on the plate once.
(6) Resterilise the loop and cool it as above and take the culture from one end of the plate to other end. Repeat the procedure 2-3 times.
(7) Incubate the plates in inverted position at 37°C for 24 hours.

(B) Pour Plate Techniques:

(1) Melt the nutrient agar in the tubes by heating them in water bath.
(2) Cool the tubes at 50°C and hold at this temperature.
(3) Take 1 ml of sample and mix it with agar medium by gently rotating the tube between your palms.
(4) Pour the contents of the tube into the petriplates and allow to solidify.
(5) Incubate the plates in inverted position at 37°C for 24 hours.

(C) Spread Plate Technique:

(1) Prepare nutrient agar plate as above.
(2) Serially dilute the given culture as follows.

(3) Arrange sterile test tubes in a test tube stand and label the tubes as 10^{-1}, 10^{-2}, 10^{-3} till 10^{-7} dilutions.

(4) Aseptically pipette out 9 ml sterile saline solution (0.85% NaCl) to each tube.

(5) Transfer 1 ml of the bacterial culture to the first tube and mix by vortexing (Dilution 10^{-1}).

(6) Take 1 ml from the first tube using a sterile pipette and transfer to the second tube and mix by vortexing (Dilution 10^{-2}).

(8) Repeat the same procedure to get 10^{-3} – 10^{-7} dilutions.

(9) Place 0.1 ml from the each dilution in separate nutrient agar plates.

(10) Sterilize the 'L' rod with alcohol and then by flaming it.

(11) Cool the rod and place the rod gently on the surface of the agar. Spread the sample over the surface of the agar medium by rotating the petriplate.

(12) Incubate the plates in inverted position at 37°C for 24 hours.

STAINING OF MICROORGANISMS

The microbial cells are very small and transparent in nature, so it is very difficult to observe them under microscope. To make the microorganisms visible and also to understand their structure and chemical nature, staining is used. Thus staining is the method of artificially colouring microorganism to allow it to visualise under the microscope.

Dyes or stains are synthetic chemicals with respect to their colour property in the anionic and cationic form, they can be divided into acid dyes and basic dyes. An acid dye is the salt of a coloured acid while a basic dye is the salt of a coloured base. Acid fuchsin and Erythrosin are acid dyes, while methylene blue and crystal violet are the basic dyes. Acid dyes have stronger tendency to combine with the cytoplasm whereas basic dyes have greater affinity for the nuclear region of the cell. In the bacteriological studies, basic dyes are generally used.

Depending on the purpose and type of organism to be studied, the various techniques of staining is available, they are as follows:

(A) *Simple Staining:* The simple staining technique contain only one dye that allow visualisation of gross, morphological characteristics of cells like shape, size, cell arrangement, etc.

Materials Required: 24 hours old bacterial cultures, methylene blue stain, staining tray, glass slides, Inoculation loop, spirit lamp, tissue paper, microscope, etc.

Procedure:

(1) Take a clean, dry glass slide.

(2) Take a loopful of water on the centre of slide.

(3) Aseptically transfer small amount of bacterial suspension in to water droplet with inoculation loop.

(4) Spread the drop over the centre of the slide to about 1 cm diameter and allow the film to air dry.

(5) Fix the smear by passing the slides rapidly over the flame.

(6) Flood the smear with the methylene blue dye for 3 minutes.

(7) Wash the slide in a gentle stream of tap water to remove excess stain.

(8) By using tissue paper, blot excess water and allow to air dry.

(9) Observe under microscope.

(B) *Negative Staining:* Negative staining is the simple staining technique. In negative staining, instead of basic dyes, acidic dyes are used, for example nigrosin, India ink. As bacterial cells are negatively charged, the acidic dyes does not stain the microbes but stains only the background. Thus after staining, under microscope, the microbes appear transparent in a dark background. The main advantage of negative staining is that there is less morphological distortion since cells are not subjected to vigorous physical or chemical treatment.

Materials required: 24 hour old bacterial culture, Nigrosin stain (Dissolve 10 gm Nigrosin in 100 ml of boiling distilled water, add 10 drops of formaline as preservative), slides, inoculation loop, spirit lamp, microscope, etc.

Procedure:

(1) Using a sterile wire loop, transfer a loopful of bacterial suspension to the centre of a clean slide.

(2) Add a small drop of nigrosin solution on the drop of the suspension on the slide.

(3) Mix thoroughly and spread to form a thin film.

(4) Air dry; do not heat fix.

(5) Observe under microscope.

(C) *Differential Staining*: Differential staining for example gram staining is employed to visualise and differentiate between organisms. It is useful in presumptive identification of organisms before carrying out several other tests such as serological and biochemical tests.

The gram staining was developed by Dr. Hens Christian Gram, a Danish physician in 1884. It is a very useful staining method for identifying and classifying bacteria into two major groups i.e. gram positive and gram negative. In this process the fixed bacterial smear is subjected to four different reagents called crystal violet (primary stain), iodine solution (mordant), alcohol (decolourization agent) and safranin (counter stain). In grams staining, the bacteria which retain the primary stain i.e. organisms that appear dark blue or violet are called gram positive, while the bacteria that lose the primary stain and counter stained by safranin i.e appear red are called gram negative.

The difference in staining responses to the gram stain can be related to chemical and physical differences in their cell walls. The gram negative bacterial cell wall is thin, complex, multilayered structure and contains relatively a high lipid contents, in addition to protein and mucopeptides. The higher amount of lipid is readily dissolved by alcohol, resulting in the formation of large pores in the cell wall which facilitate the leakage of crystal violet-iodine complex and resulting in the decolourisation of the bacterium which later takes the counter stain and appears red. In contrast the gram positive cell walls are thick and chemically simple, composed mainly of protein and cross-linked mucopeptides. When treated with alcohol, it causes dehydration and closure of cell wall pores, thus not allowing the loss of crystal violet-iodine complex and cells remain blue.

Materials Required: 24 hour old Staphylococcus and Escherichia cultures, crystal violet. Gram's iodine solution, 95% ethanol, safranin, staining tray, droppers, inoculation loop, glass slides, blotting paper, microscope, etc.

Procedure:

(1) Make thin smears of Staphylococcus and Escherichia on separate glass slides.

(2) Let the smears air dry.

(3) Fix the smear by heat.

(4) Stain it with crystal violet for 30 seconds.

(5) Wash the slide with distilled water for a few seconds.

(6) Apply iodine solution (mordant) for 1-minute.

(7) Decolourise with 95% ethanol by adding it drop by drop on the tilted slide until no more colour flows from the smear.

(8) Wash it with distilled water.

(9) Flood the slide with safranin (counter stain) for 30 seconds.

(10) Wash with distilled water and air dry.

(11) Observe under microscope.

Result: The bacteria that appear blue are called as gram positive, while those appearing pink are called gram negative.

Precautions:

(1) Always use fresh and young cultures to avoid misleading results.

(2) During heat fixing, excessive heating should be avoided.

(3) Smears should be thin and uniform.

(D) *Endospore staining*: Under unfavourable conditions for growth, specialised structures called spores or endospores are formed within certain gram positive cells such as Bacilli, Clostridia, Sporosarcinae, etc. These spores have no metabolic activity and are resistant to heat drying, freezing, toxic compounds and radiation. The heat resistance of spores has been linked to their high content of calcium and dipicolinic acid. An endospore develops in a characteristic position within a cell i.e. central, subterminal or terminal. In favourable conditions, endospores return to its vegetative state.

The endospore have low affinity for stain but once stained they resist decolouration. For staining, it is required to soak endospore in stain for longer period in the steam.

Materials Require: 48 hours old culture of Bacillus, Malachite green, safranin, staining tray, glass slides, inoculation loop, blotting paper, microscope, etc.

Procedure:

(1) Prepare a smear of Bacillus on a clean slide. Air dry and fix with heat.

(2) Flood the smear with malachite green.

(3) Heat the slides to steaming for 5 minutes, add more stain to the smear for time to time.

(4) Wash the slide gently with running tap water until all excess stain is removed.

(5) Counter stain with safranin for 30 seconds.

(6) Wash smear with water. Blot it and air dry.

(7) Observe the slide under microscope using immersion objective.

Result: The endospores stain green and vegetative cells stain red.

(E) *Lactophenol cotton blue mounting of fungi*: Lactophenol cotton blue stains the fungal cytoplasm and provides a light blue background, against which the walls of hyphae can readily be seen. Lactophenol cotton blue is made up of *phenol*, which serves as a fungicide; *lactic acid*. which acts as a clearing agent; *cotton blue*, which stains the cytoplasm of fungus and *glycerine*, which gives a semi permanent preparation.

For rapid and routine examination of almost all types of fungi, spores and spore bearing structures are teased out on a clean slide on a drop of lactophenol cotton blue and a cover glass placed over the preparation which is then ready for microscope examination.

Material Required: Fungal culture, Lactophenol cotton blue, needles, glass slides, cover slips, 70% alcohol, microscope, etc.

Procedure:

(1) Place a drop of lactophenol cotton blue on a clean slide.

(2) Transfer a small tuft of the fungus into the drop of stain using a needle.

(3) Gently tease the material using the needles.

(4) Place a cover slip without trapping air bubbles in the stain.

(5) Observe under microscope using low power and high power objectives.

Results: The fungal cytoplasm is seen as a lightly stained blue region forming a layer inside the unstained cell wall of hypae, that is surrounded by a light blue background on the slide.

TO STUDY BIOCHEMICAL ACTIVITIES OF MICROORGANISMS

Microorganisms in general utilise complex organic compounds as nutrients. Some of these compounds are insoluble. If the microbe has to use them as substrate for growth and energy, they have to necessarily break them down to simpler substances which are soluble. Such hydrolytic reactions are carried out by the extracellular enzymes secreted by some microorganisms in their surroundings. Those organisms that are producing these specific enzymes can utilise certain complex organic substances (e.g. starch, cellulose, casein) as substrate for growth.

The ability of the microorganism to metabolise and breakdown some of the substrates can be determines by the use of a suitable nutrient medium containing the substrates and the products formed can be tested using some reagents.

(A) *Hydrolysis of starch*: Amylase is a exoenzyme produces by micro-organisms of specific genera that hydrolyses starch, a polysaccharide, into a disaccharides and monosaccharides. These disaccharides and monosaccharides enters into the microbial cell through semipermeable membrane and used by microbes as nutrients.

The ability to degrade starch is used as the criteria for determination of enzyme amylase production by microbes. In the laboratory, the production of amylase by microbes can be tested by using starch agar medium. If starch is utilised by microorganism, after adding iodine, a yellow zone around the colony is observed.

Materials Required: Starch agar medium, petridishes, bacterial cultures of E. coli, Bacillus, Streptococci.

Procedure:

(1) Pour the starch agar medium in the sterile petridishes and allow to solidify.

(2) Aseptically transfer microbial culture into the starch agar plate.

(3) Incubate the inoculated plates inverted at 37°C for 48 hours.

(4) Flood the agar surface with Gram's iodine solution. Keep it for a minute and then pour off.

(5) Observe the plate carefully.

Result: Iodine reacts with starch to form a blue compound. The colourless zone surrounding colonies indicates production of extracellular amylase enzyme.

(B) *Hydrolysis of Casein:* Casein is a major protein component of milk. Hydrolysis of casein may be achieved by enzyme caseinase. Severai microorganisms have the ability to hydrolyse casein by the production of caseinase.

Materials Required: Nutrient agar medium, skim milk, sterile petridishes, microbial cultures, etc.

Procedures:

(1) Nutrient agar has been cooled to 45°C and then add 2 ml of sterile skim milk.

(2) Mix rapidly and thoroughly and pour the milk agar into a sterile petridish.

(3) Allow it to solidify.

(4) Aseptically transfer various microorganisms to different petridishes.

(5) Incubate the plates in an inverted position at 37°C for two days.

Result: A transparent zone surrounding colonies producing caseinase can be observed after growth.

(C) *IMViC Test:* The IMViC test consists of four different tests (1) Indole production, (ii) Methyl red, (iii) Voges-Proskaues and (iv) Citrate utilisation. The term IMViC stands for the first letter of the name of each test in the series, with the lower case 'i' included for ease of pronunciation. The IMViC tests were designed to differentiate gram negative intestinal bacilli (i.e. family Enterobacteriaceae) particularly *Escherichia coli* and the Enterobacter-Klebsiella group, on the basis of their biochemical properties and enzymatic reactions in the presence of specific substrates.

(1) *Indole Production Test:* Tryptophan an essential amino acid, is oxidised by some bacteria by the enzyme *tryptophanase* resulting in the formation of indole, pyruvic acid and ammonia. The indole test is performed by inoculating a bacterium into tryptone broth, the indole produced during the reaction is detected by adding Kovac's reagent (dimethylaminobenza-aldehyde) which produces a *cherry-red colour* in the top layer of the solution.

$$\text{Tryptophan} \xrightarrow{\text{Tryptophanase}} \text{Indole} + \text{Pyruvicacid} + H_3N$$

$$\text{Indole} + \text{Kovac's reagent} \xrightarrow[\text{(Butanol)}]{\text{HCl}} \underset{\text{(cherry red compound)}}{\text{Rosindole}} + H_2O$$

Materials Required: Cultures of E. coli and Enterobacter aerogenes, Tubes containing 5 ml of 1 % tryptone broth (3), Kovac's reagent, 1 ml pipette, inoculation needle.

Procedure:

(1) Inoculate the tubes of tryptone broth with one loop of culture.

(2) Incubate for 48 hours at 37°C.

(3) After 48 hours of incubation, add 1 ml of Kovac's reagent to each tube inoculated and an uninoculated comparative control tube.

(4) Shake the tubes gently after intervals for 5-10 minutes.

(5) Allow the tubes to stand to permit the reagent to come to the top.

(6) Observe the tubes for the colour in the top reagent layer.

Results: Development of cherry red colour in the top layer of the tube in E. coli is a positive test for indole production. Absence of red colouration in E. aerogenes is a negative test for indole production.

(2) *Methyl-Red and Voges-Proskauer tests:* The methyl-red (MR) and the Voges-Proskauer (VP) tests are used to differentiate two major types of facultatively anaerobic enteric bacteria that produce large amounts of acid and those that produce the neutral product *acetoin* as end product. Both tests are performed simultaneously because they are physiologically related and are performed on the same medium i.e. MR-VP broth. Opposite results are usually obtained for the MR and VP tests i.e. MR+, VP– or MR–, VP+. In these tests, if an organism produces large amount of organic acids from glucose, the medium will remain red (a positive test) after the addition of methyl red a pH indicator. In other organisms, methyl red will turn yellow (a negative test) due to the elevation of pH above 6.0 because of the enzymatic conversion of the organic acids to nonacidic end products such as ethanol.

MRVP tests are useful in the separation of *Escherichia coli* and *Enterobacter aerogenes* (both coliform bacteria) which appear virtually identical that are used as indicators of the sanitary quality of water, foods, etc.

Materials Required: MR-VP broth, cultures of E.coli and Enterobacter aerogenes, Methyl red pH indicator, VP reagent I, (napthol solution), VP reagent II (40% potassium hydroxide), inoculation loop.

Procedure:

(1) Prepare MRVP broth (pH 6.9) tubes.

(2) Inoculate two MRVP tubes with E.coli and two with E. aerogenes and keep two tubes as uninoculated comparative control.

(3) Incubate all tubes at 37°C for 48 hours.

(4) Add 5 drops of methyl red pH indicator to the tube of each set (E. coli, E.aerogenes and control)

(5) Observe the change in colour for MR test.

(6) Add 12 drops of VP reagent-I and 2-3 drops of V-P reagent II to the other set of tubes.

(7) Shake the tubes gently for 30 seconds with the caps off to expose the media to oxygen.

(8) Allow the reaction to complete for 15-30 minute.

(9) Observe the tubes for change in colour for the VP test.

Observation: In the MR test, the methyl red indicator in the pH range of 4 will remain red indicating a positive test, while turning of methyl red to yellow is a negative test.

In the VP test, the development of crimson-to ruby pink (red) colour is indicating a positive VP test while no change in colour is a negative test.

Results: E.coli shows positive MR test and negative VP test, and E.aerogenes shows negative MR and positive VP test.

(3) *Citrate utilisation test:* Citrate test is used to differentiate among enteric bacteria on the basis of their ability to utilise citrate as the carbon source. An enzyme *citrase* produced by the microorganism, breaks down the citrate to oxaloacetic acid and acetic acid. These products are later converted to pyruvic acid and carbon dioxide.

$$\text{Citric acid} \xrightarrow{\text{citrase}} \text{Oxaloacetic acid} + \text{Acetic acid}$$

$$\downarrow$$

$$\text{Pyruvic acid} + CO_2$$

In citrate test, microorganisms are inoculated into an organic synthetic medium, '*Simmon's citrate agar*', where sodium citrate is the only source of carbon and energy. Bromothymol blue is used as an indicator. When the citric acid is metabolised, the CO_2 generated combines with sodium and water to form sodium carbonate an alkaline product, which changes the colour of the indicator from green to blue which indicates positive test.

Materials Required: Cultures of E.coli Enterobacter aerogenes, Simmon's citrate agar slants, inoculation loop.

Procedure:

(1) Prepare Simmon's citrate agar slants.

(2) Inoculate one slant with E.coli and another with E.aerogenes by streaking. The third tube is kept as an uninoculated comparative control.

(3) Incubate all the slants at 37°C for 48 hours.

(4) Observe the slant cultures for the growth and change in colour of the medium.

Results: In Enterobacter aerogenes inoculated slant, growth is visible and the medium colour is blue i.e. citrate positive. In E.coli inoculated slant, there is no growth and there is no change in colour i.e. citrate negative test.

ANTIBIOTIC ASSAY

Antibiotic assay is performed to check the antibiotic sensitivity of bacterial strains and minimal inhibitory concentration of antibiotic to bacterial strain.

Antibiotics are commonly used for the treatment of infectious diseases. Once the causative organism is identified for specific disease and is isolated, it is important to test the sensitivity of the organism to the effective antibiotics. In the laboratories antibiotic impregnated discs are commonly used to identify the antibiotic sensitivity of the causative organism. The effectiveness of the antibiotic in the test is based on the size of the zone of inhibition. The zone of inhibition also depends on the diffusibility of antibiotic, the size of the inoculum, type of medium and other factors. The concentration of the antibiotic at the edge of zone of inhibition represent minimal inhibitory concentration (MIC) of antibiotic. The minimal inhibitory concentration is the lowest concentration of antibiotic that exhibits the zone of inhibition on the assay plate.

Materials Required: Antibiotic discs, Nutrient agar, Bacterial cultures, petridishes, etc.

Procedure:

(A) Antibiotic sensitivity Test:

(1) Prepare nutrient agar plates

(2) Aseptically take a swab of the test cultures and inoculate surface of the nutrient agar plate completely so as to make a lawn.

(3) Allow atleast 5 minutes for the agar surface to dry.

(4) Take a forcep and sterilise the tip by dipping it in alcohol and then by flaming. Allow to cool.

(5) Carefully take the antibiotic disc and place over the agar plate at least 15 mm from the edge of the plate. Gently press disc to give a better contact with the agar.

(6) Incubate the plates in inverted position for 16-18 hours at 37°C.

(7) Observe the zone of inhibition around antibiotic disc.

(B) Determination of Minimal inhibitory concentration of antibiotic:

(1) Prepare nutrient agar plate.

(2) Select a test organism and an antibiotic from the previous experiment that showed good inhibition zone.

(3) Inoculate the nutrient agar plate with the selected test organisms as lawn culture.

(4) Allow the plate to dry for 5 minutes.

(5) Place a disc impregnated with different concentration of antibiotic (10μg to 500 μg/ml).

(6) Incubate the plates at 37°C for 12-18 hours.

(7) Measure the zone of inhibition on the plate by ruler.

(8) Tabulate the result concentration of antibiotic versus diameter of zone of inhibition.

MEASUREMENT OF CELL NUMBER IN A CULTURE BY TOTAL COUNT AND VIABLE COUNT METHOD

Microbial analysis of food, water, milk and air requires quantitative determination i.e. total population of microorganisms in these substrates. The number of cells and spores of microorganisms can be measured in the laboratory by several methods either by direct or indirect counts, total count or viable count.

Total Count: The major disadvantage of total count method is it counts both the dead as well as living cells. For total count, generally cells are measured directly by counting under the microscope, sometimes electronic cell counter is also used. Since most of the microbial cells are smaller in size and their population is much higher in the sample, it is necessary to use special chambers such as haemocytometer chamber to count the total number of cells in the sample.

The haemocytometer is used for counting the microbes in liquid suspension. It is a special microscope slide with a counting chamber 0.1 mm deep so that volume of liquid over a one square mm is 0.1 cubic mm. The counting chamber has a total of nine squares, each of 1 mm × 1 mm engraved over it. A one mm square is divided into 25 medium sized squares (0.2 mm × 0.2 mm each). The medium sized squares are further subdivided into 16 small squares (0.05 × 0.05 mm each).

Each large square has a volume of 1/10000 cm^3 or 10^{-4} cm^3. One ml of a cell suspension is put into the counting chamber, number of cells is counted and the total cell number is determined mathematically.

Materials Required: Haemocytometer, cell or spore suspension cover glass, pipette, etc.

Procedure:

(1) Place a drop of cell suspension on the engraved grid and let the preparation stand for 1-2 minutes to allow the cells to settle at the bottom.

(2) Put the cover glass over the grid carefully so that no air bubble enters between the slides and cover glass.

(3) Slide the cover glass backwards and forwards until coloured rings are visible as the two surfaces of cover glass and slide come into close contact.

(4) Count the spores or cells in the middle square which consists of 25 groups of 16 small squares, each group 0.2 mm square.

(5) For larger spores or fewer spores count spores in four corner large squares and in the middle one to have a total count of 200-250.

Results: Calculate the number of spores per ml of the suspension mathematically as follows

(i) For small spores:

Spores/ml = Number of spores counted on the middle square of the grid × 10,000 (10^4) (Each large square has a volume of 1/10 cm × 1/10 cm × 1/100 cm = 10^{-4} cm^3) × dilution factor.

(ii) For large spores:

Spores/ml = Average number of spores in five large square × $10^4/cm^3$

$$= \frac{A+B+C+D+E}{5} \times 10^4/cm^3$$

Precautions:

(1) Always count cells/spores on top and left touching middle line of the perimeter of each square and do not count spores touching the middle line at bottom and right sides.

(2) Do not overfill or underfill the chambers of hemocytometer.

Viable count: The plate count technique is one of the most routinely used procedure for the enumeration of viable cells. This method is based on the principle that when material containing bacteria is cultured, every viable bacterium develops into a visible colony on a nutrient agar medium. Therefore, the number of colonies are the same with the number of organisms present in the sample.

Materials Required: Bacterial suspension, 9 ml dilution blanks (7), sterile petridishes (12), sterile 1 ml pipettes (7), nutrient agar medium, colony counter.

Procedure: In this method, small amount of sample is diluted by serial dilution method. Then the specific volume of the diluted sample from various dilutions is added to sterile petridishes to which molten and cooled (45°C) agar medium is added. Incubate the plates in an inverted position at 37°C for 24-48 hours. After incubation period, the number of colonies developed are counted on a colony counter. Then calculate the number of microorganisms in the given sample by following formula.

$$\text{Number of cells/ml} = \frac{\text{Number of colonies}}{\text{Amount of sample plated} \times \text{dilution}}$$

TO CALCULATE THE GENERATION TIME OF E.COLI

Bacterial growth can be determined by measuring *turbidity* or *optical density* of a broth culture. The more turbid suspension, the less light will be transmitted through it. Since turbidity is directly proportional to the number of cells, this property is used as an indicator of bacterial concentration in a sample. Absorbance of light is a logarithmic value and is used to plot bacterial growth on a graph.

Materials Required: Test tubes containing 5 ml of nutrient broth (13), 24 hours old culture of E. coli, sterile 1 ml serological pipette, spectrophotometer, etc.

Procedure:

(1) Inoculate six labelled tubes (0, 2, 4, 6, 8, 16, 24) of nutrient broth each with 0.1 ml of the E. coli culture.

(2) The remaining seven uninoculated broth tubes served as blank for adjusting the spectrophotometer.

(3) Incubate all the 13 tubes at 37°C.

(4) Observe the inoculated cultures following incubation at an interval of 0, 2, 4, 6, 8, 16 and 24 hours for the amount of growth which is determined by measuring the turbidity.

Results: Plot the graph of absorbance versus the time. It is observed that with time, turbidity of sample increases.

BACTERIOLOGICAL EXAMINATION OF WATER BY MULTIPLE-TUBE FERMENTATION TEST

Multiple-tube fermentation test or most probable number (MPN) test is used to detect coliform bacteria which are used as indicator of fecal contamination. The test is performed sequentially in three stages i.e. presumptive, confirmed and completed tests.

(A) *Presumptive coliform test*: The presumptive coliforms test is used to detect and estimate coliforms population of a water sample. In this test, known volumes of water is added to lactose fermentation tubes and production of acid and gas from the fermentation of lactose is a positive test for coliform bacteria. The lactose broth used in this test is selective for the isolation of coliforms. A pH indicator such as Bromocresol purple is added to lactose broth for the detection of acid. The colour of pH indicator changes to yellow with the fermentation of acid from lactose. A statistical method is used to estimate the population of coliforms, i.e. the result obtained is expressed as the most probable number (MPN) of coliforms, by matching the test results with those provided in the statistical table.

Materials Required: Water sample, Lactose broth medium, Durham tubes, 10 ml double-strength lactose broth tubes, (5), 5 ml single strength lactose broth tubes (10), sterile 10 ml, 1ml and 0.1 ml pipettes, etc.

Procedure:

(1) Collect water sample from different sources.

(2) Label 5 double strength lactose broth tubes as "10"; the 5 single strength broth tubes as "1.0"; and 5 single strength lactose broth tube as "0.1".

(3) Aseptically inoculate each "10" labelled tubes with 10 ml of water sample.

(4) Aseptically inoculate each '1' labelled tubes with 1 ml of water sample.

(5) Aseptically inoculate each '0.1' labelled tubes with 0.1 ml of water sample.

(6) Incubate all the tubes at 37°C for 48 hours.

(7) Examine all the tubes for the production of acid (yellow colour) and gas after 24 and 48 hours of incubation.

Results: Production of acid and gas after 24 hours incubation indicates a positive presumptive test for coliform bacteria. If gas develops in tubes after 48 hours incubation, the presumptive test is doubtful and if these is no gas produced after 48 hours incubation, it shows negative presumptive test. A positive or doubtful presumptive test suggests that the water sample is polluted or non potable. These tubes are retained and used for confirmed test.

Most probable number (MPN) of bacteria present in 100 ml of water sample is read off from the MPN table by applying the combination of positive and negative results obtained in the inoculated lactose fermentation units.

(B) *Confirmed coliforms test:* This test is used to confirm the presence of coliforms in water samples showing positive or doubtful presumptive test. In the confirmed test, the samples from the positive presumptive lactose broth tubes are streaked onto a selective differential *eosin-methylene blue* (EMB) medium for coliforms. EMB medium is selective in nature because of the presence of the dye methylene blue which inhibits the growth of gram positive bacteria and allowing the growth of gram negative bacteria. EMB medium is differential in nature in that lactose fermenting bacteria gives coloured colonies due to the formation of a complex in EMB that precipitates out onto the coliform colonies while non-lactose fermenters produce colourless colonies on EMB agar.

Materials Required: EMB agar plates, 24 hour old positive lactose broth culture, inoculation loop.

Procedure:

(1) Inoculate the EMB agar plates with the positive 24 hour lactose broth culture with a sterile inoculation loop.

(2) Incubate the plate for 24-48 hours at 37°C.

(3) Observe the plates for the presence of E.coli colonies.

Results: A positive confirmed test is shown by the appearance of typical

E. coli colonies with dark centre and metallic sheen indicating that the water is non-potable.

(C) *Completed coliforms test:* Completed test is used as a confirmatory test for the presence of E.coli in water sample. In this test, lactose positive colonies from EMB agar are isolated and inoculated into a lactose broth tube and streaked on a nutrient agar plate to perform gram-staining. If there is production of acid and gas in the inoculated lactose broth and there are rod-shaped bacteria showing gram-negative reaction, these confirms the presence of E.coli in the water and is considered a positive completed test.

Material Required: 24 hours old coliform positive EMB agar culture, lactose fermentation broth tube, nutrient agar slant, gram-staining reagents, inoculation loop, etc.

Procedure:

(1) Inoculate lactose fermentation broth tube with the isolated coliform colony from the EMB agar plate using inoculation ioop.

(2) Streak nutrient agar slant with the colony from the EMB agar plate using inoculation loop.

(3) Incubate both the inoculated broth and slant at 37°C for 24 hours.

(4) Gram stain the organism found on the slant.

(5) Observe lactose fermentation broth tube for the production of acid and gas and slide for positive or negative gram reaction and cell morphology.

Results: Production of acid and gas in an inoculated lactose broth and presence of gram negative rods indicate a positive completed test and confirm the presence of coliform bacteria in the water sample.